Oceans from Space V

Oceans from Space V

Guest Editor

Vittorio Barale

MDPI

Basel • Beijing • Wuhan • Barcelona • Belgrade • Novi Sad • Cluj • Manchester

Guest Editor
Vittorio Barale
Joint Research Centre
European Commission
Ispra
Italy

Editorial Office
MDPI AG
Grosspeteranlage 5
4052 Basel, Switzerland

This is a reprint of the Special Issue, published open access by the journal *Remote Sensing* (ISSN 2072-4292), freely accessible at: https://www.mdpi.com/journal/remotesensing/special_issues/N5M29MQT28.

For citation purposes, cite each article independently as indicated on the article page online and as indicated below:

Lastname, A.A.; Lastname, B.B. Article Title. *Journal Name* **Year**, *Volume Number*, Page Range.

ISBN 978-3-7258-7631-0 (Hbk)
ISBN 978-3-7258-7632-7 (PDF)
https://doi.org/10.3390/books978-3-7258-7632-7

Contents

About the Editor

Vittorio Barale

Vittorio Barale graduated in physics in 1977 from the University of Milan, Italy, and then received both his M.S. in 1982 and Ph.D. in 1986 from the University of California, San Diego, affiliated with the Scripps Institution of Oceanography. During his graduate studies, he was also a visiting scientist at the Massachusetts Institute of Technology in Boston, MA, in 1984, and at the Goddard Space Flight Center of NASA, in Washington, DC, in 1985. After returning to Europe in 1986, he became a private consultant on marine science and technology for various institutions, companies, and international organizations. In 1990, he joined the Joint Research Center of the European Commission in Ispra, Italy, where he held a senior scientist position until his retirement in 2019. He is currently affiliated with the Tethys Research Institute in Milan, Italy. He has been an advisor for the European Space Agency, various United Nations agencies, a number of European Commission Inter-Service Working Groups, and the Scripps Alumni Leadership Team of the Scripps Institution of Oceanography. Furthermore, he has served as an invited lecturer in international schools and university courses. He has been affiliated, for variable periods, with ISPRS, PORSEC, EGU, AGU, EARSeL, CIESM, and the MEDCOAST Network. He is the Chair of the 'Oceans from Space' conference series, held every 10 years in Venice, Italy. His research interests focus mainly on the ecological assessment of marginal and enclosed seas, as well as their coastal margins, using satellite remote sensing. He has worked extensively on the phytoplankton patchiness problem, a key factor in the survival of the marine food web and the Earth's ecosystem. In recent years, he has been active in integrated coastal management and maritime spatial planning. He was the lead scientist for the development of the European Atlas of the Seas and for other initiatives dealing with marine and maritime information systems in the European Union.

Preface

The "Oceans from Space" conference series, which has been held in Venice, Italy, at 10-year intervals since 1980, aims to review the use of remote sensing from Earth's orbit in the study of the oceans. Indeed, in the last half century satellite observations have become a cornerstone of all planetary sciences and of our efforts to understand and sustainably manage the Earth. The 5th "Oceans from Space" symposium, held at the Scuola Grande di San Marco, in Venice, on 24–28 October 2022 (after being delayed for two years by the COVID-19 pandemic), focused on the most recent scientific and technological achievements, innovations, and challenges of satellite oceanography. Major new developments have been achieved in ocean observations, to the point that many aspects of modern oceanography have been revolutionized by the unprecedented capabilities offered by orbital remote sensing. No other technology allows gathering information about marine variables and processes, at suitable space and time scales, like satellite observations do. Ocean exploration and environmental trend monitoring, coupled ocean and atmosphere forecasting, marine resources management, maritime spatial planning: the list of current or potential applications is virtually endless. The present collection of papers provides a significant sample of these, originating from the latest "Oceans from Space" edition.

Vittorio Barale
Guest Editor

remote sensing

MDPI

Editorial

Half a Century of Oceans from Space: Features and Futures

Vittorio Barale

Tethys Research Institute, 20121 Milan, Italy; vittorio_barale@yahoo.it; Tel.: +39-333-1235757

Abstract: Half a century separates us from the dawning of satellite oceanography. Aircraft flights, photographs from early space missions, and data from meteorological satellites in the 1960s already provided glimpses of the future role of remote sensing in marine science. A first generation of dedicated ocean-viewing satellites followed in the 1970s. The "Oceans from Space" conference series, which convenes every ten years in Venice, Italy, started in 1980, when unprecedented data sets originated by a second generation of satellites, SEASAT, TIROS-N, and NIMBUS-7, were just beginning to be analyzed. When "Oceans from Space II" was held in 1990, no major new missions were operating. However, in the 1990s, a third generation of missions were underway, based on a longer satellite series and larger orbital platform. By the time "Oceans from Space III" was held in 2000, increasing data quality, accessibility, and usability were contributing to the growth of this young research field. "Oceans from Space IV", in 2010, came at a time when remote sensing was already in everyday use as part of the marine scientist's standard toolkit. "Oceans from Space V", delayed by the COVID pandemic until 2022, offered a scientific and technical program reflecting the astounding panorama of missions, instruments, and innovations available today.

Keywords: Oceans from Space; satellite oceanography; surface elevation; surface roughness; marine emissivity; sea surface temperature; sea surface salinity; marine reflectance; ocean color; Venice

1. Introduction

Half a century—that is how much time separates us, today, from the dawning of the early satellite oceanography concepts. Being able to conjugate and even merge together such diverse ideas seemed unreal, back then, and almost contradictory in a way. Space exploration was still in its infancy in the 1960s, a promising, exciting, unexplored avenue of research reserved for a scant group of scientists, sometimes appearing to both experts and the general public alike as a scientific and technological competition, mixed with the allure of a thrilling voyage into the unknown. Additionally, oceanography was perceived by most, including many of its practitioners, as an intriguing and even romantic affair, which should have been carried out primarily, if not exclusively, on the open sea, or at best near the coast—certainly not in the faraway offices of a Space Agency. A bit simplistic, a bit naïve maybe, but not too far from the reality of things: we will race to the moon, bringing spacecrafts, sensors, and brave men into Earth's orbit, developing, in the meantime, the scientific understanding required to conquer that new frontier, which is waiting for us in outer space. Conversely, we, a meagre group of smart adventurers, with tanned skin and hairs ruffled by the wind, will embark, literally, on our next expedition across the Pacific Ocean, hopping from island to island in search of the Holy Grail of sea floor spreading. Two different, distant worlds ....

However, already in the 1970s, when the idea of an "Oceans from Space" conference series was still in the making, that early concept of conjugating space and marine sciences was well on its way to becoming a reality, thanks to the novel ideas of a relatively small group of newborn space oceanographers: we shall demonstrate, they foretold, that the ocean CAN be explored from Earth's orbit! Yes, we shall seek to calibrate our instruments in the lab and on the water. Yes, we shall need to validate our results from buoys and ships. However, in fact, all that will be required to make sense of our data, our "imagery", is a powerful computer, which does not even need to be located close to coastal infrastructure, let alone to the piers of a seaport. Therefore, back in San Diego, the people working on the famed Scripps Pier would feel a bit seasick, looking up at that huge satellite dish being mounted on a nearby hill, where rattlesnakes had been hissing amid the cable reels from past seagoing expeditions just a short while before.

After such a prelude, the "Oceans from Space" conference series, which convenes only once every ten years, always in the city of Venice, Italy, and which will be the main focus of this paper (see Appendix A for details), finally got started in early 1980. The first edition attracted only a few tens of participants, mostly from Europe and North America. However, the atmosphere during the Symposium was one of genuine excitement: the partakers simply knew that they were breaking new scientific ground, while expectations of making revolutionary progress in many sectors of marine science were running high. There had been another gathering on the same topic, the so-called "Oceanography from Space" conference, held in Woods Hole, Massachusetts, in August 1964, to be recalled in the following, which is traditionally seen as the starting point of all things to come in satellite oceanography. A few other landmark events, also to be recalled in the following, soon came after that first one. However, Venice 1980, i.e., "Oceans from Space I", was the maiden meeting for people to discuss actual results of the first suite of space missions specifically designed to probe the oceans, in virtually all ways possible. Indeed, after much planning and years of preparation, things had started to happen, all at the same time, as usual, in the last few months of 1978.

2. Oceans from Space I and II, Venice 1980 and 1990

A bit more than a decade was required to properly set the stage for "Oceans from Space I". Data from visible, infrared and microwave sensors carried on aircraft flights and photographs from early orbital missions, as well as records from video and thermal sensors on the initial meteorological satellites, in the early 1960s, had already provided the first indications that remote sensing had something to offer to the advancement of marine sciences. However, the report of the 1964 "Conference on the Feasibility of Conducting Oceanographic Explorations from Aircraft, Manned Orbital and Lunar Laboratories"workshop, held at Woods Hole Oceanographic Institution, in Woods Hole, Massachusetts [1], was the ideal starting point of all the pioneering programs on ocean observations developed during the following decade. In the wake of this promising beginning, the so-called "Williamstown Conference", held in 1969 at Williams College, in Williamstown, Massachusetts, to discuss a space-based geodesy mission [2], and the 1972 "Conference on Sea Surface Topography from Space" held in Miami, Florida [3], further contributed to focusing the attention of oceanographers onto the advantages of space-based radars to address a number of their data requirements.

A first generation of ocean-viewing satellites [4] carrying a suite of (active) microwave sensors, Skylab in 1973 and Geos-3 in 1975, started to provide data on the structure of the sea surface, i.e., its elevation (with respect to the geoid) and roughness (as a function of winds, waves, wakes, and slicks). At the same time, measurements of emissivity in the far infrared coming from (passive) infrared sensors on meteorological satellites [5], such as the Visible and Infrared Scanning Radiometer (VISR), supplied the first assessments of sea surface temperature (while assessments of salinity, through emissivity in the microwave spectral range, were still far to come). Additionally, aircraft programs, such as those conducted over both American and European waters using the airborne Ocean Color Scanner (OCS) [6], continued to give indications about both potential and difficulties of measuring the reflectance of the surface sea in the visible and near-infrared spectral range, thereby determining its color and deriving its optical properties and constituents.

The availability of several, very promising, preliminary data, generated by a suite of diverse, cutting-edge sensors and a growing scientific understanding of the processes that they were describing (see Table 1 for a synthetic listing of Ocean Observations techniques, spectral regions, and primary and derived parameters, which summarizes the main features of satellite oceanography [7]), led to the conception of orbital remote sensors explicitly designed to look at the sea surface. Three novel spacecrafts were launched, in only a four-month interval of 1978, which collectively carried a suite of sensors covering virtually all the known ways of observing the oceans remotely from space and which would profoundly change the way ocean scientists would study the sea in the years to come. The now-famed trio comprised SEASAT, launched on 26 June; TIROS-N, which followed on 13 October (immediately after the catastrophic failure of SEASAT on 10 October, when a massive short circuit in the electrical system prematurely ended the mission); and NIMBUS-7, reaching its final orbit on 24 October. This second generation of satellites devoted to ocean observations returned a wealth of data, which definitively proved the claim of their promoters about the techniques' potential and value and which paved the way for almost all the subsequent developments in satellite oceanography.

Table 1. Ocean Observations: techniques, spectral regions, primary and derived Parameters.

	Surface Parameters	*Spectral Regions*	λ *
	(Passive Techniques: Spectrometers, Radiometers) (Active Techniques: LIDARs)		
1.	**REFLECTANCE** *1	*Visible & Near Infrared*	(*a*)
	Surface Color, Optical Properties, Water Constituents, Bathymetry, Ice Pigment(s) Concentration, Dissolved Organic Matter, Suspended Matter Phytoplankton Biomass, Productivity, Functional Types, Macro-algae		
2.	**EMISSIVITY** *2	*Thermal Infrared & Microwaves*	(*b*)
	Sea Surface Temperature (Skin temperature, Mixed Layer Temperature) Sea Surface Salinity (Surface Roughness)		
	(Active Techniques: Imaging RADARs, Scatterometers, Altimeters)		
3.	**ROUGHNESS** *3	*Microwaves*	(*c*)
	Wave Height, Wave Spectra, Internal Waves Wakes, Films, Slicks, Ice Surface Winds		
4.	**ELEVATION** *4	*Microwaves*	(*c*)
	Surface Height (with respect to the Geoid), Slope Geostrophic Currents, Eddies, Sea Level Rise Ocean Geoid, Bathymetry		

* *Common observation wavelengths λ are: (a) 380–750 nm and >750 nm; (b) typically 3.5–12.5 μm and 0.1–10 cm; (c) primarily 1.3–76.9 cm, band letter designation K_u to L.*
*1 *Measure is reflected Sunlight or return of LIDAR pulse.*
*2 *Measure is emitted Earth radiation.*
*3 *Measure is fine detail of reflected Imaging RADAR pulse, or averaged power of Scatterometer return.*
*4 *Measure is time, shape, and strength of nadir return, following a pulse emitted by RADAR Altimeter.*

The importance of this combination of events for the future of ocean observations from space cannot be overstated. The SEASAT mission in particular explicitly aimed to validate the feasibility of global ocean monitoring using satellites and to define the requirements of future operational satellite systems for ocean remote sensing [8]. The satellite carried five major instruments: a radar altimeter, to measure the spacecraft height above the ocean surface; a microwave scatterometer, to measure the wind speed and direction; a Synthetic Aperture Radar (SAR), to monitor the global surface wave field, as well as polar sea ice; the Scanning Multifrequency Microwave Radiometer (SMMR), to measure the sea surface temperature; and the Visible and Infrared Scanning Radiometer (VISR), to identify the cloud, land, and water features. TIROS-N, conversely, carried the very first Advanced Very-High-Resolution Radiometer (AVHRR), soon to become the forefather of a long series of ever-improving thermal infrared instruments dedicated to the assessment of sea surface temperature, the workhorse of satellite oceanography on a number of successive operational NOAA satellites [9]. Finally, NIMBUS-7 hosted onboard, among other sensors, a new visible radiometer labeled Coastal Zone Color Scanner (CZCS) [10], which aimed to monitor the surface patterns of water constituents—in particular the coastal plumes already imaged successfully in early photographs, taken with a held-held camera by astronauts from Earth's orbit—and conceivably measure planktonic pigments as well, while yet another SMMR was also included in the payload.

It was the availability of the results from this varied array of missions that prepared the grounds for "Oceans from Space I". At the time of the conference in 1980, the unprecedented datasets originated by the SEASAT, TIROS-N, and NIMBUS-7 suite of sensors were just beginning to be analyzed, amid the lingering problems of data availability and accessibility and cumbersome processing requirements [11]. Although in the early stages of some missions, the members of the respective experimental teams were granted periods for exclusive use of the data, most major space agencies soon adopted open data policies, by which their archives were made freely accessible to researchers around the world. Therefore, the vast majority of "Oceans from Space I" participants, although essentially still coming from North America and Europe, were fully engaged in the analysis of the data originated from these initial, multi-faceted missions, leading the way into completely uncharted scientific territory—even though most had little or no knowledge of the substantial, if isolated, advances being made at the same time in the (passive) microwave remote sensing of the oceans by the *Kosmos* (and later *Okean*) satellite series in the former Soviet Union [12]. The success of this early start of satellite oceanography meant that new missions could be justified only on the basis of their potential, but proven, cost-effective, concrete contribution to ocean sciences. The consequent painstaking assessment and time-consuming preparation of science-driven follow-up missions, satellites, and sensors resulted in a lack of major new developments in the 1980s (with the notable exception of the American GEOSAT, launched in 1985; the two Japanese Marine Observation Satellites, MOS-1A and MOS-1B, in 1987 and 1990; and the continuing *Kosmos/Okean* satellite series, the results of which were reviewed in 1978, 1982, and 1987, respectively, by the first, second, and third "All-Union Oceanographers Congress", held in the USSR to discuss the applications of satellite remote sensing in oceanography [13]).

When the "Oceans from Space II" conference took place in 1990, no major new missions were operating and the scientific community was still busy looking back at those first, original, and very promising, but short and broken time series of satellite data covering most of the world's oceans. In fact, the full processing of the enormous amount of data generated by the lot of second-generation ocean-observing sensors (with respect to the size of the data sets normally handled by oceanographers) took years to be completed. New, longer, and wider time and space dimensions were also added in those years to the concept of satellite oceanography, in order to comply with the increasing awareness that understanding climate change required precisely the kind of data only satellite oceanography could provide. In the following years, this gave rise to an ever-growing third generation of missions, based this time no longer satellite series and larger orbital platforms, the precursors of contemporary (quasi) operational global observing systems, akin to those used for meteorological forecasting.

Thus, the 1990s saw several new developments, arising at an increasing pace and involving more and more space agencies over the world [14]. To name just a few, the main ones only, first came the European ERS-1 in 1991, carrying a Radar Altimeter, Wind Scatterometer, C-band SAR, Microwave Radiometer (MR), and Along Track Scanning Radiometer (ATSR), to be followed by ERS-2 in 1995, carrying essentially the same payload, with the additional Global Ozone Monitoring Experiment (GOME), a nadir-scanning ultraviolet and visible spectrometer, and an improved ATSR-2. Then, new altimeters were deployed on the American–French Topex-Poseidon in 1992, and a C-band SAR on the Canadian RADARSAT-1 in 1995. A group of national space agencies contributed to the manifold suite of sensors on the Japanese satellite series composed of an Earth Resources Satellite (JERS-1), launched in 1992; the first Advanced Earth Observing Satellite (ADEOS-I) in 1996, also carrying a number of sensors capable of observing the oceans using a wide range of techniques; and the Tropical Rainfall Measuring Mission (TRMM) in 1997. Finally, a whole decade after the final demise of the CZCS in 1986, and just a few months after the sudden failure of the Ocean Color and Temperature Scanner (OCTS) and the French Polarization and Directionality of the Earth's Reflectance (POLDER)—both carried on the ADEOS-I platform—through an industry/government partnership came the long-awaited visible/near-infrared radiometer Sea-viewing Wide Field-of-view Sensor (SeaWiFS), developed and launched in 1997 by a private company under contract by the American space agency, which retained responsibility for the data collection, processing, calibration, validation, archive, and distribution [15].

3. Oceans from Space III and IV, Venice 2000 and 2010

By the time "Oceans from Space III" was held in 2000, increasing data quality, accessibility and usability from the third generation of oceanographic space missions were contributing to the growth spurt of this young research field. The need for new, unprecedented commitments, in terms of both programs and funding, had become evident not only to the scientific community, but also in the political circles where goals and priorities of public investments were set. Climate change and its startling consequences, in both the environmental as well as socio-economic realms, were upon us at that time, and action seemed urgent. Thus, at the turn of the century, and of the millennium, a new generation of multi-sensor platforms—the TERRA and AQUA satellites [16], part of the American Earth Observing System (EOS); the European ENVISAT [17]; and the Japanese short-lived ADEOS-II [18]—were placed into Earth's orbit between 1999 and 2002, essentially to prolong the existing, if still preliminary, time series of climate-related data. The payload of TERRA/AQUA and ENVISAT, in particular, included sensors destined to give continuity to global ocean color assessments, the Moderate Resolution Imaging Spectroradiometer (MODIS) and the Medium Resolution Imaging Spectrometer (MERIS), together with new tracking systems, altimeters, advanced SAR, and infrared radiometers.

Within the first decade of the new century, (quasi) operational tools started to be provided to meteorological and environmental services by specialized missions, such as those of the SeaWinds scatterometer aboard QuikSCAT, launched in 1999 to quickly replace the loss of NSCAT on ADEOS-I; or the American WINDSAT on the Coriolis satellite, launched in 2002; or again the European Advanced SCATterometer (ASCAT) on the Metop-A satellite, launched in 2006 [19]. Similarly, the joint American–German two-satellite Gravity Recovery and Climate Experiment (GRACE) mission started in 2002 (and continued with GRACE-FO, Follow On, in 2018), while the European Gravity Field and Steady-State Ocean Circulation Explorer (GOCE) was launched in 2009, in order to map, in unprecedented detail, the Earth's gravity field [20]. The German twin Earth Observation satellites TerraSAR-X and TanDEM-X were deployed, respectively, in 2007 and 2010, to carry out SAR global monitoring and build up a worldwide and homogeneous Digital Elevation Model. Monitoring of the variations in the extent and thickness of polar ice was pursued by the American ICESat, launched in 2003, for a long-term altimetry mission (and followed by ICESat-2 in 2018); and by the European CryoSat, which was lost in a launch failure in 2005 but then replaced by CryoSat-2 in 2010 [21]. Furthermore, American–European cooperative altimetry missions were continued by the Jason satellite series, i.e., Jason-1 in 2001, Ocean Surface Topography Mission OSTM/Jason-2 in 2008, and Jason-3 in 2016 [22]. The French POLDER visible radiometer series also continued in 2002, with POLDER-2, and then again in 2004 with POLDER-3 [23], while a second Canadian RADARSAT-2, carrying another C-band SAR, was launched in 2007.

Still around the turn of the century, new space agencies from emerging countries also started to pursue independent ocean observation programs, in order to support both national scientific communities and growing blue economy aspirations. India launched its own OceanSat-1 in 1999 and OceanSat-2 in 2009, while Oceansat-3, carrying a suite of sensors composed of an Ocean Color Monitor (OCM-3), Sea Surface Temperature Monitor (SSTM), and K_u-band scatterometer (SCAT-3), was finally launched in 2022 [24]. Even more significantly, China began to deploy three satellite series named *Haiyang* (HY)—"ocean" in Chinese—carrying a suite of visible, infrared, and microwave sensors, both active and passive [25]. The HY-1 series, designed, primarily, to measure the ocean color and sea surface temperature, was started in 2002 by HY-1A, followed by 1B in 2007, and later by 1C in 2018 and 1D in 2020 (satellites 1E and 1F are supposed to continue this series). The HY-2 series, devoted to marine environmental dynamics, with all-weather and round-the-clock observations of wave height, sea surface height, wind, and temperature, was started in 2011 by HY-2A, followed by 2B in 2018, and later continued by 2C and 2D in 2021 (satellites 2E, 2F, 2G, and 2H are supposed to continue this series). The HY-3 series, devoted to ocean surveillance, composed of satellites 3A, 3B, 3C, and 3D, was planned for 2022, but has yet to get underway.

"Oceans from Space IV", held in 2010, came at a time when most, if not all, of the above was already in everyday use as part of the marine scientist's standard toolkit, while a score of novel ideas and technical developments were coming of age. New mission concepts were emerging, such as that of substituting the classical global, periodical monitoring from polar orbit with continuous observations of a single region from geostationary orbit, as was performed for northeast Asia by the South Korean Geostationary Ocean Color Imager (GOCI), launched in 2010 and followed by GOCI-II in 2020 [26]. Sensor technology advances also allowed the introduction of new items to the classical list of oceanographic parameters measured by orbital remote sensors. In late 2009, the Soil Moisture and Ocean Salinity (SMOS) European mission brought into orbit the Microwave Imaging Radiometer with Aperture Synthesis (MIRAS), which was capable of measuring changes in the salinity of seawater by observing variations in the microwave emission coming up from its surface. This was soon to be followed by the joint American and Argentinian mission Aquarius/*Satélite de Aplicaciones Científicas* (SAC)-D in 2011, and by the American follow-up Soil Moisture Active Passive (SMAP) mission in 2015, also operating passive microwave radiometers capable of assessing sea surface salinity [27].

In the decade following "Oceans from Space IV", a new multi-sensor satellite series, providing continuity to particular observation fields, started to become operational. A long string of American missions began in 2011 with the Suomi National Polar-orbiting Partnership (NPP), and continued in 2017 with the Joint Polar satellite System (JPSS-1)/NOAA-20, and in 2022 with the JPSS-2/NOAA-21, all carrying onboard, among other instruments for atmospheric sounding, the Visible Infrared Imaging Radiometer Suite (VIIRS) [28]. The Global Change Observation Mission—Climate 1 (GCOM-C1), nicknamed *Shikisai*—"color" in Japanese—was launched in 2017 as part of a Japanese project for the long-term observation of Earth's environmental changes [29], carrying the visible/infrared radiometer Second-generation Global Imager (SGLI) and the Advanced Microwave Scanning Radiometer 2 (AMSR2). The European Sentinel satellite series also got underway in the same decade, as part of the Copernicus program [30], with the launches of Sentinel-1 (A and B) in 2014 and 2016, in order to provide all-weather, day and night SAR imaging; of Sentinel-3 (A and B) in 2016 and 2018, both carrying the Ocean and Land Colour Instrument (OLCI), the Sea and Land Surface Temperature Radiometer (SLSTR), and the Synthetic Aperture Radar Altimeter (SRAL); and then of Sentinel-6 *Michael Freilich* (S6MF) in 2020, to provide continuity in high-precision altimetric sea level measurements, mapping previously unresolvable features over both open and coastal waters.

Furthermore, during the same years, a score of international specialty missions has been widening the spectrum of possibilities for constant, advanced oceanographic monitoring. In this category, the Indian–French altimetry mission SARAL/ALtiKa was launched in 2013 [31]. The Chinese–French Oceanography Satellite (CFOSAT) was launched in 2018, to conduct wind wave measurements with the Surface Waves Investigation and Monitoring (SWIM) instrument [32]. And additionally, the American–French Surface Water and Ocean

Topography (SWOT) latest altimetry mission, with Canadian and British contributions, was launched in late 2022 [33].

4. Oceans from Space V, Venice 2022

"Oceans from Space V"—which took place in Venice like its preceding editions, but, due to the COVID pandemic, had to be postponed twice from its original date in October 2020 to October 2022, and also somewhat downsized—offered a scientific and technical program reflecting the astounding panorama of missions, instruments, and innovations available today [34]. The conference formula was designed to provide an overview of the path followed so far by satellite oceanography and the tools currently available, as well as plans for tomorrow. Table 2 provides a summary of the main scientific and technical themes chosen for this purpose. More information on the conference and its contents can be obtained as indicated in the Supplementary Materials endnote.

Table 2. "Oceans from Space V" scientific and technical themes.

No.	Themes	Sessions
1	**Missions, Satellites, and Sensors**	4
	Technical Issues, Instruments, and Data Synergies	
	Novel Analyses, Sea Surface Salinity, and Acidification	
	Sensor Arrays, Multi-Mission/Sensor, and Cube/Small-Sats	
2	**Models, Assimilation, and Cal/Val**	2
	Modelling and Data Assimilation (in Global Models)	
	Assimilation of Model Data for New Space Products	
	Uncertainties, Cal/Val, and In Situ Technologies	
3	**Regional and Planetary Issues**	1
	Ocean Basins, Arctic Ocean, and Antarctica	
	Continental Waters, Marginal Basins, and Sea Level Rise	
	Relating Earth Oceans and Oceans on Exoplanets	
4	**Surface Processes, Coastal Issues, and Extreme Events**	3
	Ocean–Atmosphere Exchange and Coupled Modelling	
	Wind, Waves, and Total Surface Current Velocity	
	Global Coastal Ocean, Near-Coastal Waters, and Hazards	
5	**Bio–Geo–Chemical Issues and Pollution Processes**	1
	Aquatic Carbon from Space	
	Biodiversity and Floating Vegetation in the Global Ocean	
	Pollution (Plastics, Hydrocarbons, and Runoff) and Debris	
6	**Society, Policy, and Economics**	1
	Science and Society, and Management and Policy	
	Coastal and Marine Ecosystem Services, and Economics	
	Emerging Countries and Early Career Researchers	
(a)	**"Venice Syndrome" (Coastal Areas at Risk)**	2
(b)	**COVID-19 Pandemic: Oceans' Responses**	1/3 *
(c)	**Coupling Space Science and Citizen Science**	1/3 *
(d)	**UN Decade of Ocean Science (2021–2030)**	1/3 *

* Themes b, c, and d were all covered during the Symposium's Closing Session.

The scope and breadth of the Symposium Program offer a better impression of the current field of satellite oceanography and its likely future developments than any listing of missions, satellites, and sensors currently available or planned. It highlights the added value of many complementary systems, techniques, and their combined views, addressing the multidisciplinary character of the scientific direction that future missions will take. This broad spectrum of topics includes new sensing techniques and results for water color, surface wind, all kinds of surface and internal waves, ice, floating vegetation, temperature, salinity, and pollution, in both open and coastal waters, with multi-mission combinations, and includes policy applications as well as citizen science. Additionally, it also comprises a special session on global sea level rise and the future of Venice, in the hope that future conferences, in 2030 and later, will continue to be possible at the same location. A sample of the papers presented, i.e., the winners of the Best and Distinguished Contributions Awards, are listed in Appendix B, together with the winners of the other *ad personam* Awards issued in 2022 and throughout the conference series.

With Theme 1 (Missions, Satellites, and Sensors), the program focused on ocean-oriented missions, satellites, and sensors. Several specialized sessions were devoted to the latest technical developments, instruments, data analyses, and synergies. Novel issues and principles, such as polarimetry and lidar applications, sea surface salinity, or acidification assessments were also foreseen. Lastly, this first Theme included the use of satellite constellations, multi-mission and multi-sensor approaches, and cube-sats and small-sats arrays (devoted, e.g., to global climate research).

Numerical models, data assimilation, and calibration/validation issues constituted Theme 2 (Models, Assimilation, and Cal/Val) of the program. This comprised modelling and assimilation at large, and, in particular, the assimilation of ocean data into global models; the assimilation of numerical model data in order to derive new products from space; uncertainties and validation; in situ technologies; and new calibration approaches.

Theme 3 (Regional and Planetary Issues) encompassed regional and planetary issues: ocean basins, the Arctic Ocean, and Antarctica. Continental waters, marginal/enclosed seas, and local sea level rise, as well as interfacing research on Earth's oceans and the oceans on exoplanets, were also considered.

The processes taking place at the surface of the ocean or at its coastal boundaries, as well as extreme events recurring as a consequence of climate change, were grouped into Theme 4 (Surface Processes, Coastal Issues, and Extreme Events). Ocean–atmosphere exchange and coupled modelling fell into this category, together with ocean surface observations for assessing wind and waves, or for monitoring the total surface current velocity (i.e., Doppler oceanography). Furthermore, the global coastal ocean, spanning across the land-to-open-ocean boundary, plus coastal waters at large (wetlands, deltas, estuaries, and areas with saline intrusions), coastal hazards, and extremes (e.g., inshore vs. offshore sea level rise, surges, erosion, and sediment transport) were also included.

Theme 5 (Bio–Geo–Chemical Issues and Pollution Processes) dealt with bio–geo–chemical issues, pollution agents, and processes. The aquatic-carbon-from-space item aimed to link biology to the physics of the air–sea interface, while that on biodiversity called for a vision for assessing the diversity of life beyond the simple and popular measures of biomass/carbon. Floating vegetation in the global ocean meant observations of the changing distribution of Sargassum, its causes and future projections, and observations of other microvegetation species in other oceans. Marine pollution (including coastal/river runoff, hydrocarbons, and plastics) and marine debris monitoring from high-resolution microwave and optical sensors completed this Theme.

Finally, Theme 6 (Society, Policy, and Economics) brought about a connection to strategies and management, linking what can be observed and the next generation of science questions in need of being addressed (e.g., water quality, harmful algal blooms, carbon management, and water resources) to how these could provide support for decision makers. Science and society, valuations of coastal and marine ecosystems (i.e., ecosystem services), and economic issues were considered. Furthermore, Theme 6 focused on involving early career researchers, new contributors such as satellite sensor engineers/technologists, and scientists from emerging countries.

An additional set of particular, targeted topics concluded this series of Themes: (a) global sea level changes and the future of Venice, which parallel those of the many other urban areas in the Mediterranean Sea and indeed around the world, that are at risk from analogous hazards; (b) the oceans' responses to the COVID-19 pandemic, as seen from space; (c) the emerging potential of coupling space science and citizen science for water quality monitoring in particular; and (d) the United Nations (UN) Decade of Ocean Science for Sustainable Development (2021–2030), to which "Oceans from Space V" offered a key opportunity for articulating the space applications' contribution to "ocean science that is fit for purpose" [35].

5. Conclusions

The task of providing a comprehensive review of satellite oceanography is truly prohibitive, as it would be very hard for anybody to be fully aware of all the pyrotechnic developments that have led us to the current state of affairs. Nevertheless, the present synthetic—probably rather incomplete—historical outline of its progress, as witnessed by the partakers in the "Oceans from Space" conference series, attempts to give a general impression of this multi-faceted subject.

Satellite observations have become a cornerstone of all planetary sciences. No other technology allows for information gathering at the proper space and time scales like orbital remote sensing does, while the list of its current or potential applications is virtually endless. Sustainable environmental management, in particular, relies on the provision of information services via ad hoc communication, and of knowledge exchange between scientific community and user community at large. A key role in this interaction is played by specialized conventions, which can focus the awareness of both know-how providers and their customers. In the marine sector, the "Oceans from Space" conference series is one of the most celebrated and widely attended events, offering, with a decadal timeframe, a thorough outline of the state-of-the-art in satellite oceanography, an overview of the current research on the global scene, and a forum for debating topical issues.

"Oceans from Space V" was the latest in this series of discussions on the past, present, and future successes, problems, and requirements of using satellites for studying the oceans. It has been shown that research in Ocean—and indeed Earth System—Science has become increasingly multidisciplinary, as well as focused on societal benefits. It demands the collaboration of experts not only in traditional subjects, but also in social sciences, informational technology, and policy. The "Oceans from Space" conference series provides an ideal opportunity to review the use of spaceborne measurements, promote interdisciplinary education, stimulate new collaborations, and involve in this process participants with varied backgrounds, including the next generation of early career scientists.

Supplementary Materials: Additional supporting information can be downloaded from the website https://www.oceansfromspacevenice2020.org/ (accessed on 2 August 2023), where past, present, as well as future developments about the "Oceans from Space" conference series are summarized, together with the "Oceans from Space V" facts and figures, recent news, and a final version of the Symposium Proceedings.

Funding: This research received no external funding.

Data Availability Statement: No new data were created or analyzed in this study. Data sharing is not applicable to this article.

Acknowledgments: Although signed by a single Author, the present paper is based on private communications provided by a number of colleagues—too numerous to be listed here, let alone added as co-authors—who contributed to, previewed and then approved the resulting historical reconstruction. This endeavor would not have been possible without their contribution. The Author is also in debt with the "Oceans from Space V" attendees, who delivered their presentations at the Symposium held in Venice, Italy, on 24–27 October 2022, and who participated in its various discussion sessions. Further, special thanks are due to J.F.R. Gower, initiator of the "Oceans from Space" conference series, and to L. Alberotanza, local host of all conferences held so far in Venice, as well as to the members of the "Oceans from Space V" Scientific Committee—composed by J. Benveniste, ESA, Frascati (I); P. Bontempi, University of Rhode Island, Narragansett RI (USA); E. Boss, School of Marine Sciences, University of Maine, Orono ME (USA); V. Brando, ISMAR CNR, Roma (I); C. Donlon, ESA, Noordwijk (NL); M. Gade, Institut für Meereskunde, Universität Hamburg, Hamburg (D); L. Lorenzoni, NASA, Washington DC (USA); F. Melin, JRC EC, Ispra (I); F. Muller-Karger, University of South Florida, St Petersburg FL (USA); E. Obligis, EUMETSAT, Darmstadt (D)—who provided advice for the compilation of this Preface, and for the selection of the representative collection of papers constituting this "Oceans from Space V" Special Issue. The invaluable contribution of the "Oceans from Space V" Secretariat staff, in particular G. Canali and his colleagues at the NSA Group, Roma (I), and of the funding individuals and organizations listed here below, is also gratefully acknowledged. Support to the "Oceans from Space" conference series has been provided, over the years, first and foremost by its international participants, through their Registration Fees. A number of Space Agencies and Scientific Institutions provided additional financial assistance to each conference (see Appendix A for details). In particular, Oceans from Space V—which prompted the preparation of this paper, and the present Special Issue, made possible by funding from MDDPI Remote Sensing—has been sponsored by the Joint Research Centre of the European Commission (JRC EC); the European Space Agency (ESA); the European Organization for the Exploitation of Meteorological Satellites (EUMETSAT); the US National Aeronautics and Space Administration (NASA), through a grant to the Scientific Committee on Oceanic Research (SCOR), managed by the International Ocean Colour Coordinating Group (IOCCG); and by the *Isituto di Scienze Marine*, of the Italian *Consiglio Nazionale delle Ricerche* (ISMAR CNR). Furthermore, the Tethys Research Institute (TRI) and the aforementioned NSA Group provided additional in-kind support.

Conflicts of Interest: The author declares no conflict of interest.

Appendix A

The first event in the "Oceans from Space" conference series, was held in Venice, Italy, on 26–30 May 1980, as a "COSPAR/SCOR/IUCRM Symposium". These acronyms refer to international entities, which at the time were, and in some cases still are, rather active on the international scene. The Committee on Space Research (COSPAR) was created in 1958 by the International Council of Scientific Unions (ICSU), predecessor of the International Science Council (ISC)—a non-governmental organization that unites scientific bodies across the social and natural sciences [36]—to promote scientific research in space. The Scientific Committee on Oceanic Research (SCOR) is an interdisciplinary body of the ISC. It was established in 1957 by the ICSU/ISC, coincident with the International Geophysical Year of 1957–1958, to help address interdisciplinary science questions related to the ocean. And finally, the Inter-Union Commission on Radio Meteorology (IUCRM) was created in 1959, as a constituent association of the International Union of Geodesy and Geophysics (IUGG), which was one of the 20 scientific Unions grouped within the ICSU/ISC. It ceased activities in 1982.

In the late 1970′s, COSPAR had suggested holding a joint event with SCOR, as a major review of space oceanography. Since this meeting fitted well with a series of colloquia organized by IUCRM, these three bodies joined forces to sponsor the very first "Oceans from Space". The Symposium was hosted by the *Università Cà Foscari* of Venice, Italy, in the halls of its *Aula Magna* at *Cà Dolfin*. The location proved to be so much appreciated by participants, that all subsequent events of the conference series have always been held in the city of Venice, although at different venues. Additional support to the first conference edition was also provided by the Venice-based *Istituto per lo Studio delle Grandi Masse* (ISDGM), of the Italian *Consiglio Nazionale delle Ricerche* (CNR). With the help of the local organizers, an effort was made to complement the scientific and technical sessions with a rich social program, a characteristic that later became a distinguishing trait of the whole conference series. "Oceans from Space I" was chaired by J. Gower, then at the Institute of Ocean Sciences (IOS), Sidney, Canada. The Symposium Proceedings, which he edited for Springer [11], were published in 1981, and soon became a celebrated *manifesto* of the whole satellite oceanography community.

The first event ended with the commitment to reconvene once again the same community, in the same place, possibly in a sensible time frame, for a progress review. But eventually "Oceans from Space II" did not take place until 1990. The decadal periodicity has since become another distinguishing trait of the conference series. Both IOS and ISDGM/CNR, together with some of the Unions and Associations in the original group of sponsors, continued to support in various ways the conference series. A new entry in this group was the ICSU/ISC-affiliated International Union of Geodesy and Geophysics (IUGG), participating through the International Association for the Physical Sciences of the Oceans (IAPSO)—which was organized in its present form in 1967, to promote the study of the oceans and its interactions at the sea floor, coastal, and atmospheric boundaries using mathematics, physics, and chemistry. The *Scuola Grande di San Giovanni Evangelista*—a monumental complex with noteworthy examples of Gothic, Renaissance and Venetian Baroque art and architecture, which has been the seat of the lay brotherhood by the same name since 1261—was chosen as the new venue for the second Symposium, held on 21–25 May 1990. Once again, J. Gower acted as Chairman and convenor of the conference. Unfortunately, although participation was even larger than that of the first edition, no proceedings of this event were ever published.

"Oceans from Space III" was held on 9–13 October 2000, in the halls of the *Scuola Grande di San Teodoro*—the seat of the traders' lay brotherhood, devoted to the Patron Saint of the Merchant Guild of Venice since the mid 1600s and originally the heavenly protector of Venice itself—in the very heart of Venice, near the *Ponte di Rialto*. About 250 scientists, representing 25 countries from around the world, attended the various sessions and took part in the many events of the Symposium. New sponsors joined IOS and ISDGM/CNR in backing the event. The Joint Research Centre (JRC) of the European Commission (EC) became deeply involved in organizing and financing the Symposium, and later maintained the same role in all its subsequent editions. The Italian *Istituto Centrale per la Ricerca scientifica e tecnologica Applicata al Mare* (ICRAM) provided additional support. Furthermore, the European Association of Remote Sensing Laboratories (EARSeL) and, most prominently, the European Space Agency (ESA) also made their appearance in the sponsors' roster. These were later joined by the US Office of Naval Research (ONR) and National Aeronautics and Space Administration (NASA). The chairmanship of the event was enlarged to include V. Barale from JRC/EC and L. Alberotanza, then the Director of ISDGM/CNR, together with J. Gower (see Figure A1). The conference proceedings, collecting two-page Extended Abstracts from all the contributions, were published in 2000, in conjunction with the event, as a JRC Technical Report [37]. A subset of these Extended Abstracts, reformulated as more detailed Research Letters, were also published in 2004 as a Special Issue of the International Journal of Remote Sensing [38].

Ten years later, "Oceans from Space IV", held on 26–30 April 2010, moved back to the *Scuola Grande di San Giovanni Evangelista* with a new conference formula, which rendered it one of the memorable events in the mind of many participants. Oral sessions were limited to 20 keynote addresses delivered by invited speakers (whose lectures were collected and published prior to the conference in a volume edited by Springer [39]), while all the submitted contributions were presented in a sequence of multi-disciplinary poster sessions (the Extended Abstracts of which were, once again, published as a JRC Scientific and Technical Report [40], in conjunction with the event). This formula eliminated the need for parallel sessions, allowing all participants to have a complete overview of all presentations. Early-morning Special Topic Seminars and late-afternoon Panels and Plenary Discussions, focused on topics of particular interest, completed the program. The technical sessions were complemented by daily art exhibits by *Maestro* E. Balliano, a celebrated representative of the contemporary Futurist movement, whose painting "*Le Pagine nell'Acqua*", created purposely for the Symposium, was reproduced in a limited, signed, and numbered series of prints offered to all attendees. Participation in the event was somewhat lower than expected due to a persistent economic crisis, the unavoidable competition of other conferences, and, perhaps primarily, to the untimely eruption of the Icelandic volcano *Eyjafjallajökull*, which threw European air traffic into complete chaos during the weeks preceding the conference and caused a number of last-minute cancellations. In spite of the attendance losses, the Symposium still attracted 170 participants from several EU countries; Ukraine; Russia; the US and Canada; countries from central and South America; African countries; India, China, Japan, and Australia. Support was offered by IOS, ISDGM/CNR, JRC/EC, and the Tidal Forecasting and Early Warning Centre (TFEWC) of the Venetian municipality. ESA also renewed its sponsorship, together with the US ONR, NASA, and National Oceanic and Atmospheric Administration (NOAA). Once again, chairmanship of the event was provided by V. Barale, L. Alberotanza and J. Gower.

Figure A1. Co-chairs of the "Oceans from Space III, IV and V" Symposiums: Luigi Alberotanza (left), Vittorio Barale (center), and Jim Gower (right). In the background, a partial view of the façade of the *Scuola Grande di San Marco*, where the fifth edition of the Symposium was held in 2022.

After being carefully planned and scheduled to take place at the same decadal interval, "Oceans from Space V" had to be postponed twice, first from its original date in 2020 to 2021, and then again to 2022, due to the continuing COVID-19 pandemic. When it was finally held on 24–28 October 2022, in the halls of the *Scuola Grande di San Marco* (the seat of the Fraternity of Saint Mark since 1260, rebuilt in 1485 after a fire, with elements of Renaissance classicism and a masterful front of Byzantine flavor), both its technical and social programs had to be somewhat downsized, in order to accommodate the reduced number of participants and health protection measures that needed to be implemented. Furthermore, the lingering effects of the COVID-19 plague and the war in eastern Europe, generated by the 2022 Russian invasion of Ukraine, again caused a number of last-minute cancellations, before as well as during the conference itself. In spite of this reduced attendance, the Symposium still attracted about 100 participants, primarily from Europe and North America, with smaller representations from South America, Africa, and Asia. The main sponsors of the event were JRC/EC and the newly established *Istituto di Scienze Marine* (ISMAR) of the Italian CNR, together with the ESA and European Organization for the Exploitation of Meteorological Satellites (EUMETSAT). NASA also renewed its sponsorship through a grant to the SCOR, managed by the International Ocean Colour Coordinating Group (IOCCG). The Italian Tethys Research Institute (TRI), MDPI Remote Sensing, and NSA Group, which manned the Symposium Secretariat, provided additional in-kind support. The chairmanship of the event was still provided by V. Barale, L. Alberotanza, and J. Gower. The NSA Group published the conference Proceedings [34] as a draft version prior to the conference and in their final form in 2023, again as a collection of Extended Abstracts. A number of representative contributions were selected during the conference itself for publication as full-length papers in the present "Oceans from Space V" Special Issue of the MDPI Remote Sensing journal.

Appendix B

In the last three editions of the Symposium, the "Oceans from Space" Scientific Committee assigned *Fero da Pròra* Awards to recognize the sometimes obscure, but fundamental, work of those colleagues who, over the years, have played a major role in the satellite oceanography community (Table A1). The *Fero da Pròra*, or "prow iron" in Venetian dialect, seen on every *gondola* (see Figure A2), is a counter-weight balancing the *gondoliere* on the stern, thus allowing the vessel to sail smoothly with any number of passengers.

Table A1. "Oceans from Space V" *Fero da Pròra* Awards.

Year	*Fero da Pròra* Award Recipients	Spec.
2000	**Jim Gower** (Ret.) Institute for Ocean Sciences, Sidney (CAN)	
2010	**Tom Allan** (Ret.) Satellite Observing Systems, Godalming (UK)	In Memoriam
2010	**Roberto Frassetto** (Ret.) *Consiglio Nazionale delle Ricerche*, Venice (I)	
2010	**Dennis K. Clark** NOAA NESDIS, Silver Spring, MD (USA)	
2010	**Ezio Balliano** Vercelli (I)	For Artistic Contribution
2022	**Giuseppe Zibordi** (Ret.) Joint Research Centre, EC, Ispra (I)	
2022	**Christina K. Cross & Victoria K. Cross** Florida Atlantic University, Fort Pierce, FL (USA)	Youngest Participant Ever *

* High-school students C.K. Cross and V.K. Cross were co-authors of an award-winning poster.

Symposium participants of the the last three editions of the "Oceans from Space" received, in their conference kit, a ballot to be used for the evaluation of every contribution, oral or poster, on a scale from 1 (poor) to 5 (excellent), on the basis of either the presentation itself or of the Extended Abstract available in the Proceedings. After compilation of all the entries, on the last day of the Symposium, a Best Contribution Award and 10 Distinguished Contribution Awards were issued during the Closing Session. The results of this exercise for "Oceans from Space V" are shown in Table A2.

Table A2. "Oceans from Space V" Best Contribution (BC) and Distinguished Contributions (DC).

	Author(s) & Title	
1	**M.M. Martinez** *, L.A Ruiz-Etcheverry, B. Picard, M. Saraceno, and C. Guinet "Analysis of hydrographic data collected by Southern Elephant Seals in the Argentine Continental Shelf"	BC
2	**C. Gommenginger** *, A. Martin, A. Egido, K. Hall, P. Martin-Iglesias, and T. Casal et al. "Imaging small-scale ocean dynamics at interfaces of the Earth System with the SEASTAR Earth Explorer 11 mission candidate"	DC
3	**P. Bontempi** * "The value of regional and planetary ocean information for Earth's blue economy"	DC
4	**C. Donlon** * "Oceans from Space: Achievements of the European Space Agency over the last decade and plans for the next decade"	DC
5	**D.T. Sandwell** *, J.A. Goff, J. Gevorgian, H. Harper, S.S. Kim, Y. Yu, and B. Tozer et al. "Global Seafloor Mapping from Ships, Satellites, and Geological Information for High Resolution Ocean Models: SYNBATH"	DC
6	**L. Lorenzoni** * "Looking at the past to pave the future: *NASA*'s OBB perspective on priorities for aquatic ecosystems and bio-geo-chemistry"	DC
7	**A. Moiseev** *, F. Collard, J.A. Johannessen, and B. Chapron "Total surface current radial velocity from Sentinel-1 SAR Doppler shift observations"	DC
8	V.K. Cross, **C.K. Cross** *, M. McCoy, and T. Moore "Validation and Comparison of the HawkEye CubeSat Sensor to Current Ocean Color Satellites over AERONET-OC Sites"	DC
9	**G. Umgiesser** * "Venice and Acqua Alta in the perspective of Climate Change"	DC
10	**K. Ruddick***, M. Beck, A. Bialek, V. Brando, A. Cattrijsse, J. Concha, A. Corizzi et al. "WATERHYPERNET: Automated in situ measurements of hyperspectral water reflectance for satellite validation. . .and more"	DC
11	**D. Ciani** *, M-H. Rio, B. Buongiorno Nardelli, S. Guinehut, E. Charles, H. Etienne et al. "Ocean circulation from the synergy of altimeter and oceanic tracers observations"	DC

* Presenting and/or corresponding Author, who received the Contribution Award.

In line with its longstanding tradition of coupling science and art, "Oceans from Space" has promoted high-quality publishing (i.e., the photographic volume on "*Venice, places and history*" by D. Reato, re-published by White Star as a special edition for "Oceans from Space III") and visual arts expressions (i.e., the painting "*Le Pagine nell'Acqua*" by E. Balliano, commissioned for "Oceans from Space IV"). On the occasion of "Oceans from Space V", a poetry contest was set up among all participants. A jury composed of the conference Co-Chairs, and of other volunteers willing to evaluate the submitted entries, chose the *Haiku* shown in Table A3 as the best poem of the Symposium.

Table A3. "Oceans from Space V" Poetry Contest: winning composition.

"Oceans from Space V" Poetry Contest
A Whirling Haiku *by Phoebe Hudson* ***Turbulent mixing*** ***of wind, waves and ideas.*** ***Swirling discussion.***

Figure A2. "Oceans from Space V" Closing Session slide introducing the *Fero da Pròra* Awards.

References

1. Ewing, G.C. Oceanography from Space. In Proceedings of the Conference on the Feasibility of Conducting Oceanographic Explorations from Aircraft, Manned Orbital and Lunar Laboratories, Woods Hole, MS, USA, 24–28 August 1964; WHOI Ref. No. 65-10; Woods Hole Oceanographic Institution: Woods Hole, MS, USA, 1965.
2. Kaula, W.M. The Terrestrial Environment: Solid Earth and Ocean Physics. In *Report of the Williams College Conference on Solid Earth and Ocean Physics, Williamstown, MS, USA, 11–21 August 1969*; NASA Contractor Report CR-1579; National Aeronautics and Space Administration: Langley, VA, USA, 1970.
3. Apel, J.R. (Ed.) Sea surface topography from space. In *Vols. 1 & 2. NOAA Technical Reports: ERL No. 228 and AOML No. 7*; National Oceanic and Atmospheric Administration: Boulder CO, USA, 1972.
4. Apel, J.R. Three Decades of Satellite Oceanography: The View from on High. In *Space Remote Sensing of Subtropical Oceans, Proceedings of the COSPAR Colloquium on Space Remote Sensing of Subtropical Oceans (SRSSO), Taiwan, 12–17 September 1995*; Liu, C.T., Ed.; COSPAR Colloquia Series; Pergamon: Oxford, UK, 1997; Volume 8, pp. 11–19. [CrossRef]
5. Minnett, P.J.; Alvera-Azcárate, A.; Chin, T.M.; Corlett, G.K.; Gentemann, C.L.; Karagali, I.; Li, X.; Marsouin, A.; Maturi, E.; Santoleri, R.; et al. Half a century of satellite remote sensing of sea-surface temperature. *Remote Sens. Environ.* **2019**, *233*, 111366.
6. Sørensen, B.M. *The North Sea Ocean Color Scanner Experiment 1977, Final Report*; Commission of the European Communities, Joint Research Centre & Directorate General XII: Ispra, Italy; Brussels, Belgium, 1979; p. 126.
7. Robinson, I.S. *Measuring the Oceans from Space. The Principles and Methods of Satellite Oceanography*; Springer-Praxis Books in Geophysical Sciences; Springer: Berlin/Heidelberg, Germany; New York, NY, USA, 2004.
8. Born, G.H.; Dunne, J.A.; Lame, D.B. Seasat Mission Overview. *Science* **1979**, *204*, 1405–1406. [PubMed]
9. Schwalb, A. *The TIROS-N/NOAA A-G Satellite Series. NOAA Technical Memorandum NESS 95*; National Oceanic and Atmospheric Administratio, National Environmental Satellite Service: Washington, DC, USA, 1979; p. 75.
10. Hovis, W.A.; Clark, D.K.; Anderson, F.; Austin, R.W.; Wilson, W.H.; Baker, E.T.; Ball, D.; Gordon, H.R.; Mueller, J.L.; El-Sayed, S.Z.; et al. Nimbus-7 Coastal Zone Color Scanner: System Description and Initial Imagery. *Science* **1980**, *210*, 60–63. [PubMed]
11. Gower, J.F.R. (Ed.) *Oceanography from Space*; Marine Science; Springer: Boston, MA, USA, 1981; Volume 13. [CrossRef]
12. Nelepo, B.A.; Terekhin, Y.V.; Kosnirev, V.K.; Khmirov, B.E. *Sputnikovaya Hydrophysica (Satellite Hydrophysics)*; Nauka Publishing House: Moscow, Russia, 1983.
13. Mitnik, L.M.; Victorov, S.V. (Eds.) *Radiolokatsiya Poverkhnosti Zemli iz Kosmosa (Radar Sensing of the Earth's Surface from Space)*; Hydrometeoizdat Publishing House: Leningrad, Russia, 1990.
14. Wilson, W.S.; Fellous, J.L.; Kawamura, H.; Mitnik, L.M. A History of Oceanography from Space. In *Remote Sensing of the Marine Environment*; Ryerson, R.A., Ed.; Manual of Remote Sensing; American Society for Photogrammetry and Remote Sensing: Bethesda, MD, USA, 2006; Volume 6, pp. 1–31.

15. McClain, C.R.; Feldman, G.C.; Hooker, S.B. An overview of the SeaWiFS project and strategies for producing a climate research quality global ocean bio-optical time series. *Deep. Sea Res.* **2004**, *5*, 5–42.
16. NASA Earth Observing System. Available online: https://eospso.nasa.gov (accessed on 26 May 2023).
17. ESA Earth Online. Available online: https://earth.esa.int/eogateway/missions/envisat (accessed on 26 May 2023).
18. JAXA Advanced Earth Observing Satellite II "Midori II" (ADEOS-II). Available online: https://global.jaxa.jp/projects/sat/adeos2/index.html (accessed on 26 May 2023).
19. Guo, Q.; Xu, X.; Zhang, K.; Li, Z.; Huang, W.; Mansaray, L.R.; Liu, W.; Wang, X.; Gao, J.; Huang, J. Assessing Global Ocean Wind Energy Resources Using Multiple Satellite Data. *Remote Sens.* **2018**, *10*, 100. [CrossRef]
20. Visser, P.N.A.M. Gravity field determination with GOCE and GRACE. *Adv. Space Res.* **1999**, *23*, 771–776. [CrossRef]
21. Kwok, R.; Kacimi, S.; Webster, M.A.; Kurtz, N.T.; Petty, A.A. Arctic snow depth and sea ice thickness from ICESat-2 and CryoSat-2 freeboards: A first examination. *J. Geophys. Res. Ocean.* **2020**, *125*, e2019JC016008. [CrossRef]
22. International Altimetry Team. Altimetry for the future: Building on 25 years of progress. *Adv. Space Res.* **2021**, *68*, 319–363. [CrossRef]
23. POLDER. Available online: https://en.wikipedia.org/wiki/POLDER (accessed on 28 May 2023).
24. Oceansat. Available online: https://en.wikipedia.org/wiki/Oceansat (accessed on 16 May 2023).
25. Haiyang (Satellite). Available online: https://en.wikipedia.org/wiki/Haiyang_(satellite) (accessed on 16 May 2023).
26. KOSC Korean Ocean Satellite Center. Available online: https://kosc.kiost.ac.kr/index.nm?lang=en&contentId=82 (accessed on 25 May 2023).
27. Vinogradova, N.; Lee, T.; Boutin, J.; Drushka, K.; Fournier, S.; Sabia, R.; Stammer, D.; Bayler, E.; Reul, N.; Gordon, A.; et al. Satellite Salinity Observing System: Recent Discoveries and the Way Forward. *Front. Mar. Sci.* **2019**, *6*, 243. [CrossRef]
28. NASA. Earth Data. Visible Infrared Imaging Radiometer Suite (VIIRS). Available online: https://www.earthdata.nasa.gov/sensors/viirs (accessed on 30 May 2023).
29. JAXA. Global Change Observation Mission—Climate "Shikisai" (GCOM-C). Available online: https://global.jaxa.jp/projects/sat/gcom_c/ (accessed on 30 May 2023).
30. Copernicus. Europe's Eyes on Earth. Available online: https://www.copernicus.eu/en (accessed on 30 May 2023).
31. Steunou, N.; Desjonquères, J.D.; Picot, N.; Sengenes, P.; Noubel, J.; Poisson, J.C. AltiKa Altimeter: Instrument Description and In-Flight Performance. *Mar. Geod.* **2015**, *38* (Suppl. 1), 22–42. [CrossRef]
32. Hauser, D.; Dong, X.; Aouf, L.; Tison, C.; Castillan, P. Overview of the CFOSAT Mission. In Proceedings of the IEEE International Geoscience and Remote Sensing Symposium (IGARSS) Conference, Beijing, China, 10–15 July 2016.
33. Fu, L.L.; Alsdorf, D.; Rodriguez, E.; Morrow, R.; Mognard, N.; Lambin, J.; Vaze, P.; Lafon, T. The SWOT (Surface Water and Ocean Topography) Mission: Spaceborne Radar Interferometry for Oceanographic and Hydrological Applications. In Proceedings of the OceanObs'09, Venice, Italy, 21–25 September 2009.
34. Barale, V.; Gower, J.F.R.; Alberotanza, L. (Eds.) *Proceedings "Oceans from Space V", Venice 2022*; NSA GROUP: Rome, Italy, 2022; p. 248.
35. The Ocean Decade. The Science We Need for the Ocean We Want. Available online: https://oceandecade.org (accessed on 25 May 2023).
36. International Science Council. Available online: https://council.science (accessed on 18 May 2023).
37. Barale, V.; Gower, J.F.R.; Alberotanza, L. (Eds.) *Oceans from Space, Venice 2000, Abstracts*; EUR 19661 EN; Publication Office of the European Union: Luxemburg, 2010; p. 282.
38. Barale, V. Special Issue Oceans from Space, Venice 2000. *Int. J. Remote Sens.* **2004**, *25*, 1255–1543.
39. Barale, V.; Gower, J.F.R.; Alberotanza, L. (Eds.) *Oceanography from Space, Revisited*; Springer: Dordrecht, The Netherlands; Heidelberg, Germany; London, UK; New York, NY, USA, 2010; p. 374.
40. Barale, V.; Gower, J.F.R.; Alberotanza, L. (Eds.) *Proceedings (Extended Abstracts Collection) "Oceans from Space" Venice 2010*; EUR 24324 EN; Publication Office of the European Union: Luxemburg, 2010; p. 265.

remote sensing

MDPI

Review

Overview of Operational Global and Regional Ocean Colour Essential Ocean Variables Within the Copernicus Marine Service

Vittorio E. Brando [1,*], Rosalia Santoleri [1], Simone Colella [1], Gianluca Volpe [1], Annalisa Di Cicco [1], Michela Sammartino [1], Luis González Vilas [1], Chiara Lapucci [1], Emanuele Böhm [1], Maria Laura Zoffoli [1], Claudia Cesarini [1], Vega Forneris [1], Flavio La Padula [1], Antoine Mangin [2], Quentin Jutard [2], Marine Bretagnon [2], Philippe Bryère [2], Julien Demaria [2], Ben Calton [3], Jane Netting [3], Shubha Sathyendranath [3], Davide D'Alimonte [4], Tamito Kajiyama [4], Dimitry Van der Zande [5], Quinten Vanhellemont [5], Kerstin Stelzer [6], Martin Böttcher [6] and Carole Lebreton [6]

1 Consiglio Nazionale delle Ricerche, Istituto di Scienze Marine (CNR-ISMAR), 00133 Rome, Italy; rosalia.santoleri@cnr.it (R.S.); simone.colella@cnr.it (S.C.); gianluca.volpe@cnr.it (G.V.); annalisa.dicicco@cnr.it (A.D.C.); michela.sammartino@cnr.it (M.S.); luis.gonzalezvilas@artov.ismar.cnr.it (L.G.V.); chiara.lapucci@cnr.it (C.L.); emanuele.bohm@cnr.it (E.B.); marialaura.zoffoli@cnr.it (M.L.Z.); vega.forneris@cnr.it (V.F.); flavio.lapadula@cnr.it (F.L.P.)

2 ACRI-ST S.A.S., 06904 Sophia-Antipolis, France; antoine.mangin@acri-st.fr (A.M.); quentin.jutard@acri-st.fr (Q.J.); marine.bretagnon@acri-st.fr (M.B.); philippe.bryere@acri-st.fr (P.B.); julien.demaria@acri-st.fr (J.D.)

3 Plymouth Marine Laboratory (PML), Plymouth PL1 3DH, UK; bac@pml.ac.uk (B.C.); jann@pml.ac.uk (J.N.); ssat@pml.ac.uk (S.S.)

4 Aequora, 8200-567 Lisbon, Portugal; davide.dalimonte@aequora.org (D.D.); tamito.kajiyama@aequora.org (T.K.)

5 Royal Belgian Institute of Natural Sciences (RBINS), 1000 Brussels, Belgium; dvanderzande@naturalsciences.be (D.V.d.Z.); qvanhellemont@naturalsciences.be (Q.V.)

6 Brockmann Consult GmbH, 21029 Hamburg, Germany; kerstin.stelzer@brockmann-consult.de (K.S.); martin.boettcher@brockmann-consult.de (M.B.); carole.lebreton@brockmann-consult.de (C.L.)

* Correspondence: vittorio.brando@cnr.it

Abstract: The Ocean Colour Thematic Assembly Centre (OCTAC) of the Copernicus Marine Service delivers state-of-the-art Ocean Colour core products for both global oceans and European seas, derived from multiple satellite missions. Since 2015, the OCTAC has provided global and regional high-level merged products that offer value-added information not directly available from space agencies. This is achieved by integrating observations from various missions, resulting in homogenized, inter-calibrated datasets with broader spatial coverage than single-sensor data streams. OCTAC enhanced continuously the basin-level accuracy of essential ocean variables (EOVs) across the global ocean and European regional seas, including the Atlantic, Arctic, Baltic, Mediterranean, and Black seas. From 2019 onwards, new EOVs have been introduced, focusing on phytoplankton functional groups, community structure, and primary production. This paper provides an overview of the evolution of the OCTAC catalogue from 2015 to date, evaluates the accuracy of global and regional products, and outlines plans for future product development.

Keywords: Ocean Colour; operational oceanography; essential ocean variables; regional products; sentinel-2; sentinel-3; environmental reporting

1. Introduction

Awareness of the role that the ocean plays in the climate, environment, economy, and more generally the entire society has increased over the past decades [1,2]. At the European level, this has given birth to the Copernicus Marine Environment Monitoring Service (CMEMS), which constitutes one of the six pillar services of the Copernicus program [3]. CMEMS was established in 2015, building on the experience gained through a series of

European projects from 2004 through 2015 (MERSEA, MyOcean, and MyOcean2). Currently, CMEMS is the European provider of operational information (both observations and model outputs) about the global ocean and the European regional seas [4,5].

Within CMEMS, the Ocean Colour Thematic Assembly Centre (OCTAC) provides state-of-the-art Ocean Colour (OC) core products for the global ocean and the European seas based on multiple satellite missions [6–8]. The OCTAC serves users across the scientific and operational oceanography communities, commercial providers focused on the use of marine resources, and public agencies focused on environmental monitoring, with interests in data across oceanic, shelf, and coastal waters. Depending on their applications, these users require different spatial resolutions (i.e., 1 to 4 km in open ocean, 300 m over the shelf, and down to 10s of meters in coastal waters) [9,10]. To meet these needs, the global and regional higher-level combined OCTAC products generate added-value information not readily available from space agencies. Since 2015, the OCTAC has continued to improve the accuracy at the basin level of existing essential ocean variables (EOVs), i.e., chlorophyll-a concentration (CHL), inherent optical properties (IOPs), as well as the radiometry in itself [5,11]. EOVs are key parameters for understanding the spatiotemporal variability of the ocean's physical and biological compartments and are required for inclusion in climate models and projections [11]. Given that the variability in the phytoplankton community structure and the composition of the dissolved and particulate matter occurring across oceanic basins cause significant optical differences [12–14], the regional algorithms differ from those available for global applications because they are specifically derived to reflect the bio-optical characteristics of each European sea [6,13,15]. Blended CHL datasets are produced for all basins applying the appropriate algorithms across the open ocean and coastal waters depending on the water types [6,7,9]. From 2019 onwards, new EOVs related to phytoplankton functional and size groups, community structure, and primary production (PP) were introduced [5,11].

The present review will provide (i) a summary of the operational OC products and datasets across the different spatial resolutions and their evolution from 2015 to date; (ii) an overview of the uncertainty associated with selected variables; (iii) examples of the use of products for operational monitoring and reporting; and (iv) a description of the planned and foreseen product evolutions.

2. Product Overview

Within the CMEMS operational oceanography framework, data are produced both in near-real time (NRT) and as reprocessed multiyear (MY) data delivered as daily consistently projected Level 3 (L3) datasets, as well as monthly average and daily "gap-free" Level 4 (L4) products to overcome cloud cover in subsequent oceanographic analyses [3–5]. The daily L4 datasets are retrieved using optimal interpolation or variants of the DINEOF (data interpolating empirical orthogonal functions) procedure [7,16,17]. The daily NRT products are available by the end of the day following the satellite data acquisition. The daily MY products are produced within 8 to 12 days of acquisition. Since 2015, OCTAC has delivered global and regional OC products covering the CMEMS regions: Global (GLO), Arctic (ARC), and North-East Atlantic (ATL) regions, and Baltic (BAL), Black (BLK), and Mediterranean (MED) seas (Figure 1, Tables 1 and 2).

In 2015, the NRT regional products were based on single-sensors data; between 2016 and 2018, the multisensor datasets were introduced across the whole catalogue (Figure 1). Such datasets are based on harmonized multisensor time series of remote sensing reflectance (R_{rs}) acquired by different OC satellites, significantly increasing the spatial coverage of daily observations) [6,7,9]. These products are available at 1 km spatial resolution for European seas, and at 4 km resolution for the global ocean. Since 2020, the MY processing chains have become fully consistent with the NRT multisensor processors, for all basins. Hence, the only difference between NRT and MY datasets lies in the upstream input data: the Level 2 (L2) granules processed with consolidated auxiliary data (hindcast meteorological and ephemerides data, usually available a few days after their acquisition) are used to produce the consistent and quality checked MY time series.

In May 2021, higher spatial resolutions were added to the catalogue with the OLCI (Ocean and Land Colour Instrument) datasets at 300 m resolution combining Copernicus Sentinel-3 A and B, as well as the Copernicus Sentinel-2 MSI (MultiSpectral Instrument) datasets at 100 m (Figure 1). The Sentinel 2 MSI datasets are produced for the European coastal waters in a 20 km strip from the coastline, while the OLCI datasets are available at 300 m for all European regional seas and in the global product over a 200 km strip from the coastline (Figure 2). In 2022, the OCTAC catalogue was fully reorganized to reduce the number of products and datasets, so that each product now contains up to five datasets:

(i) Plankton—with the phytoplankton chlorophyll concentration (CHL), phytoplankton size classes (PSC) and phytoplankton functional types (PFT);
(ii) Primary Production—integrated productivity within the euphotic zone (PP);
(iii) Reflectance—with the spectral remote sensing reflectance (R_{rs});
(iv) Transparency—with the diffuse attenuation coefficient of light at 490 nm (Kd490), Secchi depth (ZSD—an indicator of water transparency), turbidity (TUR), and the suspended particulate matter (SPM);
(v) Optics—including the inherent optical properties (IOPs), such as absorption and backscattering by particulate and dissolved matter.

As of December 2024, the OCTAC catalogue is composed of 38 OC Products and 214 datasets, across the multisensor, Sentinel-3, and Sentinel-2 data streams (Tables 1 and 2).

The OCTAC operational production is shared among European research centers and private companies to ensure a distribution of the necessary expertise across data streams (Table 2). The development, refinement, and implementation of the processing chains is based on Copernicus funding as well as the uptake of state-of-art algorithms and approaches developed by the space agencies, large collaborative projects, and the OC community.

The OCTAC Catalogue (Tables 1 and 2) includes two complementary global reprocessed products from the Copernicus-GlobColour [7] and OC-CCI (Ocean Colour Climate Change Initiative) [8,9]. These are the main and only two existing operational initiatives providing global long-term daily observations of L3 OC products based on a multisensor approach with 4 km resolution. The Copernicus-GlobColour operational processor ensures consistency of MY and NRT products, with periodical updates when new upstream data from NASA (National Aeronautics and Space Administration) or ESA (European Space Agency)/EUMETSAT (European Organisation for the Exploitation of Meterological Satellites) are available, or following processing chain evolutions [7]. On the other hand, OC-CCI targets climate quality consistency with minimal inter-sensor bias [8,9], with the unavoidable cost of being unable to reach this consistency with an NRT production. Since OC-CCI V5, the CCI algorithms have been applied to delayed-time NRT data to produce interim climate data records (ICDR) for the Copernicus Climate Change Service (C3S) within a month of acquisition, though these should not be considered climate grade. The two products therefore feature different and complementary characteristics, serving various user needs. A downstream service offering NRT products might choose the GC MY to identify anomalies in yesterday's NRT, while a study investigating long-term subtle changes or deriving historical measures for later use may opt for OC-CCI or Copernicus-GlobColour.

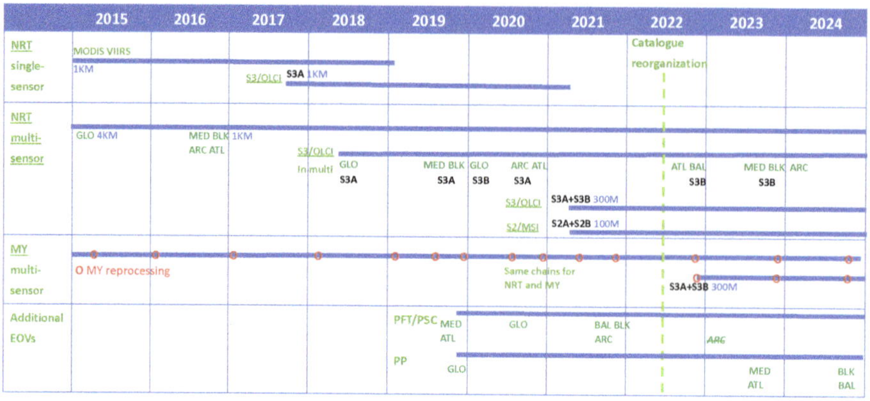

Figure 1. Overview of the OCTAC catalogue evolutions of the single-sensor and multisensor global and regional OC products from 2015 to 2024. The blue lines mark the timelines of each product type; covered basins are marked in green and listed under each line; satellite sensors are marked in black; spatial resolution of products/datasets is marked in blue. The red dots mark the dates of the MY reprocessing.

Table 1. Overview of OCTAC products at various spatial resolutions: 1 and 4 km multisensor datasets, merged Sentinel-3 OLCI A + B datasets at 4 km and 300 m, and merged Sentinel-2 MSI A + B datasets at 100 m.

CMEMS Region	Multi-Sensor 1 km (Regions), 4 km (ARC, GLO)				Sentinel-3 OLCI A + B 300 m (Regions and GLO)/4 km (GLO)				Sentinel-2 MSI A + B 100 m			
	NRT		MY		NRT		MY		NRT		MY	
	L3	L4	L3	L4	L3	L4	L3	L4	L3	L4	L3	L4
Arctic Ocean	-	-	✓	✓	✓	✓	✓	✓	✓	✓	-	-
NE Atlantic Ocean	✓	✓	✓	✓	✓	-	✓	-	✓*	✓*	-	-
Baltic Sea	-	-	✓	✓	✓	✓	✓	✓	✓	✓	-	-
Black Sea	✓	✓	✓	✓	✓	✓	✓	✓	✓	✓	-	-
Mediterranean Sea	✓	✓	✓	✓	✓	✓	✓	✓	✓	✓	-	-
Global	✓	✓	✓	✓	✓	✓	✓	✓	-	-	-	-
Global (C3S/OC-CCI)			✓	✓								

* Due to size of the files, the Sentinel-2 based products for the North-East Atlantic are produced over the Iberia–Biscay–Ireland (IBI) and North-West Shelf (NWS) areas and are provided in tiles linked to UTM zones.

Table 2. Listing of the OCTAC products in December 2024.

Region	L3/L4	NRT/MY	Product Name	DOI	Production Unit
GLO	L3	NRT	OCEANCOLOUR_GLO_BGC_L3_NRT_009_101	https://doi.org/10.48670/moi-00278	ACRI-ST
GLO	L4	NRT	OCEANCOLOUR_GLO_BGC_L4_NRT_009_102	https://doi.org/10.48670/moi-00279	ACRI-ST
GLO	L3	MY	OCEANCOLOUR_GLO_BGC_L3_MY_009_103	https://doi.org/10.48670/moi-00280	ACRI-ST
GLO	L4	MY	OCEANCOLOUR_GLO_BGC_L4_MY_009_104	https://doi.org/10.48670/moi-00281	ACRI-ST
GLO	L3	MY	OCEANCOLOUR_GLO_BGC_L3_MY_009_107	https://doi.org/10.48670/moi-00282	BC/PML *
GLO	L4	MY	OCEANCOLOUR_GLO_BGC_L4_MY_009_108	https://doi.org/10.48670/moi-00283	BC/PML *
ATL	L3	NRT	OCEANCOLOUR_ATL_BGC_L3_NRT_009_111	https://doi.org/10.48670/moi-00284	ACRI-ST
ATL	L3	MY	OCEANCOLOUR_ATL_BGC_L3_MY_009_113	https://doi.org/10.48670/moi-00286	ACRI-ST

Table 2. *Cont.*

Region	L3/L4	NRT/MY	Product Name	DOI	Production Unit
ATL	L4	NRT	OCEANCOLOUR_ATL_BGC_L4_NRT_009_116	https://doi.org/10.48670/moi-00288	ACRI-ST
ATL	L4	MY	OCEANCOLOUR_ATL_BGC_L4_MY_009_118	https://doi.org/10.48670/moi-00289	ACRI-ST
ARC	L3	NRT	OCEANCOLOUR_ARC_BGC_L3_NRT_009_121	https://doi.org/10.48670/moi-00290	CNR
ARC	L4	NRT	OCEANCOLOUR_ARC_BGC_L4_NRT_009_122	https://doi.org/10.48670/moi-00291	CNR
ARC	L3	MY	OCEANCOLOUR_ARC_BGC_L3_MY_009_123	https://doi.org/10.48670/moi-00292	CNR
ARC	L4	MY	OCEANCOLOUR_ARC_BGC_L4_MY_009_124	https://doi.org/10.48670/moi-00293	CNR
BAL	L3	NRT	OCEANCOLOUR_BAL_BGC_L3_NRT_009_131	https://doi.org/10.48670/moi-00294	CNR
BAL	L4	NRT	OCEANCOLOUR_BAL_BGC_L4_NRT_009_132	https://doi.org/10.48670/moi-00295	CNR
BAL	L3	MY	OCEANCOLOUR_BAL_BGC_L3_MY_009_133	https://doi.org/10.48670/moi-00296	CNR
BAL	L4	MY	OCEANCOLOUR_BAL_BGC_L4_MY_009_134	https://doi.org/10.48670/moi-00308	CNR
MED	L3	NRT	OCEANCOLOUR_MED_BGC_L3_NRT_009_141	https://doi.org/10.48670/moi-00297	CNR
MED	L4	NRT	OCEANCOLOUR_MED_BGC_L4_NRT_009_142	https://doi.org/10.48670/moi-00298	CNR
MED	L3	MY	OCEANCOLOUR_MED_BGC_L3_MY_009_143	https://doi.org/10.48670/moi-00299	CNR
MED	L4	MY	OCEANCOLOUR_MED_BGC_L4_MY_009_144	https://doi.org/10.48670/moi-00300	CNR
BLK	L3	NRT	OCEANCOLOUR_BLK_BGC_L3_NRT_009_151	https://doi.org/10.48670/moi-00301	CNR
BLK	L4	NRT	OCEANCOLOUR_BLK_BGC_L4_NRT_009_152	https://doi.org/10.48670/moi-00302	CNR
BLK	L3	MY	OCEANCOLOUR_BLK_BGC_L3_MY_009_153	https://doi.org/10.48670/moi-00303	CNR
BLK	L4	MY	OCEANCOLOUR_BLK_BGC_L4_MY_009_154	https://doi.org/10.48670/moi-00304	CNR
ARC	L3	NRT	OCEANCOLOUR_ARC_BGC_HR_L3_NRT_009_201	https://doi.org/10.48670/moi-00061	BC-RBINS
BAL	L3	NRT	OCEANCOLOUR_BAL_BGC_HR_L3_NRT_009_202	https://doi.org/10.48670/moi-00079	BC-RBINS
NWS	L3	NRT	OCEANCOLOUR_NWS_BGC_HR_L3_NRT_009_203	https://doi.org/10.48670/moi-00118	BC-RBINS
IBI	L3	NRT	OCEANCOLOUR_IBI_BGC_HR_L3_NRT_009_204	https://doi.org/10.48670/moi-00107	BC-RBINS
MED	L3	NRT	OCEANCOLOUR_MED_BGC_HR_L3_NRT_009_205	https://doi.org/10.48670/moi-00109	BC-RBINS
BLK	L3	NRT	OCEANCOLOUR_BLK_BGC_HR_L3_NRT_009_206	https://doi.org/10.48670/moi-00086	BC-RBINS
ARC	L4	NRT	OCEANCOLOUR_ARC_BGC_HR_L4_NRT_009_207	https://doi.org/10.48670/moi-00062	BC-RBINS
BAL	L4	NRT	OCEANCOLOUR_BAL_BGC_HR_L4_NRT_009_208	https://doi.org/10.48670/moi-00080	BC-RBINS
NWS	L4	NRT	OCEANCOLOUR_NWS_BGC_HR_L4_NRT_009_209	https://doi.org/10.48670/moi-00119	BC-RBINS
IBI	L4	NRT	OCEANCOLOUR_IBI_BGC_HR_L4_NRT_009_210	https://doi.org/10.48670/moi-00108	BC-RBINS
MED	L4	NRT	OCEANCOLOUR_MED_BGC_HR_L4_NRT_009_211	https://doi.org/10.48670/moi-00110	BC-RBINS
BLK	L4	NRT	OCEANCOLOUR_BLK_BGC_HR_L4_NRT_009_212	https://doi.org/10.48670/moi-00087	BC-RBINS

* Global ocean OC-CCI reprocessed multisensor data produced by PML (2015–2023) and by BC in 2023–2025.

2.1. Upstream OC Data Streams

Over the years, the upstream data shifted from OC science missions—i.e., SeaWiFS (Sea-viewing Wide Field-of-view Sensor), MERIS (MEdium Resolution Imaging Spectrometer) and MODIS (Moderate Resolution Imaging Spectroradiometer)—towards operational missions (Figure 3). As well as the two OLCIs on the Copernicus Sentinel-3 A and B and three VIIRSs (Visible Infrared Imaging Radiometer Suites) on NOAA's (National Oceanic and Atmospheric Administration) SNPP (Suomi-national polar-orbiting partnership), NOAA-20 and NOAA-21, the two MSI (MultiSpectral Instrument) sensors on the Copernicus Sentinel-2 A and B satellite are ingested for the coastal products due to their finer spatial resolution, although they were initially designed for terrestrial applications and have a revisit time of 3–5 days (Figure 3). In 2015, all NRT regional products were based on MODIS and VIIRS; between 2016 and 2018 these datasets were then replaced by multisensor datasets [6,7,9] (Figure 1). The number of sensors contributing to the multisensor product time series changed over the years from one sensor (SeaWiFS) from 1997 to 2002, up to six sensors (MODIS-Aqua, VIIRS-SNPP/NOAA-20/NOAA-21, and OLCI-Sentinel-3A/B) from late 2022 to the moment of writing (Figure 3).

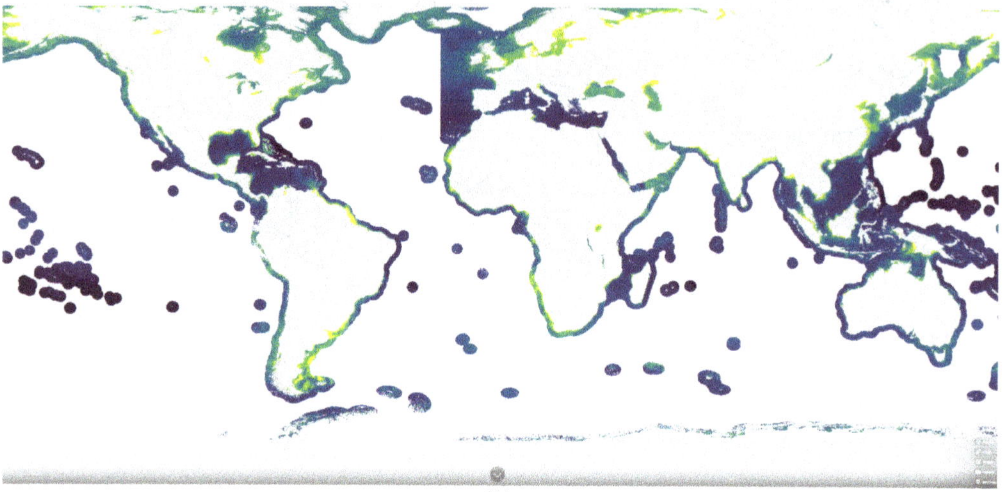

(**A**)

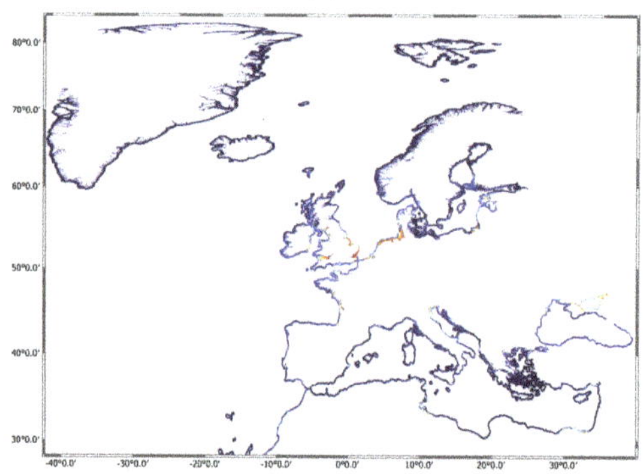

(**B**)

Figure 2. Spatial coverage of the Sentinel-3 OLCI 300 m and Sentinel-2 MSI 100 m datasets. (**A**) All European regional seas and a 200 km strip from the coastline in the global product for Sentinel-3 OLCI. (**B**) A 20 km strip from the coastline for the European coastal waters covered in 5 days with Sentinel-2 MSI.

2.2. Merging Strategies and Atmospheric Correction

Within OCTAC, specific strategies to merge data from sensors with different sets of central wavelengths and spectral response functions are adopted in the processing chains. For the OC-CCI multisensor production, atmospheric correction is performed independently for each sensor, and the merging is performed for the calibrated reflectances, after band shifting to the reference sensor bands [8,9,18,19]. In particular, MERIS, MODIS, VIIRS, and OLCI data were processed to L2 with the POLYMER algorithm [20,21] while L2 data downloaded from NASA were used for SeaWiFS. For the Mediterranean and Black Sea regional products, the method developed within OC-CCI has been adapted to rely on L2 data distributed by space agencies [6]. The Copernicus-GlobColour products for GLO and ATL are based on the L2 data distributed by the agencies with multisensor merging and flagging strategies detailed in Garnesson et al. [7].

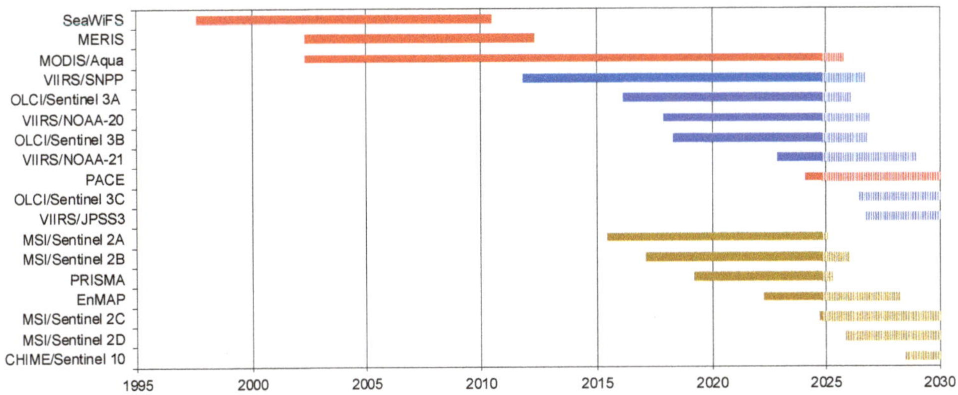

Figure 3. OC sensors and high-resolution imagers adopted upstream in OCTAC processing chains. Timelines of legacy, and current and forthcoming (approved and planned) sensors are displayed (source CEOS): red identifies science OC missions, blue identifies operational OC missions, and brown identifies high-resolution/land imagers.

Combining Sentinel-3 A and B (Sentinel-3 OLCI datasets), as well as combining Sentinel-2 A and B (Sentinel-2 MSI datasets), does not require band-shifting or inter-sensor bias correction because the two companion A and B sensors are assumed to be fully consistent. The Sentinel-3 OLCI datasets for the global and regional products are based on L2 reflectances distributed by EUMETSAT [22], except for the Baltic Sea, where the EUMETSAT L2 processor often yields inaccurate R_{rs} spectra with low and even negative values [23–25]. Thus for the atmospheric correction of the OLCI L1 granules in the Baltic Sea, the OLCI neural network swarm [26] was used until 2023, and then POLYMER [20,21] was selected based on a round-robin comparison of several algorithms [25].

The Sentinel-2 MSI products are available every 3–5 days at each location for a 20 km strip from the coastline of the European coasts, characterized by diverse atmospheric conditions and fast-changing water types in space and time (Figure 2B). As an operational L2 reflectance product for water applications is not available within Copernicus for Sentinel-2 MSI [27], two algorithms, C2RCC (version 1.0, normal NN) [28,29] and ACOLITE/DSF [30], are used as the baseline atmospheric correction approach to deal with both optically complex and clear waters under the very challenging atmospheric conditions of the nearshore environments [31]. These two methods are highly complementary because ACOLITE/DSF makes no assumption about the water reflectance, thus achieving good results even for unexpected water types (e.g., dredging plumes, very concentrated algal blooms, etc.). Instead, C2RCC constrains the water reflectance to correspond to the training data, always retrieving R_{rs} spectra that look like water. This extra information/constraint on water reflectance, embedded within the C2RCC approach, provides greater retrieval power in the most challenging circumstances (sunglint, highly absorbing waters) but at the expense of imposing a solution for R_{rs} that may not correspond to reality. The C2RCC to ACOLITE/DSF pixel-based switching is performed by means of the comparison of the R_{rs} (560) and R_{rs} (865) spectral bands (as provided by the C2RCC processor).

2.3. *Retrieval Algorithms in the Global and Regional Processing Chains*

2.3.1. Chlorophyll Algorithms

This section provides an overview of the CHL algorithms implemented in the operational processing chains for the global and regional products across the three spatial resolutions. All processing chains generate blended CHL datasets by ensuring that the most appropriate algorithms are applied across the water types that occur in the open ocean and coastal waters. The selection of the algorithms for CHL retrieval and the merging schemes were carried out based on the optical characteristics of each basin and round-robin procedures. To generate the NRT and MY CHL for the OLCI and the multisensor datasets, the regional bio-optical algorithms are consistently applied for each basin. For the S2 MSI datasets, the same processing chain is applied across all European waters to generate NRT CHL.

For the Global Ocean, CHL concentration estimated within the Copernicus-GlobColour as a daily multisensor merged dataset, where CHL values are individually computed for each sensor using a blended algorithm and then combined [7]. For oligotrophic waters, the product relies on the CI algorithm [32], while for mesotrophic and coastal waters the OC5 algorithm [33] was tuned for each sensor. The OC5 and CI blending uses the same approach as NASA's implementation of the CI algorithm [32], with a transition between 0.15 to 0.2 mg m^{-3} to ensure a smooth merging.

For the Global Ocean, the CHL values for the OC-CCI products are calculated by blending algorithms based on the water-types utilising the same OC-CCI R_{rs} described above [8,9]. For v6.0, the blending of the OCI algorithm (as implemented by NASA, itself a combination of CI and OC4 [32]), the OCI2 algorithm (an updated OCI parameterization), the OC2 algorithm and the OCx algorithm [8,19].

For the Arctic and Atlantic Oceans, the regional CHL algorithm adopted until 2022 was OC5CCI—i.e., a variation of OC5 [33]—developed by IFREMER and PML [34]. To this end, an OC5CCI look-up table was specifically generated for application over OC-CCI daily merged R_{rs}. The resulting OC5CCI algorithm was tested and selected after a calibration exercise and sensibility analysis of the existing algorithms (OC3, OC4, OCI, OC5CI, OC5, OC5CCI) that included a round-robin quantitative performance assessment against in situ data [34]. Following a catalogue reorganization and change of production responsibility, since 2023 the Atlantic Ocean product has been retrieved with the Copernicus-GlobColour processor described above [7], while for the Arctic a new regional algorithm was developed [35], and applied to OLCI and OC-CCI R_{rs}.

The water-column in the Arctic region has particular characteristics, namely high and heterogeneous distribution of colored dissolved organic matter (CDOM) due to freshwater inputs that reach different ARC sectors, which limits the performance of global CHL algorithms [36,37]. When acquiring satellite data in polar regions, additional challenges arise due to low solar zenith angles, frequent ice coverage and high aerosol content [37] that usually introduce high uncertainties in retrieving water-leaving radiance. Since 2023, CHL has been retrieved by a new regional algorithm, seasonal spatially adjusted for the Arctic Ocean (CHL-SeSARC) [35], developed using supervised machine learning techniques and trained with a compilation of in situ databases for the Arctic waters from 1998 to 2018. In the proposed pan-Arctic CHL algorithm, the use of the longitude of the pixel center and the day of the year enables accounting for the regional particularities and spatial heterogeneity within the ARC and the seasonal variability of the bulk phytoplankton community and/or the associated uncertainties in atmospheric correction.

In the Mediterranean Sea, the blended CHL product is based on two regional algorithms: the MedOC4, an updated version of the regionally parameterized maximum band ratio [6] for clear waters, and the ADOC4 algorithm [38] for optically complex waters. From 2020, the determination of the water type accounts specifically for waters with high CHL concentration due to phytoplankton blooms (e.g., Gulf of Lions) or mixing (e.g., Alborán Sea) that can be erroneously identified as Case II waters [39].

In the Black Sea, the retrieval of the CHL concentration is based on a merging scheme [40] designed for two different regional algorithms exhibiting lower and higher optical complexity. These are, respectively, a band-ratio algorithm based on two wavelengths (490 and 555 nm) [41], and a multilayer perceptron (MLP) neural net based on R_{rs} values at three wavelengths (490, 510, and 555 nm) that features interpolation capabilities helpful to fit data non-linearities [40]. In 2019, this merging scheme substituted the regional band ratio approach by Kopelevich et al. [42].

In the Baltic Sea, CHL is derived from the MLP neural net developed under the umbrella of the BiOMaP program of JRC/EC [15,43,44]. The BAL product is based on an ensemble algorithm that combines the CHL retrievals from individual MLPs based on different R_{rs} spectral subsets to address the optical complexity of the basin and to account for the temporal and spatial variation of uncertainties introduced by the atmospheric correction [25,45]. In 2020, this ensemble approach substituted the previous operational regional algorithm based on the recalibration of the OC4v6 with in situ data [46].

The coastal products based on Sentinel 2 MSI introduced in 2021 are produced for a 20 km strip from the coastline in the coastal waters of the ARC, NWS, BAL, IBI, MED, and BLK regions (Figure 2B). For these products, the same processing chain is applied across all European waters to address the fast-changing water types in space and time by combining different algorithms for the CHL concentration retrieval. The CHL datasets are generated by merging two complementary algorithms following the approach of Lavigne et al. [47]: the OC3 empirical blue-green bands ratio algorithm [48], and the Gons [49,50] semi-analytical algorithm. The OC3 algorithm was selected for application over low-to-moderate biomass waters and over clear-to-moderately turbid waters. The Gons algorithm was chosen for application over moderate-to-high-biomass waters and for turbid coastal waters. The operational limits of the CHL algorithms are determined on the basis of the optical conditions of the considered pixels, using the quality control routines developed by Lavigne et al. [47] adapted to the Sentinel-2 bands. Within this framework, pixels are flagged in waters with a turbidity level of approximately 10 FNU or higher and CHL lower than 5 mg m^{-3}, because the uncertainties associated with CHL retrieval in such water types would be too high.

2.3.2. Phytoplankton Type Variables

The phytoplankton type variables were introduced in the OCTAC catalogue from 2019 for the global ocean and all regional seas using global and regionally tuned methods [39,51–56]. The phytoplankton size classes (PSCs) and phytoplankton functional types (PFTs) are expressed as CHL concentration (mg m^{-3}). Both for the global ocean and regional seas, PFTs include diatoms, dinoflagellates, green algae, prokaryotes, and haptophytes (except for BAL). For GLO and ATL, the prochlorococcus group is also distributed, while cryptophytes are provided only for MED and BAL. PSCs consist of three main size groups, micro-, nano- and pico-phytoplankton, based on Sieburth et al.'s [57] size classification and Vidussi et al.'s [58] approach founded on the relationships between diagnostic pigments, taxonomic groups, and their most common dimensions. For BLK, only PSCs are distributed.

For both algorithm calibration and validation, the in situ Chl-a concentration of each group was quantified through diagnostic pigment analysis (DPA) [58] and its implementations and refinements [51,59–61]. The DPA was updated for GLO and ATL [54,55], and regionalized for MED, BAL, and BLK [39,53,56]. For GLO and ATL, the algorithm [49,50] was initially implemented using OLCI reflectance in the visible spectrum (bands comprised between 400 and 681 nm) using an empirical orthogonal function (EOF) approach and then extended to the multisensor datasets. The regional algorithms for PFT and PSC retrieval for MED, BAL, and BLK [39,53,56] rely on empirical functions based on statistical relationships between the in situ contribution of each group (PFT or PSC) and the corresponding log10-transformed in situ CHL concentrations (that are applied to each of the regional CHL datasets).

2.3.3. Inherent Optical Properties

The operational processing chains for the global and regional products across the three spatial resolutions have implemented different approaches for retrieval of the main IOPs. The coefficients for the absorption by phytoplankton (aph), the absorption by dissolved and detrital matter (adg), and the backscattering by particulate matter (bbp) are provided at reference wavelengths.

For MED, BLK, BAL, and ARC multisensor and OLCI datasets, the algorithm used to produce the aph(443), adg(443), and bbp(443) is the quasi-analytical algorithm (QAA V6 [62,63]), also used in the context of the band-shifting procedures [18]. For all the Sentinel 2 MSI products for European coastal waters, the bbp coefficient is spectrally dependent and is also estimated using the QAA V6 [62,63].

For GLO and ATL since 2023, the adg(443) and bbp(443) are estimated from a semi-analytical model based on Kd490 and R_{rs} [64,65], replacing the retrieval carried out with the Garver–Siegel–Maritorena (GSM01) bio-optical model [66] implemented in the Copernicus-GlobColour processor described above [7].

2.3.4. Primary Production

Primary Production data products were added to the catalogue in 2019. As of the December 2024 version, PP is distributed for the GLO, ATL, MED, BLK, and BAL (Figure 1). The GLO and ATL version is based on the Antoine and Morel algorithm [67], and uses OC products—merged CHL, PAR (photosynthetically active radiation [68]), sea surface temperature from OSTIA (operational sea surface temperature and ice analysis) and mixed layer depth from model reanalysis (GLORYS12V1) [69]).

The regional PP datasets for MED, BLK, and BAL are based on an updated version of the bio-optical model by Morel [70], incorporating the regional CHL retrievals [39]. This model uses outputs from the atmospheric model by Tanré et al. [71], which allows the estimation of the photosynthetic radiation at the sea surface and its attenuation through the water column. With a parameterization of the main physiology processes, the model allows the computation of the primary productivity starting from algal biomass concentration. The empirical approach developed by Morel and Berthon [72] establishes relationships between the pigment concentration in the upper layer, the integrated content across the entire euphotic zone, and the shape of the vertical pigment profile. As a result, this model enables the linkage of satellite-derived pigment concentrations with vertical pigment distributions. For the regional products, the atmospheric model was replaced by the revised version of the multispectral ocean atmosphere spectral irradiance model (OASIM [73]). This updated OASIM model provides daily estimations of the direct and diffuse irradiance over the ocean with 5 nm spectral resolution (400–700 nm) and 4 km spatial resolution. Moreover, the empirical approach by Morel and Berthon [72] to associate a pigment vertical distribution with a satellite pigment concentration has been refined for MED through the specific utilization of a Mediterranean Sea in situ dataset (MedBiOp, [6]).

3. Uncertainty of OCTAC Products

The validation of the satellite products is carried out by pairwise comparison against in situ reference observations using a common methodology defined and agreed within CMEMS [74]. Since those distributed by OCTAC are all multisensor daily products, the temporal collocation criteria are more relaxed than those of the L2 matchup analyses (e.g., [75]) and allow the inclusion of any in situ observations up to 24 h. As for the spatial matching, the median values are extracted from a $n \times n$ satellite data pixels (with n varying according to the product spatial resolution), centered on the in situ measurement location only in the presence of at least 50% valid values and a coefficient of variation smaller than 20% [75]. The quality assessment is mainly based on an inter-comparison with in situ data gathered from publicly available datasets (e.g., [76]) and/or collected from the production units (e.g., MedBiOp, [6]). MY and NRT are considered together as a homogeneous time series for the assessment. Many uncertainties are linked to these in situ data (e.g., instruments quality, methodologies, water depth of sample compared to surface satellite observation, and time of observations [74,75]). Hence, the estimated accuracy numbers (EANs [74]) used to compare satellite and in situ observations (Table 3) are based on a regression of type 2 (with a reasonable assumption of the same weight for observation and in situ) to compute the determination coefficient (r^2), slope, and intercept

(S, I) and completed by the root mean square distance (RMSD), the center-pattern root mean square distance (cRMSD), and the bias.

The validation metrics for all datasets for the OCTAC products are reported in quality information documents (QuIDs) that are updated with every operational release. The QuIDs for all products can be retrieved from the CMEMS portal following the links reported in Table 2.

Table 3. Metrics used to compare the estimated (satellite-based) dataset $X^E_{i,i=1.N}$ to a reference (in situ) dataset $X^M_{i,i=1.N}$. For log-normally distributed variables (such as Chl), both datasets are log-transformed prior to computing the metrics.

Name	Definition
Estimated dataset mean ($\overline{X}^E$)	$\overline{X}^E = \frac{1}{N}\sum_{i=1}^{N} X_i^E$
Reference dataset mean ($\overline{X}^M$)	$\overline{X}^M = \frac{1}{N}\sum_{i=1}^{N} X_i^M$
Type-2 slope (S)	$S = \frac{\sum_{i=1}^{N}\left(x_i^E-\overline{x}^E\right)^2 - \sum_{i=1}^{N}\left(x_i^M-\overline{x}^M\right)^2 + \left[\left\{\sum_{i=1}^{N}\left(x_i^E-\overline{x}^E\right)^2 - \sum_{i=1}^{N}\left(x_i^M-\overline{x}^M\right)^2\right\}^2 + 4\left\{\sum_{i=k}^{N}\left(x_k^E-\overline{x}^E\right)\left(x_k^M-\overline{x}^M\right)\right\}^2\right]^{\frac{1}{2}}}{2\sum_{i=k}^{N}\left(x_k^E-\overline{x}^E\right)\left(x_k^M-\overline{x}^M\right)}$
Type-2 intercept (I)	$I = \overline{X}^E - S \cdot \overline{X}^M$
Determination coefficient (r^2)	$r^2 = \frac{\left[\sum_{i=1}^{N}\left(x_i^E-\overline{x}^E\right)\left(x_i^M-\overline{x}^M\right)\right]^2}{\sum_{i=1}^{N}\left(x_i^E-\overline{x}^E\right)^2 \sum_{i=1}^{N}\left(x_i^M-\overline{x}^M\right)^2}$
Root mean square difference (RMSD)	$RMSD = \sqrt{\frac{\sum_{i=1}^{N}\left(x_i^E-x_i^M\right)^2}{N}}$
Center-pattern root mean square difference (cRMSD)	$cRMSD = \sqrt{\frac{\sum_{i=1}^{N}\left\{\left[x_i^E-\left(\sum_{j=1}^{N} x_j^E\right)\right]-\left[x_i^M-\left(\sum_{k=1}^{N} x_k^M\right)\right]\right\}^2}{N}}$

Uncertainty Associated to Chlorophyll Datasets

As an example of the OCTAC validation effort, Table 4 provides a summary of the matchup metrics for all CHL datasets for the global and regional products across the three spatial resolutions. For GLO and ATL, the EANs values show a good relationship between in situ HPLC measurements (from 1997 to present) and CHL retrieved with the GlobColour approach [7]). For daily, the statistics show a good correlation: r^2 of 0.75 (0.74 at the Atlantic level) associated to an optimal regression line 1:1 (0.94 on the Atlantic).

These statistics, based on several thousands of in situ observations covering both coastal and clear ocean, demonstrate the quality of this product for many applications. The interpolated product shows a slight degradation but r^2 still reaches 0.71, meaning that it is also of applicative interest, for instance, for model assimilation purposes. The Atlantic interpolated product at 1 km shows a similar r^2 of 0.72. The OLCI specific EANs suffer from a very limited number of matchups and should improve over time. The r^2 is good at more than 0.7, but the slope is high because of a slight overestimation at higher values.

For the OC-CCI global product [8,9,19], the CHL results show a strong correlation (r^2 = 0.88) with low error (RMSD = 0.23 and cRMSD = 0.23) and low bias (−0.022) for more than 30,000 matched in situ observations. Based on the high quality of the product, and in particular the very low bias, the OC-CCI CHL product is within the GCOS target requirement of 5% accuracy, thus suggesting that the OC-CCI program is meeting the GCOS target for the ECV climate quality criteria.

For MED, the CHL validation of the multisensor datasets show good relationships between in situ measurements and CHL retrieval with the regional algorithm [6], although for in situ values larger than 0.3 mg m^{-3} there is a slight dispersion increase. The EANs show low biases (i.e., 0.0017 and −0.029 for daily and daily-interpolated, respectively, Table 3) with r^2 values of 0.79 and 0.78 for daily and daily-interpolated, respectively.

Table 4. Summary of the OCTAC validation metrics for CHL datasets. All symbols are defined in Table 3.

Region	CHL Dataset	N	Slope	Intercept	r^2	RMSD	cRMSD	Bias
GLO (GC)	MULTI MY L3 daily 4 km	17,019	1.00	0.05	0.75	0.340	0.340	0.050
	MULTI MY L4 interpolated 4 km	36,438	0.99	0.00	0.71	0.370	0.370	0.010
	OLCI MY L3 4 km	669	1.32	0.21	0.68	0.395	0.388	0.078
	OLCI MY L3 300 m	288	1.35	0.27	0.71	0.417	0.376	0.180
GLO (OC-CCI)	MULTI MY L3 daily 4 km	34,221	0.925	−0.026	0.88	0.226	0.225	−0.022
ATL	MULTI MY L3 daily 4 km	4621	0.94	0.07	0.74	0.350	0.34	0.080
	MULTI MY L4 interpolated 4 km	10,397	0.94	0.04	0.72	0.360	0.36	0.050
	OLCI MY L3 1 km	72	1.21	0.14	0.83	0.261	0.25	0.073
	OLCI MY L3 300 m	35	1.54	0.24	0.78	0.324	0.281	0.161
ARC	MULTI MY L3 4 km	323	0.67	−0.04	0.68	0.268	0.267	0.015
	OLCI MY L3 300 m	21	0.64	0.06	0.75	0.215	0.193	0.641
BAL	MULTI MY L3 1 km	2070	1.09	−0.21	0.31	0.375	0.335	−0.168
	OLCI MY L3 300 m	460	0.83	0.01	0.32	0.271	0.262	−0.071
BLK	MULTI MY L3 1 km	1154	0.63	0.09	0.28	0.480	0.367	0.042
MED	MULTI MY daily L3 1 km	742	0.97	−0.02	0.79	0.250	0.250	0.002
	MULTI MY L4 interpolated 1 km	1819	0.91	−0.11	0.78	0.258	0.256	−0.029
	MULTI MY L4 interpolated-only 1 km	1084	0.87	−0.16	0.78	0.263	0.259	−0.050
All zones	MSI NRT daily 100 m	700	0.90	0.26	0.48	0.549	0.492	0.257
BAL	MSI NRT daily100 m	188	0.86	0.15	0.22	0.478	0.471	0.085
NWS	MSI NRT daily 100 m	289	1.03	0.22	0.12	0.557	0.508	0.245
IBI	MSI NRT daily 100 m	120	0.94	0.40	0.03	0.582	0.432	0.374
MED	MSI NRT daily 100 m	103	1.15	0.54	0.64	0.608	0.446	0.341

For BLK, the performances of the daily CHL retrieved with the regional merging scheme [40] yield a r^2 = 0.39 and a bias = 0.17 due to the extremely complex waters of the basin and the limited number of matchups. With interpolated data, the correlation is worse (r^2 = 0.28), but the bias is better (0.042). Dispersion of the data is evident for the entire CHL range [39].

For BAL, in view of its optical complexity, the matchup window is limited to 6 h [25,45]. The assessment of the CHL retrieved with the regional ensemble approach [25,45] shows an r^2 of 0.312 and RPD and APD of −2.8 and 66.1% for the MY multisensor datasets and an r^2 of 0.324 and RPD and APD of 4 and 51.2% for the NRT ad MY OLCI 300 m datasets. The EANs for the OLCI results are consistent with the MY multisensor datasets even with a lower number of matchups available from 2016 to date (460 vs. 2070). These matchup statistics may appear unsatisfactory, but represent an adequate performance for the CDOM-dominated optically complex waters of the Baltic Sea and an improvement of those reported in [25,45] based on a more limited in situ dataset.

The validation of CHL retrieval for ARC based on the new machine learning pan-Arctic regional algorithm [35] shows an r^2 of 0.681 and a RMSD of 0.268 for the MY multisensor datasets and an r^2 of 0.75 and RMSD of 0.215 for the NRT and MY OLCI 300 m datasets. The OLCI results should be interpreted with caution as they may be influenced by the lower number of matchups (21 vs. 323) and presented only to provide a preliminary indication of performance.

For the validation of the CHL datasets in the HROC coastal products based on the merging approach developed for European waters [47–50], the matchups between satellite and in situ observations when assessed over four regions together (BAL, NWS, IBI, MED), follow the 1:1 line (slope = 0.90) with a slight overestimation across the range <1 to 10 mg m^{-3}. A wider dispersion of points is observed (r^2 = 0.48), which can be expected because of the use of the broad spectral bands particularly in the blue region (443 nm and

490 nm) for CHL estimation. Additionally, the median time difference of 70 min between in situ and satellite measurements in dynamic coastal zones also contributes to the higher dispersion. For each of the four single regions (i.e., BAL, NWS, IBI, MED) the matchup statistics are based on a limited number of observations. Furthermore, it should be noted that matchup results for CHL in the ARC and BLK region are not present in this matchup analysis due to the scarcity of suitable in situ records available from 2020 onwards; hence these datasets are released for community evaluation.

4. Contributions to Environmental Reporting

Within CMEMS, OCTAC also contributes to ocean monitoring through the distribution of specific operational indicators delivering information on the state, variability, and change of CHL for all regions. To ensure state-of-the-art ocean monitoring in real time, the ocean monitoring indicator (OMI) framework requires that the indicator time series, their visualization, a description, and additional documents such as product and quality information are updated regularly in an operational mode [5]. As an illustration of the contributions to operational ocean monitoring, Figure 4 presents the Mediterranean Sea CHL trend analysis, derived from two operational OMIs (1997–2023) of satellite CHL based on the CMEMS L4 product.

The trend analysis (Figure 4A) shows that the basin is undergoing a general biomass decrease, in particular starting from 2011. Surface ocean warming [77,78] translates into stronger thermal stratification of the water column for increasingly extended periods. In turn, this implies a progressive nutrient decline into the upper mixed layer, which likely forces the phytoplankton vertical distribution with a deep CHL maximum (DCM) persisting in recent years for longer periods than ever before. From the remote sensing point of view, this has the effect of reducing the phytoplankton biomass resident time in the upper layer where the satellite sensors can effectively observe them. Therefore, the trend of $-0.73 \pm 0.65\%$ per year should be considered as an upper limit as part of this contribution could have more simply been undetected by OC remote sensing.

Spatially, the CHL trend is not uniform, with only a few areas characterized by positive values: in the Alboran Sea, Sicily Channel, and SE of Crete (Figure 4B). The rest of the basin is characterized by a negative trend with higher magnitude in the western region. This is in line with the hypothesis of decreasing nutrient availability in the upper mixed layer. In fact, on one side, the eastern basin is already characterized by a DCM-dominated phytoplankton vertical distribution structure [79] with less impact over the remote sensing observations. Furthermore, Pisano et al. [77] reported that the area most affected by the general warming trend in the Mediterranean Sea is the western sector, where the reduction in phytoplankton biomass in spring was recently documented by combining autonomous observations from BioGeoChemical-Argo floats, satellite-based, and marine ecosystem modeling [80].

As an example of the CMEMS contributions to Sustainable Development Goal (SDG) reporting, Figure 5 presents the 1998–2023 time series of the potential eutrophication (PE) for European waters based on the OC regional products. The SDG reporting for 14.1.1a Level 2 sub-indicator for European countries, which measures the index of coastal eutrophication, is carried out operationally by CMEMS in a harmonized, consistent, and integrated manner using satellite-derived CHL-a data to generate a single variable indicator [81,82]. The methodology for reporting on indicator 14.1.1a builds on the UNEP (United Nations Environment Programme) progressive monitoring approach based on both globally and nationally derived data and supplemental data to report on SDG indicators [82]. For each year, a satellite-based map of potential eutrophic areas in the European Seas is generated by comparing the per-pixel CHL-a data from the MY regional products in the reporting year with the corresponding CHL-a climatological 90th percentile (P90) established for a 20-year baseline (1998–2017) [81]. Then, the PE time series of PE potential eutrophication is calculated by performing for each year a spatial average of the PE map, weighted by pixel area over the exclusive economic zones (EEZs) of each European country [81].

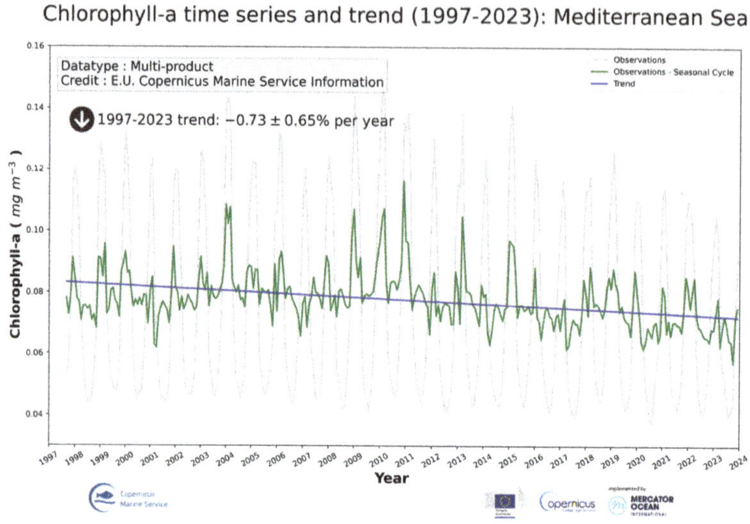

(A)

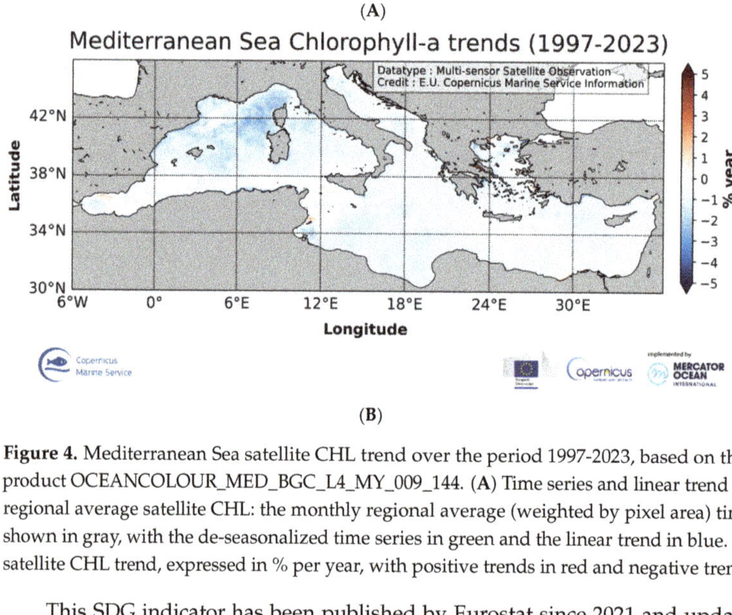

(B)

Figure 4. Mediterranean Sea satellite CHL trend over the period 1997-2023, based on the CMEMS product OCEANCOLOUR_MED_BGC_L4_MY_009_144. (**A**) Time series and linear trend of monthly regional average satellite CHL: the monthly regional average (weighted by pixel area) time series is shown in gray, with the de-seasonalized time series in green and the linear trend in blue. (**B**) Map of satellite CHL trend, expressed in % per year, with positive trends in red and negative trends in blue.

This SDG indicator has been published by Eurostat since 2021 and updated every year [83–86]; the data presented in Figure 5 are publicly available on the Eurostat data browser [87]. Due to the full reprocessing of the underlying satellite products, some of the reported values differ from those reported in previous years (e.g., [81,83]), while the overall picture remains consistent. The data computed in 2024 showed minor changes for the SDG eutrophication indicator for the Mediterranean and Black Sea, while for the Baltic countries some of the values changed significantly. The values for the Atlantic countries changed for data for 2022 due to the consolidation of the MY time series. For several countries, the SDG indicator at the EEZ level was often nil or never exceeded 1% of the EEZ area (Figure 5). Some notable deviations from the CHL climatology are evident, e.g., the high PE values observed for four Baltic countries (Lithuania, Latvia, Poland, and Sweden) in 2008 capturing the extended spring bloom reported for the central and southern Baltic Sea [45]. From 2012 onwards, all European countries yielded a eutrophication index lower than

2%, consistent with the findings based on ensemble analyses of bio-geochemical models for all European seas [45], in situ and satellite data for the Atlantic and Baltic regions (e.g., [45,88–91]), and the CMEMS OMIs as shown for the Mediterranean sea (Figure 4).

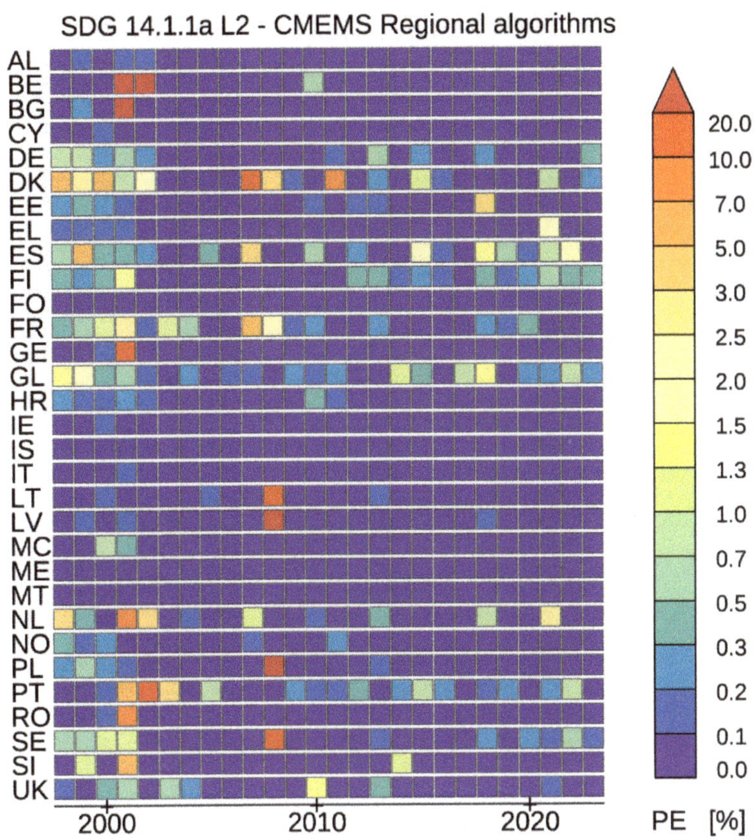

Figure 5. Time series (1998–2023) of SDG 14.1.1a Level 2 sub-indicator for European countries. The potential eutrophication values for European waters are based on CMEMS OC regional products aggregated over the EEZ for each country. AL: Albania, BE: Belgium, BG: Bulgaria, CY: Cyprus, DE: Germany, DK: Denmark, EE: Estonia, EL: Greece, ES: Spain, FI: Finland, FO: Faroe Islands, FR: France, GE: Georgia, GL: Greenland, HR: Croatia, IE: Ireland, IS: Iceland, IT: Italy, LT: Lithuania, LV: Lat- via, MC: Monaco, ME: Montenegro, MT: Malta, NL: Netherlands, NO: Norway, PL: Poland, PT: Portugal, RO: Romania, SE: Sweden, SI: Slovenia, UK: United Kingdom.

5. Future Evolutions

To ensure the state of the art of OC global and regional products, OCTAC will continue to focus on increasing the number and the accuracy of the EOVs included in the CMEMS catalogue, aiming to include most of, if not all, the EOVs that can be retrieved from OC radiometry. This will entail the uptake within the operational processing chains of algorithms and approaches developed within CMEMS and by the OC community.

In 2025–2028, the introduction of new products and/or resolution in the catalogue will be based on the uptake of outcomes from Copernicus Marine Service Evolution projects [92] and some recent internal development activities:

- Introduction of multi-resolution SPM/TUR/CHL products based on harmonized S2 and S3 products generated using multi-resolution data interpolation techniques algorithms [93–95]. The same approach will also improve the daily L4 products for the Sentinel-2 coastal high-resolution products.

- New datasets for particulate and dissolved organic carbon will be added to the catalogue for open and coastal waters and managing the transitions between the two domains [64,65,96]. Basically, POC (particulate organic carbon) estimates result in a combination of different algorithms as a function of the optical water classes [64,96]. For the retrieval of DOC (dissolved organic carbon), the main contributor to organic carbon over open ocean water, the algorithm proposes an innovative approach considering a temporal window to account for the fate of the organic matter with a neural-net [65,96].
- Update of the phytoplankton functional types (PFTs) retrieval algorithms in the catalogue to build a more consistent time series over the OC archive based on a recalibration of the OLCI product [54,55,97].
- The generation of gap-filled R_{rs} fields using the DINEOF technique from which all subsequent biogeochemical parameters will be retrieved [98]. The introduction of gap-filled R_{rs} and IOPs datasets in L4 products will be carried out in the regional multisensor datasets for BLK, MED and BAL.

Furthermore, to enhance the accuracy of the global and regional products, new efforts will be dedicated to:

- Update ATL products based on the Copernicus-GlobColour processor [7] for the coastal waters in the North Sea to support the OSPAR (OSlo PARis convention) requirements for eutrophication assessment. This will also entail updating the OC5 algorithms for CHL and SPM retrieval [33] to be in line with the latest version of the sensor reprocessing.
- Update of the strategies to merge data from sensors with different sets of central wavelengths and spectral response functions in the processing chains. In particular, the reference sensor will be changed to OLCI.
- Extend the applicability over water types of the blended approaches for CHL retrieval in the optically complex waters in ARC, BAL, BLK, and MED, also further incorporating machine learning approaches.
- Uptake of the new L2 operational reflectance product dedicated to water applications for the Sentinel-2 MSI to be released by ESA, and regionalization of the CHL algorithms at the basin level for the Sentinel-2 MSI products to be consistent with the OLCI and multi-resolution products for each European sea.
- Full reprocessing of the MY time series to incorporate major changes to the upstream satellite data carried out by the space agencies, and the improvements listed above.

The continuous and sustained operational data stream of both observational classes currently in use (i.e., OC sensors and high-resolution imagers) is foreseen to continue beyond 2030. In 2025–2028, OCTAC will thus carry out dedicated assessments and studies to prepare the introduction of future missions and data streams in the catalogue. The assessment and uptake of the NASA PACE (Plankton, Aerosol, Cloud, Ocean Ecosystem) science mission will serve to prepare for the future exploitation of the Copernicus Sentinel 10 CHIME (Copernicus Hyperspectral Imaging Mission for the Environment) hyperspectral data currently planned for launch in 2028 and 2030 (CHIME-A and CHIME-B, Figure 3) and the Sentinel-3 Next Generation AOLCI that will be launched in the 2032–2035 time frame. After 2028, a further change in the resolutions will be the uptake of (sub-)hourly datasets for the European basins based on geostationary data by incorporating EUMETSAT MSG (Meteosat Second Generation) and MTG GEO-OC (Meteosat Second Generation Geostationary Ocean Colour Product) data-streams as well as multiple OLCI and VIIRS overpasses. This will enable to match the sub-daily time scales of the operational modelling effort within CMEMS, thus strengthening the potential for data assimilation.

Author Contributions: Conceptualization, V.E.B. and R.S.; methodology, V.E.B., T.K., D.D., K.S., D.V.d.Z., C.L. (Chiara Lapucci), Q.V., M.L.Z., G.V., A.D.C., L.G.V., A.M., Q.J., M.B. (Marine Bretagnon), P.B., M.S. and S.S.; software, M.B. (Martin Böttcher), Q.V., S.C., G.V., V.F., F.L.P., L.G.V., J.D., M.B. (Marine Bretagnon), Q.J., J.N. and B.C.; validation, S.C., L.G.V., V.E.B., A.D.C., K.S., D.V.d.Z., C.L. (Carole Lebreton), M.L.Z., M.B. (Marine Bretagnon), J.D., Q.J., M.S., S.S. and J.N.; formal analysis, M.L.Z., G.V., A.D.C., S.C., L.G.V. and S.S.; investigation, T.K., D.D., D.V.d.Z., Q.V., M.L.Z. and S.S.; resources, M.B. (Martin Böttcher), V.F. and F.L.P.; data curation, M.B. (Martin Böttcher), C.L. (Carole Lebreton), J.N., J.D., V.F., and F.L.P.; writing—original draft preparation, V.E.B., E.B., C.L. (Chiara Lapucci), G.V., and A.D.C.; writing—review and editing, D.D., K.S., D.V.d.Z., M.L.Z., M.S., Q.J. and B.C.; visualization, V.E.B., S.C. and K.S.; project administration, V.E.B., R.S., E.B. and C.C.; funding acquisition, V.E.B. and R.S. All authors have read and agreed to the published version of the manuscript.

Funding: This work has been performed in the context of the Copernicus Marine Service (2015–2021: 77-CMEMS-TAC-OC, 2021–2024: 21001L2-COP-TAC OC-2200).

Data Availability Statement: This study has been conducted using E.U. Copernicus Marine Service Information. All products are available on the Copernicus Marine Service portal at http://marine.copernicus.eu (accessed on 22 October 2024) as detailed in Table 2.

Acknowledgments: We are grateful to all collaborators who contributed to the development, refinement, and implementation of the algorithms and processing chains during the last decade. The data providers are acknowledged for the in situ data used for the algorithms calibration and validation, as detailed in each of the quality information documents and the published papers. We thank Vittorio Barale and the four anonymous reviewers for their valuable comments on the manuscript.

Conflicts of Interest: Authors Antoine Mangin, Quentin Jutard, Marine Bretagnon, Philippe Bryère, and Julien Demaria were employed by the company ACRI-ST S.A.S. Authors Davide D'Alimonte and Tamito Kajiyama were employed by the company Aequora. Authors Kerstin Stelzer, Martin Böttcher and Carole Lebreton were employed by the company Brockmann Consult GmbH. The remaining authors declare that the research was conducted in the absence of any commercial or financial relationships that could be construed as a potential conflict of interest.

References

1. von Schuckmann, K.; Le Traon, P.-Y.; Smith, N.; Pascual, A.; Djavidnia, S.; Gattuso, J.-P.; Grégoire, M.; Nolan, G.; Aaboe, S.; Aguiar, E.; et al. Copernicus Marine Service Ocean State Report 3. *J. Oper. Oceanogr.* **2019**, *12* (Suppl. S1), S1–S123. [CrossRef]
2. Von Schuckmann, K.; Le Traon, P.Y.; Menna, M.; Martellucci, R.; Notarstefano, G.; Mauri, E.; Gerin, R.; Pacciaroni, M.; Bussani, A.; Pirro, A.; et al. Copernicus Ocean State Report, Issue 6. *J. Oper. Oceanogr.* **2022**, *15* (Suppl. S1), s1–s220.
3. Letraon, P.; Ali, A.; Fanjul, E.A.; Aouf, L.; Axell, L.; Aznar, R.; Ballarotta, M.; Behrens, A.; Benkiran, M.; Bentamy, A.; et al. The Copernicus marine environmental monitoring service: Main scientific achievements and future prospects. *Mercat. Ocean. J.* **2017**, *56*, 101.
4. Le Traon, P.Y.; Reppucci, A.; Fanjul, E.A.; Aouf, L.; Behrens, A.; Belmonte, M.; Bentamy, A.; Bertino, L.; Brando, V.E.; Kreiner, M.B.; et al. From Observation to Information and Users: The Copernicus Marine Service Perspective. *Front. Mar. Sci.* **2019**, *6*, 234. [CrossRef]
5. Le Traon, P.; Abadie, V.; Ali, A.; Behrens, A.; Staneva, J.; Hieronymi, M.; Krasemann, H. The Copernicus Marine Service from 2015 to 2021: Six years of achievements. *Mercat. Ocean. J.* **2021**. [CrossRef]
6. Volpe, G.; Colella, S.; Brando, V.E.; Forneris, V.; Padula, F.L.; Di Cicco, A.; Sammartino, M.; Bracaglia, M.; Artuso, F.; Santoleri, R. Mediterranean ocean colour Level 3 operational multi-sensor processing. *Ocean Sci.* **2019**, *15*, 127–146. [CrossRef]
7. Garnesson, P.; Mangin, A.; Fanton d'Andon, O.; Demaria, J.; Bretagnon, M. The CMEMS GlobColour chlorophyll a product based on satellite observation: Multi-sensor merging and flagging strategies. *Ocean Sci.* **2019**, *15*, 819–830. [CrossRef]
8. Sathyendranath, S.; Brewin, R.; Brockmann, C.; Brotas, V.; Calton, B.; Chuprin, A.; Cipollini, P.; Couto, A.; Dingle, J.; Doerffer, R.; et al. An Ocean-Colour Time Series for Use in Climate Studies: The Experience of the Ocean-Colour Climate Change Initiative (OC-CCI). *Sensors* **2019**, *19*, 4285. [CrossRef]
9. Sathyendranath, S.; Brewin, R.J.W.; Jackson, T.; Mélin, F.; Platt, T. Ocean-colour products for climate-change studies: What are their ideal characteristics? *Remote Sens. Environ.* **2017**, *203*, 125–138. [CrossRef]
10. Dekker, A.G.; Pinnel, N.; Gege, P.; Briottet, X.; Peters, S.; Turpie, K.R.; Sterckx, S.; Costa, M.; Giardino, C.; Brando, V.E.; et al. *Feasibility Study for an Aquatic Ecosystem Earth Observing System Version 1.2*; Committee on Earth Observation Satellites (CEOS) and Commonwealth Scientific and Industrial Research Organization: Canberra, Australia, 2018.

11. GOOS. Global Ocean Observing System, Essential Ocean Variables Specification Sheet: Ocean Colour. 2018. Available online https://www.goosocean.org/index.php?option=com_oe&task=viewDocumentRecord&docID=19959 (accessed on 22 October 2024).
12. Qin, Y.; Brando, V.E.; Dekker, A.G.; Blondeau-Patissier, D. Validity of SeaDAS Water Constituents Retrieval Algorithms in Australian Tropical Coastal Waters. *Geophys. Res. Lett.* **2007**, *34*, L21603. [CrossRef]
13. Volpe, G.; Santoleri, R.; Vellucci, V.; Ribera d'Alcalà, M.; Marullo, S.; D'Ortenzio, F. The colour of the Mediterranean Sea: Global versus regional bio-optical algorithms evaluation and implication for satellite chlorophyll estimates. *Remote Sens. Environ.* **2007**, *107*, 625–638. [CrossRef]
14. Szeto, M.; Werdell, P.J.; Moore, T.S.; Campbell, J.W. Are the world's oceans optically different? *J. Geophys. Res.* **2011**, *116*, C00H04. [CrossRef]
15. D'Alimonte, D.; Zibordi, G.; Kajiyama, T.; Berthon, J.-F. Comparison between MERIS and regional high-level products in European seas. *Remote Sens. Environ.* **2014**, *140*, 378–395. [CrossRef]
16. Beckers, J.; Rixen, M. EOF Calculations and Data Filling from Incomplete Oceanographic Datasets. *J. Atmos. Ocean.* **2003**, *20*, 1839–1856. [CrossRef]
17. Volpe, G.; Nardelli, B.B.; Colella, S.; Pisano, A.; Santoleri, R. Operational Interpolated Ocean Colour Product in the Mediterranean Sea. In *New Frontiers in Operational Oceanography*; Chassignet, E., Pascual, A., Tintoré, J., Verron, J., Eds.; CreateSpace: Scotts Valley, CA, USA, 2018; pp. 227–244.
18. Mélin, F.; Sclep, G. Band shifting for ocean color multi-spectral reflectance data. *Opt. Express* **2015**, *23*, 2262–2279. [CrossRef]
19. Jackson, T.; Sathyendranath, S.; Groom, S.; Calton, B. ESA Ocean Colour Climate Change Initiative Product User Guide for v6.0 Dataset. 2002. Available online: https://docs.pml.space/share/s/fzNSPb4aQaSDvO7xBNOCIw (accessed on 22 October 2024).
20. Steinmetz, F.; Deschamps, P.-Y.; Ramon, D. Atmospheric correction in presence of sun glint: Application to MERIS. *Opt. Express* **2011**, *19*, 9783–9800. [CrossRef]
21. Steinmetz, F.; Ramon, D. Sentinel-2 MSI and Sentinel-3 OLCI consistent ocean colour products using POLYMER. In Proceedings of the Remote Sensing of the Open and Coastal Ocean and Inland Waters, Honolulu, HI, USA, 24–25 September 2018; Volume 10778, p. 107780E.
22. EUMETSAT. Sentinel-3 OLCI L2 Report for Baseline Collection OL_L2M_003. 2021. Available online: https://user.eumetsat.int/s3/eup-strapi-media/Sentinel_3_OLCI_L2_report_for_baseline_collection_OL_L2_M_003_2_B_c8bbc6d986.pdf (accessed on 22 October 2024).
23. Tilstone, G.H.; Pardo, S.; Simis, S.G.H.; Qin, P.; Selmes, N.; Dessailly, D.; Kwiatkowska, E. Consistency between satellite ocean colour products under high coloured dissolved organic matter absorption in the baltic sea. *Remote Sens.* **2022**, *14*, 89. [CrossRef]
24. Zibordi, G.; Kwiatkowska, E.; Mélin, F.; Talone, M.; Cazzaniga, I.; Dessailly, D.; Gossn, J.I. Assessment of OLCI-A and OLCI-B radiometric data products across European seas. *Remote Sens. Environ.* **2022**, *272*, 112911. [CrossRef]
25. Vilas, L.G.; Brando, V.E.; Di Cicco, A.; Colella, S.; D'alimonte, D.; Kajiyama, T.; Attila, J.; Schroeder, T. Assessment of ocean color atmospheric correction methods and development of a regional ocean color operational dataset for the Baltic Sea based on Sentinel-3 OLCI. *Front. Mar. Sci.* **2024**, *10*, 1256990. [CrossRef]
26. Hieronymi, M.; Müller, D.; Doerffer, R. The OLCI Neural Network Swarm (ONNS): A Bio-Geo-Optical Algorithm for Open Ocean and Coastal Waters. *Front. Mar. Sci.* **2017**, *4*, 140. [CrossRef]
27. Doxani, G.; Vermote, E.; Roger, J.-C.; Gascon, F.; Adriaensen, S.; Frantz, D.; Hagolle, O.; Hollstein, A.; Kirches, G.; Li, F.; et al. Atmospheric Correction Inter-Comparison Exercise. *Remote Sens.* **2018**, *10*, 352. [CrossRef] [PubMed]
28. Doerffer, R.; Schiller, H. The MERIS Case 2 water algorithm. *Int. J. Remote Sens.* **2007**, *28*, 517–535. [CrossRef]
29. Brockmann, C.; Doerffer, R.; Peters, M.; Stelzer, K.; Embacher, S.; Ruescas, A. Evolution of the C2RCC neural network for Sentinel 2 and 3 for the retrieval of Ocean Colour products in normal and extreme optically complex waters. In Proceedings of the Living Planet Symposium, Prague, Czech Republic, 9–13 May 2016; p. 740.
30. Vanhellemont, Q. Adaptation of the dark spectrum fitting atmospheric correction for aquatic applications of the Landsat and Sentinel-2 archives. *Remote Sens. Environ.* **2019**, *225*, 175–192. [CrossRef]
31. Van der Zande, D.; Vanhellemont, Q.; Stelzer, K.; Lebreton, C.; Dille, A.; dos Santos, J.C.; Böttcher, M.; Vansteenwegen, D.; Brockmann, C. Improving operational ocean color coverage using a merged atmospheric correction approach. *Remote Sens. Ocean. Sea Ice Coast. Waters Large Water Reg.* **2023**, *12728*, 12–29. [CrossRef]
32. Hu, C.; Lee, Z.; Franz, B. Chlorophyll a algorithms for oligotrophic oceans: A novel approach based on three-band reflectance difference. *J. Geophys. Res.* **2012**, *117*, C01011. [CrossRef]
33. Gohin, F.; Druon, J.-N.; Lampert, L. A five channel chlorophyll concentration algorithm applied to SeaWiFS data processed by SeaDAS in coastal waters. *Int. J. Remote Sens.* **2002**, *23*, 1639–1661. [CrossRef]
34. Pardo, S.; Jackson, T.; Netting, J.; Calton, B.; Howey, B. Quality information document for OC TAC Products Atlantic and Arctic Observation Products. *Copernicus Marine Service*. 2022. Available online: https://catalogue.marine.copernicus.eu/documents/QUID/CMEMS-OC-QUID-009-111to114-121to124.pdf (accessed on 22 October 2024).
35. Zoffoli, M.L.; Volpe, G.; Brando, V.E.; Pitarch, J.; Gonzalez Vilas, L.; Colella, S. QUID (Quality Information Document) for Arctic Sea Observation Products. *Copernicus Marine Service*, release 2.0. 2023. Available online: https://documentation.marine.copernicus.eu/QUID/CMEMS-OC-QUID-009-121to124.pdf (accessed on 22 October 2024).

36. Goncalves-Araujo, R.; Rabe, B.; Peeken, I.; Bracher, A. High colored dissolved organic matter (CDOM) absorption in surface waters of the central-eastern Arctic Ocean: Implications for biogeochemistry and ocean color algorithms. *PLoS ONE* **2018**, *13*, e0190838. [CrossRef]
37. IOCCG. *Ocean Colour Remote Sensing in Polar Seas*; Babin, M., Arrigo, K., Bélanger, S., Forget, M.-H., Eds.; IOCCG Report Series, No. 16; International Ocean Colour Coordinating Group: Dartmouth, NS, Canada, 2015.
38. D'Alimonte, D.; Zibordi, G. Phytoplankton determination in an optically complex coastal region using a multilayer perceptron neural network. *IEEE Trans. Geosci. Remote Sens.* **2003**, *41*, 2861–2868. [CrossRef]
39. Colella, S.; Brando, V.E.; Di Cicco, A.; D'Alimonte, D.; Forneris, V.; Bracaglia, M. *Quality Information Document for Ocean Colour Mediterranean and Black Sea Observation Product Release 4.0*; Mercator Ocean International: Toulouse, France, 2024. [CrossRef]
40. Kajiyama, T.; D'Alimonte, D.; Zibordi, G. Algorithms merging for the determination of Chlorophyll-a concentration in the Black Sea. *IEEE Geosci. Remote Sens. Lett.* **2018**, *16*, 677–681. [CrossRef]
41. Zibordi, G.; Mélin, F.; Berthon, J.-F.; Talone, M. In situ autonomous optical radiometry measurements for satellite ocean color validation in the Western Black Sea. *Ocean Sci.* **2015**, *11*, 275–286. [CrossRef]
42. Kopelevich, O.; Burenkov, V.; Sheberstov, S.; Vazyulya, S.; Kravchishina, M.; Pautova, L.; Silkin, V.; Artemiev, V.; Grigoriev, A. Satellite monitoring of coccolithophore blooms in the Black Sea from ocean color data. *Remote Sens. Environ.* **2014**, *146*, 113–123. [CrossRef]
43. Zibordi, G.; Berthon, J.-F.; Mélin, F.; D'Alimonte, D. Cross-site consistent in situ measurements for satellite ocean color applications: The BiOMaP radiometric dataset. *Remote Sens. Environ.* **2011**, *115*, 2104–2115. [CrossRef]
44. D'Alimonte, D.; Zibordi, G.; Berthon, J.-F.; Canuti, E.; Kajiyama, T. Bio-Optical Algorithms for European Seas: Performance and Applicability of Neural-Net Inversion Schemes. Technical Report JRC66326, JRC-IES Scientific and Technical Reports. 2011. Available online: https://data.europa.eu/doi/10.2788/56321 (accessed on 22 October 2024).
45. Brando, V.E.; Sammartino, M.; Colella, S.; Bracaglia, M.; Di Cicco, A.; D'Alimonte, D.; Kajiyama, T.; Kaitala, S.; Attila, J. Phytoplankton Bloom Dynamics in the Baltic Sea Using a Consistently Reprocessed Time Series of Multi-Sensor Reflectance and Novel Chlorophyll-a Retrievals. *Remote Sens.* **2021**, *13*, 3071. [CrossRef]
46. Pitarch, J.; Volpe, G.; Colella, S.; Krasemann, H.; Santoleri, R. Remote sensing of chlorophyll in the Baltic Sea at basin scale from 1997 to 2012 using merged multi-sensor data. *Ocean Sci.* **2016**, *12*, 379–389. [CrossRef]
47. Lavigne, H.; Van Der Zande, D.; Ruddick, K.; Cardoso dos Santos, J.; Gohin, F.; Brotas, V.; Kratzer, S. Quality-control tests for OC4, OC5 and NIR-red satellite chlorophyll-a algorithms applied to coastal waters. *Remote Sens. Environ.* **2021**, *255*, 112237. [CrossRef]
48. O'Reilly, J.E.; Werdell, P.J. Chlorophyll algorithms for ocean color sensors-OC4, OC5 & OC6. *Remote Sens. Environ.* **2019**, *229*, 32–47. [CrossRef]
49. Gons, H.J. Optical teledetection of chlorophyll a in turbid inland waters. *Environ. Sci. Technol.* **1999**, *33*, 1127–1132. [CrossRef]
50. Gons, H.J.; Rijkeboer, M.; Ruddick, K.G. Effect of a waveband shift on chlorophyll retrieval from MERIS imagery of inland and coastal waters. *J. Plankton Res.* **2005**, *27*, 125–127. [CrossRef]
51. Brewin, R.J.W.; Sathyendranath, S.; Jackson, T.; Barlow, R.; Brotas, V.; Airs, R.; Lamont, T. Influence of light in the mixed-layer on the parameters of a three-component model of phytoplankton size class. *Remote Sens. Environ.* **2015**, *168*, 437–450. [CrossRef]
52. Brewin, R.J.W.; Ciavatta, S.; Sathyendranath, S.; Jackson, T.; Tilstone, G.; Curran, K.; Airs, R.L.; Cummings, D.; Brotas, V.; Organelli, E.; et al. Uncertainty in Ocean-Color Estimates of Chlorophyll for Phytoplankton Groups. *Front. Mar. Sci.* **2017**, *4*, 104. [CrossRef]
53. Di Cicco, A.; Sammartino, M.; Marullo, S.; Santoleri, R. Regional Empirical Algorithms for an Improved Identification of Phytoplankton Functional Types and Size Classes in the Mediterranean Sea Using Satellite Data. *Front. Mar. Sci.* **2017**, *4*, 126. [CrossRef]
54. Xi, H.; Losa, S.N.; Mangin, A.; Soppa, M.A.; Garnesson, P.; Demaria, J.; Liu, Y.; Hembise Fanton d'Andon, O.; Bracher, A. Global retrieval of phytoplankton functional types based on empirical orthogonal functions using CMEMS GlobColour merged products and further extension to OLCI data. *Remote Sens. Environ.* **2020**, *240*, 111704. [CrossRef]
55. Xi, H.; Losa, S.N.; Mangin, A.; Garnesson, P.; Bretagnon, M.; Demaria, J.; Soppa, M.A.; Hembise Fanton d'Andon, O.; Bracher, A. Global chlorophyll a concentrations of phytoplankton functional types with detailed uncertainty assessment using multi-sensor ocean color and sea surface temperature satellite products. *J. Geophys. Res. Ocean.* **2021**, *126*, e2020JC017127. [CrossRef]
56. Brando, V.E.; Gonzalez Vilas, L.; Di Cicco, A.; Sammartino, M.; Colella, S.; D'Alimonte, D.; Kajiyama, T.; Kaitala, S.; Attila, J. Ocean Colour Production Centre—Baltic Sea Observation Products (Quality Information Document, QUID). 2023. Available online: https://catalogue.marine.copernicus.eu/documents/QUID/CMEMS-OC-QUID-009-131to134.pdf (accessed on 22 October 2024).
57. Sieburth, J.M.; Smetacek, V.; Lenz, J. Pelagic ecosystem structure: Heterotrophic compartments of the plankton and their relationship to plankton size fractions. *Limnol. Oceanogr.* **1978**, *23*, 1256–1263. [CrossRef]
58. Vidussi, F.; Claustre, H.; Manca, B.B.; Luchetta, A.; Marty, J.C. Phytoplankton pigment distribution in relation to upper thermocline circulation in the eastern Mediterranean Sea during winter. *J. Geophys. Res.* **2001**, *106*, 19939–19956. [CrossRef]
59. Uitz, J.; Claustre, H.; Morel, A.; Hooker, S.B. Vertical distribution of phytoplankton communities in open ocean: An assessment based on surface chlorophyll. *J. Geophys. Res.* **2006**, *111*, C08005. [CrossRef]
60. Brewin, R.J.W.; Sathyendranath, S.; Hirata, T.; Lavender, S.J.; Barciela, R.M.; Hardman-Mountford, N.J. A three-component model of phytoplankton size class for the Atlantic Ocean. *Ecol. Model.* **2010**, *221*, 1472–1483. [CrossRef]

61. Hirata, T.; Hardman-Mountford, N.J.; Brewin, R.J.W.; Aiken, J.; Barlow, R.; Suzuki, K.; Isada, T.; Howell, E.; Hashioka, T. Noguchi-Aita, M.; et al. Synoptic relationships between surface Chlorophyll-a and diagnostic pigments specific to phytoplankton functional types. *Biogeosciences* **2011**, *8*, 311–327. [CrossRef]
62. Lee, Z.P.; Carder, K.L.; Arnone, R.A. Deriving inherent optical properties from water color: A multiband quasi-analytical algorithm for optically deep waters. *Appl. Opt.* **2002**, *41*, 5755–5772. [CrossRef]
63. Lee, Z.P.; Carder, K.L.; Arnone, R.A. *Update of the Quasi-Analytical Algorithm (QAA_v6)*; IOCCG Software Report; International Ocean Colour Coordinating Group (IOCCG): Dartmouth, NS, Canada, 2014.
64. Jorge, D.S.F.; Loisel, H.; Jamet, C.; Dessailly, D.; Demaria, J.; Bricaud, A.; Maritorena, S.; Zhang, X.; Antoine, D.; Kutser, T.; et al. A Three-Step Semi Analytical Algorithm (3SAA) for Estimating Inherent Optical Properties over Oceanic, Coastal, and Inland Waters from Remote Sensing Reflectance. *Remote Sens. Environ.* **2021**, *263*, 112537. [CrossRef]
65. Bonelli, A.G.; Loisel, H.; Jorge, D.S.F.; Mangin, A.; d'Andon, O.F.; Vantrepotte, V. A new method to estimate the dissolved organic carbon concentration from remote sensing in the global open ocean. *Remote Sens. Environ.* **2022**, *281*, 113227. [CrossRef]
66. Maritorena, S.; Siegel, D.A. Consistent merging of satellite ocean color data sets using a bio-optical model. *Remote Sens. Environ.* **2005**, *94*, 429–440. [CrossRef]
67. Antoine, D.; Morel, A. Oceanic primary production: 1. Adaptation of a spectral light-photosynthesis model in view of application to satellite chlorophyll observations. *Glob. Biogeochem. Cycles* **1996**, *10*, 43–55. [CrossRef]
68. Frouin, R.; Franz, B.A.; Werdell, P.J. The SeaWiFS PAR product. In *Algorithm Updates for the Fourth SeaWiFS Data Reprocessing*; Hooker, S.B., Firestone, E.R., Eds.; NASA/TM-2003-206892; NASA Goddard Space Flight Center: Greenbelt, MD, USA, 2003; Volume 22, pp. 46–50.
69. CMEMS. *Global Ocean Physics Reanalysis GLORYS12V1 Product*; CMEMS: Ramonville-Saint-Agne, France, 2022. [CrossRef]
70. Morel, A. Light and marine photosynthesis: A spectral model with geochemical and climatological implications. *Prog. Oceanogr.* **1991**, *26*, 263–306. [CrossRef]
71. Tanré, D.; Herman, M.; Deschamps, P.Y.; De Leffe, A. Atmospheric modeling for space measurements of ground reflectances, including bidirectional properties. *Appl. Opt.* **1979**, *18*, 3587–3594. [CrossRef]
72. Morel, A.; Berthon, J.F. Surface pigments, algal biomass profiles, and potential production of the euphotic layer: Relationships reinvestigated in view of remote-sensing applications. *Limnol. Oceanogr.* **1989**, *34*, 1545–1562. [CrossRef]
73. Gregg, W.W.; Casey, N.W. Skill assessment of a spectral ocean-atmosphere radiative model. *J. Mar. Syst.* **2009**, *76*, 49–63. [CrossRef]
74. Hernandez, F.; Smith, G.; Baetens, K.; Cossarini, G.; Garcia-Hermosa, I.; Drévillon, M.; Maksymczuk, J.; Melet, A.; Régnier, C.; Schuckmann, K.V. Measuring performances, skill and accuracy in operational oceanography: New challenges and approaches. In *New Frontiers in Operational Oceanography*; Chassignet, E., Pascual, A., Tintoré, J., Verron, J., Eds.; GODAE OceanView: Mallorca, Spain, 2018; pp. 759–796. [CrossRef]
75. Concha, J.A.; Bracaglia, M.; Brando, V.E. Assessing the Influence of Different Validation Protocols on Ocean Colour Match-up Analyses. *Remote Sens. Environ.* **2021**, *259*, 112415. [CrossRef]
76. Valente, A.; Sathyendranath, S.; Brotas, V.; Groom, S.; Grant, M.; Jackson, T.; Chuprin, A.; Taberner, M.; Airs, R.; Antoine, D.; et al. A compilation of global bio-optical in situ data for ocean colour satellite applications–version three. *Earth Syst. Sci. Data* **2022**, *14*, 5737–5770. [CrossRef]
77. Pisano, A.; Marullo, S.; Artale, V.; Falcini, F.; Yang, C.; Leonelli, F.E.; Santoleri, R.; Buongiorno Nardelli, B. New Evidence of Mediterranean Climate Change and Variability from Sea Surface Temperature Observations. *Remote Sens.* **2020**, *12*, 132. [CrossRef]
78. Marullo, S.; Serva, F.; Iacono, R.; Napolitano, E.; di Sarra, A.; Meloni, D.; Monteleone, F.; Sferlazzo, D.; De Silvestri, L.; de Toma, V.; et al. Record-breaking persistence of the 2022/23 marine heatwave in the Mediterranean Sea. *Environ. Res. Lett.* **2023**, *18*, 114041. [CrossRef]
79. Volpe, G.; Nardelli, B.B.; Cipollini, P.; Santoleri, R.; Robinson, I.S. Seasonal to interannual phytoplankton response to physical processes in the Mediterranean Sea from satellite observations. *Remote Sens. Environ.* **2012**, *117*, 223–235. [CrossRef]
80. Li, M.; Organelli, E.; Serva, F.; Bellacicco, M.; Landolfi, A.; Pisano, A.; Marullo, S.; Shen, F.; Mignot, A.; van Gennip, S.; et al. Phytoplankton spring bloom inhibited by marine heatwaves in the North-Western Mediterranean Sea. *Geophys. Res. Lett.* **2024**, *51*, e2024GL109141. [CrossRef]
81. Brando, V.E.; Pardo, S.; Sathyendranath, S.; Howey, B.; Land, P.; Jackson, T.; Santoleri, R.; Sammartino, M.; Colella, S.; von Schuckmann, K.; et al. Potential eutrophication of European waters using satellite derived chlorophyll following the UN Sustainable Development Goal 14 framework. *J. Oper. Oceanogr.* **2022**, *15* (Suppl. S1), s83–s91. [CrossRef]
82. UNEP. *Understanding the State of the Ocean: A global Manual on Measuring SDG 14.1.1, SDG 14.2.1 and SDG 14.5.1*; UNEP: Nairobi, Kenya, 2021.
83. Eurostat. *Sustainable Development in the European Union. Monitoring Report on Progress Towards the SDGs in an EU Context (Cat. No: KS-03-21-096-EN-N)*; Eurostat: Luxembourg, 2021; ISBN 978-92-76-30698-6. [CrossRef]
84. Eurostat. *Sustainable Development in the European Union. Monitoring Report on Progress Towards the SDGs in an EU Context—2022 Edition (KS-09-22-019-EN-N)*; Eurostat: Luxembourg, 2022. [CrossRef]
85. Eurostat. *Sustainable Development in the European Union. Monitoring Report on Progress Towards the SDGs in an EU Context—2023 Edition (KS-04-23-184-EN-N)*; Eurostat: Luxembourg, 2023. [CrossRef]

86. Eurostat. *Sustainable Development in the European Union. Monitoring Report on Progress Towards the SDGs in an EU Context—2024 Edition (Cat: KS-05-24-071-EN-N)*; Eurostat: Luxembourg, 2024. [CrossRef]
87. Eurostat. *Marine Waters Affected by Eutrophication. Eurostat Data Browser—Online Data Code: Sdg_14_60 (Last Accessed on 22 October 2024)*; Eurostat: Luxembourg, 2024. [CrossRef]
88. Friedland, R.; Macias, D.M.; Cossarini, G.; Daewel, U.; Estournel, C.; Garcia-Gorriz, E.; Grizzetti, B.; Grégoire, M.; Gustafson, B.; Kalaroni, S.; et al. Effects of nutrient management scenarios on marine eutrophication indicators: A Pan-European, multi-model assessment in support of the Marine Strategy Framework Directive. *Front. Mar. Sci.* **2021**, *8*, 596126. [CrossRef]
89. Axe, P.; Clausen, U.; Leujak, W.; Malcolm, S.; Ruiter, H.; Prins, T.; Harvey, E.T.; OSPAR Coimmission. *Eutrophication Status of the OSPAR Maritime Area. Third Integrated Report on the Eutrophication Status of the OSPAR Maritime Area*; OSPAR Commission: London, UK, 2017; p. 165. ISBN 978-1-911458-34-0.
90. Gohin, F.; Van der Zande, D.; Tilstone, G.; Eleveld, M.A.; Lefebvre, A.; Andrieux-Loyer, F.; Blauw, A.N.; Bryère, P.; Devreker, D.; Garnesson, P.; et al. Twenty years of satellite and in situ observations of surface chlorophyll-a from the northern Bay of Biscay to the eastern English Channel. Is the water quality improving? *Remote Sens. Environ.* **2019**, *233*, 111343. [CrossRef]
91. Prins, T.; Enserink, L. Concentrations of Chlorophyll-a in the Greater North Sea, Celtic Seas and Bay of Biscay and Iberian Coast. In *The 2023 Quality Status Report for the Northeast Atlantic*; OSPAR Commission: London, UK, 2023; Available online: https://oap.ospar.org/en/ospar-assessments/quality-status-reports/qsr-2023/indicator-assessments/chl-a-concentrations (accessed on 22 October 2024).
92. CMEMS. *Copernicus Marine Service Evolution Projects*; CMEMS: Ramonville-Saint-Agne, France, 2024; Available online: https://marine.copernicus.eu/about/research-development-projects (accessed on 24 October 2024).
93. Alvera-Azcárate, A.; Van der Zande, D.; Barth, A.; Dille, A.; Massant, J.; Beckers, J.-M. Generation of super-resolution gap-free ocean colour satellite products using DINEOF. *EGUsphere* **2024**. [CrossRef]
94. Alvera-Azcárate, A.; van der Zande, D.V.; Barth, A.; dos Santos, J.F.C.; Troupin, C.; Beckers, J.-M. Detection of shadows in high spatial resolution ocean satellite data using DINEOF. *Remote Sens. Environ.* **2021**, *253*, 112229. [CrossRef]
95. Alvera-Azcárate, A.; van der Zande, D.V.; Barth, A.; Troupin, C.; Martin, S.; Beckers, J.-M. Analysis of 23 Years of Daily Cloud-Free Chlorophyll and Suspended Particulate Matter in the Greater North Sea. *Front. Mar. Sci.* **2021**, *8*, 707632. [CrossRef]
96. Loisel, H.; Duforêt-Gaurier, L.; Tran, T.K.; Schaffer Ferreira Jorge, D.; Steinmetz, F.; Mangin, A.; Bretagnon, M.; Hembise Fanton d'Andon, O. Characterization of the organic vs. inorganic fraction of suspended particulate matter in coastal waters based on ocean color radiometry remote sensing. In *7th Edition of the Copernicus Ocean State Report (OSR7)*; von Schuckmann, K., Moreira, L., Le Traon, P.-Y., Grégoire, M., Marcos, M., Staneva, J., Brasseur, P., Garric, G., Lionello, P., Karstensen, J., et al., Eds.; Copernicus Publications, State Planet: Göttingen, Germany, 2023; Volume 1-osr7, p. 11. [CrossRef]
97. Xi, H.; Bretagnon, M.; Losa, S.N.; Brotas, V.; Gomes, M.; Peeken, I.; Alvarado, L.; Mangin, A.; Bracher, A. Satellite monitoring of surface phytoplankton functional types in the Atlantic Ocean over 20 years (2002–2021). *State Planet* **2023**, *1*, 5.
98. Marchese, C.; Colella, S.; Brando, V.E.; Zoffoli, M.L.; Volpe, G. Towards accurate L4 Ocean Colour products: Interpolating Remote Sensing Reflectance via DINEOF. *Int. J. Appl. Earth Obs. Geoinf.* **2024**, *135*, 104270. [CrossRef]

remote sensing

MDPI

Review

Satellite Oceanography in NOAA: Research, Development, Applications, and Services Enabling Societal Benefits from Operational and Experimental Missions

Eric Bayler [1,*], Paul S. Chang [1], Jacqueline L. De La Cour [1,2], Sean R. Helfrich [1], Alexander Ignatov [1], Jeff Key [1], Veronica Lance [1], Eric W. Leuliette [1], Deirdre A. Byrne [1], Yinghui Liu [1], Xiaoming Liu [1,3], Menghua Wang [1], Jianwei Wei [1,4] and Paul M. DiGiacomo [1]

1 National Oceanic and Atmospheric Administration (NOAA), National Environmental Satellite, Data, and Information Service (NESDIS), Center for Satellite Applications and Research (STAR), College Park, MD 20740, USA; paul.s.chang@noaa.gov (P.S.C.); jacqueline.shapo@noaa.gov (J.L.D.L.C.); sean.helfrich@noaa.gov (S.R.H.); jeff.key@noaa.gov (J.K.); veronica.lance@noaa.gov (V.L.); eric.leuliette@noaa.gov (E.W.L.); deirdre.byrne@noaa.gov (D.A.B.); yinghui.liu@noaa.gov (Y.L.); xiaoming.liu@noaa.gov (X.L.); menghua.wang@noaa.gov (M.W.); jianwei.wei@noaa.gov (J.W.); paul.digiacomo@noaa.gov (P.M.D.)

2 Cooperative Institute for Satellite Earth System Studies, Earth System Science Interdisciplinary Center, University of Maryland, College Park, MD 20740, USA

3 Cooperative Institute for Research in the Atmosphere (CIRA), Colorado State University, Fort Collins, CO 80521, USA

4 Global Science and Technology, Inc., Greenbelt, MD 20770, USA

* Correspondence: eric.bayler@noaa.gov; Tel.: +1-301-683-3519

Abstract: The National Oceanic and Atmospheric Administration's (NOAA) Center for Satellite Applications and Research (STAR) facilitates and enables societal benefits from satellite oceanography, supporting operational and experimental satellite missions, developing new and improved ocean observing capabilities, engaging users by developing and distributing fit-for-purpose data, applications, tools, and services, and curating, translating, and integrating diverse data products into information that supports informed decision making. STAR research, development, and application efforts span from passive visible, infrared, and microwave observations to active altimetry, scatterometry, and synthetic aperture radar (SAR) observations. These efforts directly support NOAA's operational geostationary (GEO) and low Earth orbit (LEO) missions with calibration/validation and retrieval algorithm development, implementation, maintenance, and anomaly resolution, as well as leverage the broader international constellation of environmental satellites for NOAA's benefit. STAR's satellite data products and services enable research, assessments, applications, and, ultimately, decision making for understanding, predicting, managing, and protecting ocean and coastal resources, as well as assessing impacts of change on the environment, ecosystems, and climate. STAR leads the NOAA Coral Reef Watch and CoastWatch/OceanWatch/PolarWatch Programs, helping people access and utilize global and regional satellite data for ocean, coastal, and ecosystem applications.

Keywords: satellite; oceanography; research; development; operational

1. Introduction

Satellite remote sensing observations of the ocean, including high-latitude regions and trans-boundary coastal zones, provide significant value through synoptic measurements of geophysical properties, with combined spatiotemporal resolution generally not available from in situ observations. Exploiting satellite ocean observations serves both situational awareness needs and prediction capabilities across short- and long-term time scales. Developing and providing tools and services for data science and informatics increases the utility of ocean satellite observations through enhanced discovery, access, visualization, and comprehension. User-driven ocean and coastal applications create and extract synergistic value from these satellite observations through integrated (e.g., with in situ measurements and/or modeling data) products and derived information, serving to inform decision making for society's benefit. This discussion highlights ongoing operational and experimental satellite oceanography activities within the United States (US), specifically within the US National Oceanic and Atmospheric Administration (NOAA).

The Center for Satellite Applications and Research (STAR), located within NOAA's National Environmental Satellite, Data and Information Service (NESDIS), performs extensive science-based satellite oceanography activities, spanning innovative research, algorithm developmental and sustainment efforts, calibration/validation, application development, transitions to routine and sustained use and operations, facilitating and enabling informed decision making for societal benefit. STAR engages with diverse users and stakeholders to understand and support their needs and missions, providing fit-for-purpose ocean data and derived products that are co-designed, co-developed, and co-produced. In addition to NOAA and other US domestic users, STAR actively supports the ongoing United Nations (U.N.) Decade of Ocean Science for Sustainable Development (U.N. Ocean Decade) and the broader global community of users, particularly within developing nations.

With NOAA being an operational agency, STAR's contributions encompass the user-driven value chain: supporting identification of user needs, developing new and improved observing capabilities, exploiting observations (remote and in situ), disseminating observations and derived information for societal benefit, and soliciting feedback to enhance existing and planned products and data streams. With the increasing volume and diversity of satellite data, STAR provides valuable support through curating, translating, and integrating multi-sensor, multi-platform, and multi-parameter data that enable improved societal outcomes. Increasingly, this support means coupling environmental and social science datasets across terrestrial, aquatic, atmospheric, and cryospheric domains.

Actively supporting and contributing to domestic and international missions and global fora, STAR aims to develop and provide satellite ocean observations foundational to addressing critical societal issues. These research and development (R&D) efforts support NOAA's operational satellite programs, in both geostationary (GEO) and low Earth orbit (LEO), as well as enable NOAA efforts to leverage the broader international constellation of operational and experimental satellite missions. This support includes extensive participation and engagement with the Committee on Earth Observation Satellites (CEOS), and the Coordination Group for Meteorological Satellites (CGMS), amongst other global entities and programs.

The R&D enterprise executed by STAR targets innovation, enhancement, and exploitation of satellite earth observations across the spectrum of sensing capabilities and applications, spanning radiances, imagery, and derived parameters. These efforts enhance understanding and representation of the ocean state, ocean dynamics, and surface conditions (sea surface height (SSH), sea surface salinity (SSS), sea surface temperature (SST), roughness, sea ice, ocean color, etc.), spanning near-real time to climate temporal scales. Active (scatterometry and synthetic aperture radar (SAR)) and passive technologies and techniques are used to measure and interpret ocean surface roughness, producing developmental and operational products that extract routine and extreme ocean surface vector winds (OSVWs), e.g., tropical cyclones. Additionally, SAR is exploited to provide the capability to assess inundation impact from storm surge, as well as ocean–atmosphere and ocean–atmosphere–sea ice fluxes, factors critical to improving coupled Earth system (ocean, atmosphere, cryosphere) predictions. Efforts also target satellite observations of cryospheric parameters, such as altimetric observations of sea ice thickness and SAR observations characterizing sea ice, aiming for new parameters and higher-resolution assessments to support higher-resolution modeling and prediction. Beyond physical parameters, STAR's developments in satellite ocean color radiometry (OCR) enable ocean biological and biogeochemical measurements that support ecosystem assessments and predictions. These efforts actively contribute to collaborative international scientific coordination groups, such as the Group for High-Resolution Sea Surface Temperature (GHRSST), the International Ocean Colour Coordinating Group (IOCCG), and the Ocean Surface Topography Science Team (OSTST).

Collectively, STAR science teams support user and stakeholder needs for ocean data and information across NOAA's diverse mission. Facilitating timely uptake and effective utilization of these satellite oceanography products, extensive user engagement, training, and support activities are provided by CoastWatch/OceanWatch/PolarWatch, a NOAA Program led and managed by STAR, commonly referred to collectively as "CoastWatch". CoastWatch's structural design facilitates collaboration across NOAA, with its programmatic structure including "Regional Nodes", comprising cross-NOAA teams dedicated to addressing the specific data product and information needs of a particular geographic region and/or specific thematic interests. The OceanWatch component provides global coverage and services for diverse users, leveraging the value proposition of satellite ocean data to address the common data and information needs shared by all nations and communities relative to the ocean. STAR also operates Coral Reef Watch for NOAA, which similarly bridges the gap between satellite data providers and users for these socio-economically and ecologically important ecosystems. Timely, accurate, and sustained data products and information are essential for these at-risk regions threatened by climate change and local anthropogenic impacts.

The paper's Abbreviations Section provides the definitions for the acronyms used.

2. Satellites, Sensors, and Calibration/Validation

2.1. Instruments and Sensors

Supporting satellite missions and sensors characterizes STAR's core role within NESDIS and NOAA. This role includes providing underpinning scientific R&D expertise for implementing NOAA's operational satellite missions, as well as for leveraging complementary and gap-filling partner satellite observations. Associated science efforts address the development and refinement of satellite ocean observations across the spectrum of instruments and phenomenology, aiming to exploit passive (visible, infrared, microwave) and active (altimetry, synthetic aperture radar (SAR), and scatterometry) capabilities to span physical, biological, and biogeochemical parameters for ocean-related applications, e.g., ecological forecasting.

Complementing development of satellite ocean observation retrieval algorithms, STAR performs critical developmental and operational calibration and validation support for NOAA's GEO and LEO missions. Current NOAA satellite missions with ocean-related sensors include the GOES-R series (specifically the GOES-R series Advanced Baseline Imager (ABI) [1]), the Joint Polar Satellite System (JPSS) series (Visual Infrared Imaging Radiometer Suite (VIIRS) [2], and the Jason/Sentinel-6 series altimeters [3].

Beyond NOAA's own satellite ocean observations, STAR leverages non-NOAA partner satellite observations for operational robustness, greater coverage and resolution, and for providing observations not available from NOAA's satellites. An aim is to exploit both operational and exploratory satellite missions of domestic and international partners, from LEO and GEO platforms for more intensive coverage of non-US waters. Examples include passive microwave observations from GCOM-W1 (AMSR2) [4] and SMAP [5], scatterometry from METOP (ASCAT) [6], and synthetic aperture radar (SAR) from the RADARSAT Constellation Mission [7] and the Sentinel-1 mission series [8]. Leveraged gap-filling capabilities include sea surface salinity, sea surface temperature, ocean surface vector winds, sea ice detection and characteristics through non-optical methods, oil spill detection and monitoring, and illegal, unreported, and unregulated (IUU) fishing detection and monitoring.

2.2. Technical Developments

STAR's satellite oceanography portfolio exploits passive and active capabilities across the electromagnetic spectrum. Representative examples follow, organized by the portion of the spectrum exploited and by passive/active sensing, noting that these highlights are not intended to be an exhaustive accounting of all related STAR ocean remote sensing activities, products, and services.

2.2.1. Visible Capabilities

Within the visible portion of the electromagnetic spectrum, STAR maintains a strong R&D program for retrievals and applications of ocean and freshwater parameters derived from "ocean color", which include chlorophyll-a (Chl-a) concentration, colored dissolved organic matter (CDOM), suspended particulate matter (SPM), water diffuse attenuation coefficient parameters, and various water classification products. These and other satellite ocean color data products are crucial for ocean/inland water environmental monitoring and biological, biogeochemical, and ecological research and applications (e.g., water quality and habitat assessments, aquaculture, and fishery management). With multi-sensor merged and global gap-free ocean color data, NOAA STAR continues to improve the surveillance and forecast of harmful algal blooms (HABs), coastal eutrophication, and changes in phytoplankton dynamics, food webs, and biological productivity due to climate change. This section highlights recent progress in ocean color remote sensing for generating routine global ocean color data, as well as new products for ocean, coastal, and inland water research and applications.

Global Ocean Color Product Data

Satellite ocean color products include normalized water-leaving radiance spectra ($nL_w(\lambda)$, i.e., ocean color) [9,10], Chl-a concentration [11], Chl-a anomaly products [12], water diffuse attenuation coefficient at 490 nm ($K_d(490)$) [13] and for photosynthetically available radiation (PAR) (K_d(PAR)), SPM concentration [14], and water class [15]. Satellite-derived Chl-a and consequently Chl-a anomaly (from Chl-a) data provide continuous global estimations of ocean phytoplankton concentration (biomass) and variations, which are used for monitoring HABs, ocean/water biological productivity, and other ocean/inland water environmental processes. Diffuse attenuation coefficient $K_d(490)$ data enable the monitoring of ocean water quality and the study of ocean processes, such as thermal dynamics and phytoplankton photosynthesis. SPM data contribute to quantifying water clarity across the world ocean and inland lakes, facilitating sediment transportation modeling and ocean circulation tracing, as well as land–ocean flux and global carbon cycle studies. These ocean color products are being routinely produced from VIIRS [16] onboard the Suomi National Polar-orbiting Partnership (SNPP), NOAA-20, and NOAA-21, the Ocean and Land Colour Instrument (OLCI) [17] on the Sentinel-3A (S3A) and Sentinel-3B (S3B) satellites, and the Second-Generation Global Imager (SGLI) on the Global Change Observation Mission-Climate (GCOM-C) satellite. The discussion below reviews ocean biological and biogeochemical products routinely produced in NOAA/STAR, highlighting some new ocean color products.

Chlorophyll-a Data

Satellite Chl-a data, derived using the ocean color index (OCI) method [11,18], employs satellite-derived $nL_w(\lambda)$ for the blue, green, and red bands. Satellite Chl-a accuracy over oligotrophic oceans (i.e., low Chl-a) can be significantly improved using the OCI-based approach, greatly reducing noise and bias errors [11], as STAR has shown, using in situ optics measurements from NOAA's Marine Optical Buoy (MOBY) [19]. Consequently, the OCI Chl-a algorithm has been implemented in the NOAA MSL12 enterprise ocean color data processing system to derive Chl-a data globally from all relevant satellite sensors [20,21]. Due to sensor spectral band differences from various satellite sensors, an algorithm makes appropriate adjustments for specific sensors to produce consistent Chl-a data across the different satellite sensors [22]. Consequently, Chl-a data from VIIRS, OLCI, and SGLI can be effectively merged, producing consistent multi-sensor global Chl-a data products. Additionally, global Chl-a anomaly products have been implemented in the

NOAA MSL12 ocean color data processing system and are being routinely generated [23] for various near-real-time applications, including global HAB monitoring.

Diffuse Attenuation Coefficient $K_d(490)$

The water diffuse attenuation coefficient at 490 nm, $K_d(490)$, an important water quality parameter, characterizes light penetration/dissipation with respect to depth in oceanic and aquatic systems, which significantly impacts primary productivity [24]. STAR developed satellite $K_d(490)$ observations using combined empirical (open oceans) and semi-analytical (turbid waters) approaches for deriving consistent $K_d(490)$ data [25]. The $K_d(490)$ algorithm uses $nL_w(\lambda)$ at the blue and green bands for open oceans, while employing $nL_w(\lambda)$ at the red band to relate the particle backscattering coefficient at 490 nm to $K_d(490)$ [26]. With proper accounting for sensor spectral band variations in the $K_d(490)$ algorithm, global $K_d(490)$ data derived from various sensors can be accurately merged [27].

New Global Suspended Particulate Matter Product

Suspended particulate matter (SPM) quantity, comprising organic and inorganic, living and nonliving particulates, is estimated as the total dry weight per unit volume of water. SPM assessments for the global ocean have been difficult due to the high spatiotemporal variability that spans a dynamic range of approximately 0.01–2000 mg L^{-1}. STAR's inversion model retrieves SPM from the remote sensing reflectance ($R_{rs}(\lambda)$ or $nL_w(\lambda)$) in the near-infrared (NIR), red, green, and blue bands (NIR-RGB) [28,29], employing algorithms separately developed for "turbid" and "clear" waters, demarcated by a threshold pertinent to $R_{rs}(671)$. Validation with in situ water filtration data shows a median absolute percentage difference of approximately 35–39% overall. NOAA employs original swath projection sensor radiance (Level-1) data to routinely generate a swath projection SPM (Level-2) data product (approximately 750 m resolution) from VIIRS observations, as well as gridded (Level-3) products (2 and 9 km, daily, 8 days, and monthly). Figure 1a depicts the VIIRS-derived SPM global climatology.

New Water Class Product

Satellite-based optical water classification can distinguish water bodies with different bio-optical and biogeochemical properties. Recently, STAR devised and implemented a new water classification model [15] based on a clustering analysis of hyperspectral $R_{rs}(\lambda)$ spectra. This model globally resolves oceanic, coastal, and inland waters into 23 contiguous classes, varying from purple-blue waters to green waters and yellow waters. Distinct bio-optical and biogeochemical properties separate these water classes, such as Chl-a, SPM, $K_d(490)$, light absorption coefficient of phytoplankton ($a_{ph}(443)$), light absorption coefficient of detritus and colored dissolved organic matter (CDOM) ($a_{dg}(443)$), and light backscattering coefficient of particles ($b_{bp}(443)$). Figure 1b depicts the VIIRS-derived water class climatology (2012–2023) for the global oceans and inland waters. For open oceans, the water classes predominantly comprise Classes 1–6, while the optically more complex coastal oceans and inland waters are characteristically dominated by Classes 7–23. This water classification data, a new addition to the satellite ocean color product list, can be exploited for many applications, such as water quality, ocean ecology, and ocean color uncertainty estimation.

Global Gap-Free Ocean Color Data

Satellite-derived daily ocean color product images can often have a significant number of missing pixels in a single-sensor daily image due to cloud cover, narrow swath width, high satellite viewing angle, high solar zenith angle, contamination by high sun glint, etc. Since these missing pixels often partially or completely obscure ocean features of interest, such as oceanic mesoscale/submesoscale eddies and fronts, STAR merges daily ocean color observations from multiple sensors to reduce the missing data, using a STAR-developed innovative method to completely fill data gaps, generating gap-free ocean color products. Discussion of STAR's routinely generated global gap-free ocean color products follows.

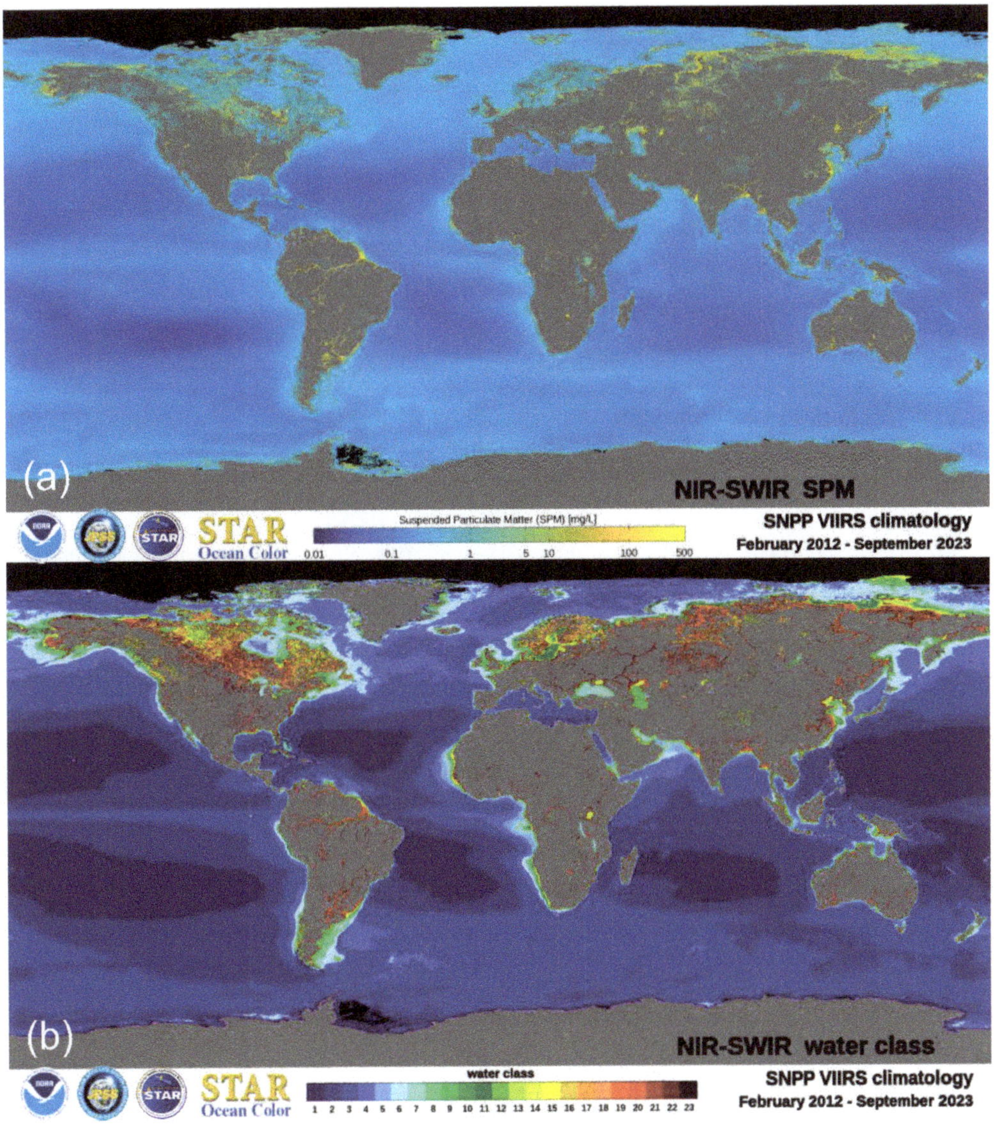

Figure 1. Climatology maps for 2012–2023 (SNPP VIIRS) for (**a**) suspended particulate matter and (**b**) water class product over global oceans and inland waters.

NOAA routinely produces global Level-3 (i.e., spatially and/or temporally aggregated from Level-2) ocean color data from multiple satellite missions, including VIIRS-SNPP, VIIRS-NOAA-20, OLCI-S3A, and OLCI-S3B. In general, a daily global image derived from VIIRS on SNPP or NOAA-20 comprises approximately 70 percent missing pixels and a daily global OLCI image from S3A or S3B comprises approximately 80 percent missing pixels [25–27]. Merging daily ocean color images from different sensors can significantly reduce the number of missing pixels. Currently, STAR routinely produces daily merged

Chl-a, $K_d(490)$, and SPM data, employing combinations of two, three, and four sensors (VIIRS and OLCI). A two-sensor merged daily image has approximately 38% more valid pixels than from a single VIIRS sensor for the same day, with three sensors adding about 12% more valid pixels and a four-sensor merged image adding approximately 8% more valid pixels beyond the three-sensor amount [26,27].

To completely fill missing pixel gaps in multi-sensor merged images, STAR employs the Data Interpolating Empirical Orthogonal Function (DINEOF) method [28,29] to reconstruct those missing pixels and generate a Level-4 product. The DINEOF method, based on Empirical Orthogonal Functions (EOFs), extracts the dominant spatial patterns (EOF modes) from a time series of satellite images to reconstruct the missing pixels. Currently, NOAA generates near-real-time daily global gap-free Chl-a, $K_d(490)$, and SPM data based on multi-sensor merged ocean color images [30], freely distributed via NOAA CoastWatch [31]. STAR provides global gap-free Chl-a, $K_d(490)$, and SPM data at 2 km and 9 km spatial resolutions. These daily gap-free ocean color data reveal large-scale and mesoscale ocean features, such as the equatorial current, the Gulf Stream, and mesoscale eddies, permitting smooth reconstruction of the evolution of mesoscale eddies, such as Loop Current rings, North Brazil Current rings, etc. [27]. Since enhanced (reduced) Chl-a values are often associated with cyclonic (anticyclonic) eddies, gap-free Chl-a data, along with satellite-observed sea level anomalies, facilitate tracking and globally studying mesoscale eddies [32]. Adding data from additional satellite sensors to merged images not only significantly increases the number of valid pixels, but also improves the quality of derived global gap-free images [27]. Notably, three-sensor gap-free images significantly enhance the definition of coastal features versus those assessed in two-sensor images [26].

To further resolve smaller-scale dynamic features, especially in coastal and inland lake regions, higher-resolution gap-free ocean color images are needed. VIIRS has a wide swath width of 3040 km, with spatial resolution gradually increasing from 0.75 km at nadir to approximately 1.5 km on the edge of the swath. Although OLCI's swath width (1270 km) is much narrower than VIIRS, OLCI has higher spatial resolution (0.3 km at nadir). Enabled by these sensor native spatial resolutions, four-sensor merged global and regional gap-free ocean color products at 2 km, 1 km, and 0.5 km spatial resolutions have been developed and tested [26]. These higher-resolution data capture more high-frequency variations in coastal oceans [26], e.g., the 2 km gap-free data images are capable of resolving fine ocean features, such as coastal eddies and filaments, with the 1 km and 0.5 km resolution images further enhancing the definition of such features [26]. Figure 2 provides examples for the ocean and inland waters of global daily 2 km gap-free Chl-a (Figure 2a), $K_d(490)$ (Figure 2b), and SPM (Figure 2c) for 1 July 2023. STAR routinely generates these gap-free ocean color products from three sensors (VIIRS-SNPP, VIIRS-NOAA-20, and OLCI-S3A). With ocean color data becoming available from more satellite sensors, more satellite data sources can be combined for higher-resolution daily global gap-free products. The community of users desiring gap-free ocean color products is large, spanning scientists "upstream" to decision-makers and "downstream" end-users across diverse research, application, management, and policy activities.

2.2.2. Infrared Capabilities

SST Data and Products

STAR has ongoing R&D efforts to develop experimental products within the infrared portion of the spectrum, with resulting significant refinement and exploitation of operational retrievals. These SST data enable NOAA and the broader community to assess ocean state and dynamics, inform ocean and weather predictions, and assess ecosystem/habitat conditions and associated implications on living marine resources.

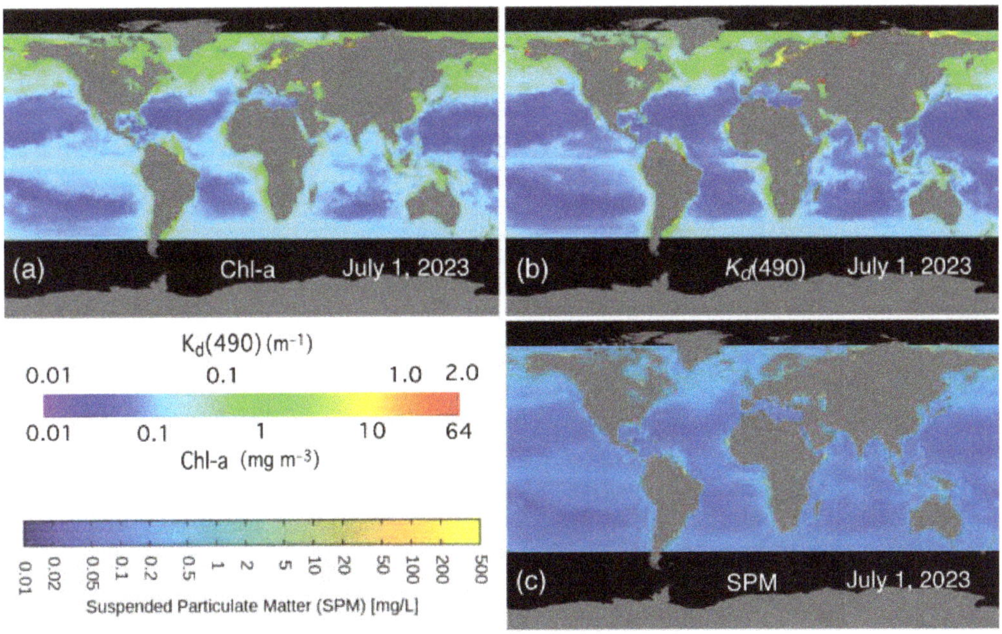

Figure 2. Three-sensor (VIIRS-SNPP, VIIRS-NOAA-20, and OLCI-S3A)-derived global daily gap-free 2 km ocean color products for (**a**) Chl-a, (**b**) K_d (490), and (**c**) SPM on 1 July 2023.

In the early 1980s, NOAA pioneered a global satellite SST product from 4 km-resolution data to produce Advanced Very High-Resolution Radiometer (AVHRR)/2 Global Area Coverage (GAC) onboard polar-orbiting NOAA-7 [33]. Over the next 20 years, NOAA's heritage LEO SST retrieval system has undergone multiple improvements [33–35]. In the 1990s, complementary SST retrievals from GEO data commenced [35,36]. In the 2000s, the scope of NOAA satellite SST capabilities significantly expanded, with changes largely driven by ongoing preparations for the next-generation US satellite systems, the JPSS, and GOES 'R' series, ultimately leading to the generation of a platform-agnostic, enterprise retrieval system for consistent SST products, the Advanced Clear Sky Processor for Ocean (ACSPO) [37–40] that supports the diverse fleet of SST sensors.

Currently, ACSPO processes data from eighteen LEO missions, including NOAA, EUMETSAT, and NASA satellites, and five GEO missions, including NOAA and Japanese Meteorological Agency (JMA) satellites (Table 1). Each of the satellites/sensors routinely provide data in several formats: original swath projection geophysical products (Level 2) and their gridded counterparts. All ACSPO products are currently provided at equal-grid 0.02° resolution. Following the GHRSST recommended standards, the Level 3 single-sensor products are available in two types: uncollated (L3U) or collated in time (L3C). ACSPO data can be accessed multiple ways, including via NOAA's Product Distribution and Access (PDA) system, NOAA CoastWatch, and NOAA's Centers for Environmental Information (NCEI), as well as EUMETSAT EUMETcast, and the NASA Physical Oceanography Distributed Active Archive Center (PO.DAAC) [41–44].

NOAA, national, and international users extensively employ ACSPO products. Consistent feedback from users notes the challenge of ingesting and processing data from multiple platforms and sensors. Users commonly request that ACSPO L3U/C products be aggregated into sensor-agnostic products, where the large number of sensors and platforms offers greater potential for comprehensive SST consistency checks and quality assurance. Achieving this goal requires ingesting SST data from various individual sensors, reconciling and merging data, thereby enabling gridded super-collated (L3S) products that employ multiple sensors, with improved global coverage and improved spatiotemporal resolution.

Table 1. Sensor data sources processed by STAR's Advanced Clear Sky Processor for Ocean (ACSPO). Sources for data ingested into the current Level 3 Super-collated Daily LEO time series and data fusion product (L3S) are in italics. Sources for the planned GEO super-collated product are noted in the GEO section.

ACSPO Enterprise Processing	Heritage Sensors	Current Generation Sensors	Next Generation Sensors to Be Added
LEO	• NOAA AVHRR GAC, NOAA-7, 9, 11, 12, 14, 15, 16, 17, 18, and 19 • EUMETSAT MetOp First Generation A, B and C AVHRR FRAC • NASA MODIS, Terra and Aqua	• JPSS VIIRS, Suomi NPP, NOAA-20 and NOAA-21	• EUMETSAT MetOp Second Generation MetImage
GEO	• NOAA GOES Imager (list of several) • EUMETSAT Meteosat Second Generation, Spinning Enhanced Visible and InfraRed Imager	• NOAA GOES-R: GOES-16, 17, and 18 Advanced Baseline Imager • JMA Himawari-8 and Himawari-9 Advanced Himawari Imager	• EUMETSAT Meteosat-Third Generation Flexible Combined Imager

NOAA first pursued developing two L3S-LEO products, aggregating mid-morning (AM) data from the MetOp-First Generation-A, B, and C satellites and aggregating afternoon (PM) data from the JPSS Suomi NPP, NOAA-20, NOAA-21, and Aqua MODIS satellites, with each product comprising one nighttime file and one daytime file. Many NOAA users also want these four files aggregated into a single daily (DY) product. Global LEO satellite coverage of any given satellite sensor depends on the observation capability (orbit, swath, observation geometry, etc.) minus anything that is not ocean SST (clouds, ice, etc.). A single satellite typically provides "good" SST values for approximately 22–26% of the global ocean, with two satellites producing approximately 35–37% "good" coverage, and three satellites giving up to approximately 40% "good" coverage. The DY product (Figure 3), which aggregates data from up to five different satellites, provides "good" SST values for approximately 65% of the global ocean. The L3S-LEO time-series begins in February 2000.

Current work focuses on super-collating SSTs from six GEO satellites into a single L3S-GEO product (Table 1), which will result in near-global coverage at hourly temporal resolution, excluding the Indian Ocean, which is not presently covered by new-generation GEO SST observations from NOAA and its partners. Ultimately, the L3S-LEO and L3S-GEO will be combined into a single global L3S product having superior coverage, exploiting LEO's high spatial resolutions and GEO's high temporal resolutions.

All ACSPO SST products have three modes: (1) near-real time (2–3 h latency) for operational users; (2) science quality delayed mode (up to 2 months' latency) having improved sensor data (Level 1b) and ancillary data input; and (3) full mission reanalysis (RAN; several years latency) [41–44]. Before archiving at NOAA and NASA, all ACSPO products undergo extensive consistency checks and quality control, as well as calibration and validation against high-quality in situ data in multi-tier NOAA SST monitoring systems, notably the SST Quality Monitor (SQUAM) [45]; Monitor of Infrared Clear-sky Radiances over Ocean for SST (MICROS) [46]; and ACSPO Regional Monitor for SST (ARMS) [47]. STAR performs calibration and validation against high-quality in situ data captured in NOAA's in situ SST Quality Monitor (iQuam) [48]. STAR monitors additional information about sensors and platforms via the NOAA Sensor Stability for SST (3S) [49] system. All anomalies identified during monitoring are promptly addressed. Comprehensive monitoring of SST products, their source radiance data, and the health of contributing sensors and platforms ensures that STAR delivers the highest-quality ACSPO SST products to users.

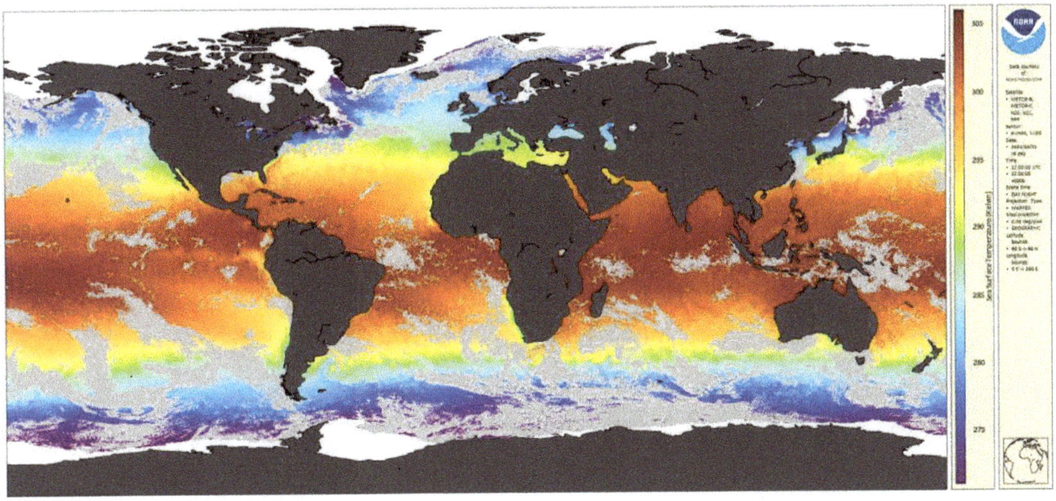

Figure 3. Global ocean LEO Level-3 super-collated (L3S-LEO) daily (DY) SST product for 1 April 2024 showing substantial global daily coverage of satellite observations (approximately 65% on average). The L3S-LEO DY time series begins in the year 2000. Gray areas indicate no SST data due to probable clouds or other quality flags and white areas represent no SST data due to probable ice.

SST Gap-Filled Analysis

Responding to user interest and need for a gap-free Level-4 (L4) product, a global 0.05° resolution (approximately 5 km) gap-free Level-4 NOAA Geo-Polar Blended (GPB) daily analysis was created at STAR [50], separate from ACSPO heritage products [23], by combining Level 2 SST data from various US, Japanese, and European LEO and GEO instruments, including ACSPO data. A multi-scale optimum interpolation methodology, approximating a Kalman filter, is employed with a data-adaptive correlation length scale to ensure a good balance between spatial feature preservation and noise reduction. The time series for this analysis product begins in 2002. Temporal extension to earlier time periods is being researched. A key user of this analysis product is NOAA's Coral Reef Watch (CRW).

Next-Generation SST at NOAA/NESDIS

NOAA scientists with expertise in SST recognize the opportunity to improve NOAA's SST data product collection, especially with regard to addressing growing user demand for various decision-making applications; consequently, in an internal report (white paper), NESDIS/STAR presented a proposal for a next-generation SST product suite [51]. Applying new computing technologies, such as AI and machine learning, to the many robust, operational SST satellite sensor data openly shared amongst the international space agency community holds the promise of creating next-generation super-high-resolution observations that exceed the native sensor resolution. Higher-spatial-resolution SST products, both near-real-time data and reanalyzed longer-term time series, are highly desirable and in demand from multiple user sectors. The outlook is optimistic that relevant institutions will promote progress in this direction.

Infrared Ice Observations

Beyond SST, STAR develops infrared geophysical products for near-real-time monitoring of sea ice conditions, lake ice conditions, and longer-term climate studies. STAR provides ice information and services to aid marine navigation and security, weather and climate prediction, and climate monitoring and change detection. Near-real-time sea ice products, used or evaluated for sea-like ice operations in support of marine navigation and assimilation in numerical environmental prediction models, include sea ice concentration, temperature, thickness, and motion. The primary instruments for infrared-based ice surface temperature, ice thickness, and ice motion are VIIRS on S-NPP, NOAA-20, and NOAA-21, and ABI on the GOES-R series satellites [52]. Ice concentration is derived from the VIIRS and the ABI [53]. All except ice motion are generated operationally. These ice products will also be generated for the MetOp-SG series of satellites with the METImage instrument. While NOAA does have satellite products for land-based snow properties, there are currently no satellite products for snow on sea ice; however, an approach using active altimetry is under investigation. Figure 4 provides some examples of STAR's infrared-based sea ice products. Current research products include blended VIIRS+AMSR2 sea ice concentration [54], sea ice leads [55,56], ice surface temperature from a single VIIRS high-resolution band [57], blended ice motion, and sea ice dynamics and deformation.

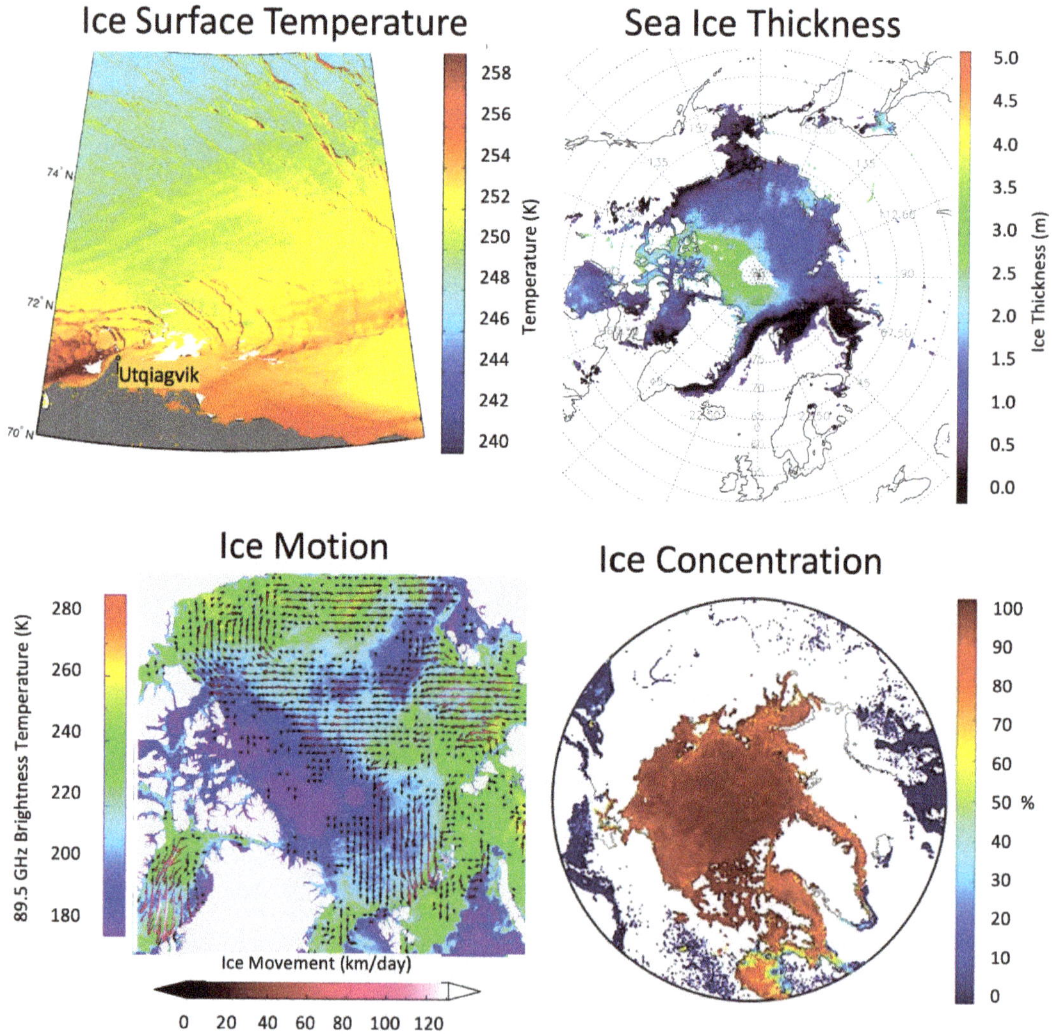

Figure 4. Near-real-time infrared-derived sea ice products. Clockwise from upper left: VIIRS sea ice surface temperature, VIIRS sea ice thickness, VIIRS + AMSR2 ice concentration, and VIIRS + AMSR2 ice motion, with the AMSR2 providing passive microwave data.

The scientific community uses sea ice climate data records (CDRs) for process studies and to better understand interactions and feedback in the climate system. STAR develops and enables robust climate data records that can be used for monitoring and assessing changes in Arctic and Antarctic climate. One NOAA CDR developed by STAR, the Extended Advanced Very High-Resolution Radiometer (AVHRR) Polar Pathfinder (APP-x), comprises a suite of 20 variables, including sea ice thickness, ice surface temperature, ice concentration, cloud properties, and radiative fluxes. APP-x covers both polar regions twice per day from 1982 to the present [58]. VIIRS data, added daily to APP-x, extends the CDR into the future. Another recently developed long-term dataset estimates ice thickness from ice age, where ice age is determined by tracking new ice parcels over a long period of time [59], a unique approach to deriving sea ice thickness that is very different from more direct altimetry-based observations.

2.2.3. Passive Microwave Capabilities

For passive microwave capabilities, STAR must leverage partner satellite observations. NOAA's formal partnership agreement with the Japan Aerospace Exploration Agency (JAXA) for timely access to data from the AMSR2 instrument on the GCOM-W1 satellite will extend to include continuity observations resulting from the launch of AMSR3 on JAXA's GOSAT-GW mission. The AMSR2 uniquely addresses several NOAA JPSS program observational requirements not met by NOAA satellites, notably microwave SST and microwave brightness temperature (MBT) imagery. STAR developed and validated several operationally generated and distributed AMSR2 ocean products, in particular sea surface wind speed (SSW), SST, total precipitable water (TPW), cloud liquid water (CLW), and MBT imagery for tropical cyclones. Recently, STAR developed an all-weather AMSR2 wind speed product, extending the utility of AMSR2 wind speed retrievals into the tropical and extratropical storm environments [60].

STAR pursues passive microwave sea ice retrievals (sea ice concentration, thickness, and motion) from AMSR2 [53] to provide observations complementary to infrared-based sea ice observations, targeting application in near-real-time monitoring of sea ice, and lake ice conditions, and for longer-term climate studies (Figure 4). Passive microwave sea ice concentration data, produced operationally from the AMSR2 and subsequent instruments, also will be generated for the Microwave Imager (MWI) instrument on the MetOp-SG series of satellites. Near-real-time passive microwave sea ice products directly support ice operations, marine navigation, assimilation in numerical ocean and atmosphere prediction models, as well as for climate data records used by the scientific community for process studies and better understanding of climate system interactions and feedback. Blended passive microwave (AMSR2) and infrared (VIIRS) products for sea ice motion and concentration have been developed but have not yet transitioned to operational production. The APP-x product suite includes a passive microwave ice concentration product. Sea ice thickness is derived from the retrieved snow–ice interface temperature and an ice growth model. Passive microwave satellite data (AMSR2) have been used to develop the first long-term basin-wide arctic climatology of dynamically and thermodynamically driven sea ice thickness effects at sub-seasonal temporal resolution [61]. Although over a decade long, this time series is not yet considered to be a CDR.

Beyond AMSR2 passive microwave observing capabilities, STAR also leverages the passive L-band observations of ESA's SMOS and NASA's SMAP missions for sea surface salinity (SSS) observations, a critical ocean state parameter. These SSS observations support NOAA operational efforts, including near-real-time global numerical modeling and seasonal-to-interannual prediction efforts [62]. STAR L-band SSS development efforts currently focus on enhancing exploitation of high-latitude SSS observations, where SSS observations are critical, but colder water impedes the geophysical retrieval. Another effort pursues exploiting satellite SSS observations as an artificial intelligence (AI) predictor for precipitation prediction at subseasonal to seasonal time scales.

2.2.4. Active Capabilities

Altimetry

STAR exploits satellite altimetry, notably for ocean surface topography efforts, which benefits numerical modeling and prediction of ocean dynamics; hurricane intensity forecasting through improved representation of ocean heat content; enhanced high wave warnings; ENSO forecasting and subsequent global water cycle implications; ocean surface current assessments that inform fishery management, fishing services, energy siting, habitat health, search and rescue, offshore operations, and incident responses; and understanding and mitigating global and regional sea level rise [63].

Routine calibration/validation activities of altimetry concentrate in two areas, near-real-time operational oceanography and the sea level climate record. The Near-Real-Time Altimeter Validation System (NRTAVS) produces comprehensive, running web-based statistical summaries of wind, wave, and sea surface height, as well as individual correction terms, which are used for rapid quality assurance. To ensure the integrity of STAR's regional and global sea level climate products, the previous tide gauge comparison system [64] was updated to detect system drifts and shifts. Additionally, a comprehensive radiometer comparison system looks for errors using atmospheric models, inter-satellite comparison, and vicarious methods [65]. Annually, STAR uses complementary sea level measurements from Argo array and the GRACE gravity missions to assess the altimetry-based sea level budget [66,67].

Through the international OSTST and in conjunction with NASA and other programs, the NOAA Jason/Sentinel-6 program supports research and development of altimetry applications, producing several innovative altimeter processing algorithms for improved sea level and sea state observations, altimetric bathymetry [68,69], inland water levels, and, for the cryosphere, sea ice thickness (freeboard). The Fully Focused Synthetic Aperture Radar (FF-SAR) algorithm [70] provides ultra-high-resolution altimetry at scales of up to 0.5 m, increasing the along-track resolution by 100 times over conventional processing methods. Widely applied, FF-SAR use includes improved altimeter range calibrations at dedicated transponders and new oceanographic observations, including swell [71]. Additionally, STAR has helped develop FF-SAR-based altimetry processors to monitor water levels of small inland water bodies, particularly for ungauged rivers and lakes in remote or undeveloped areas [72]. A more recent innovation, 2D retracking, introduces additional parameters to the processing of radar waveforms from Delay-Doppler (SAR) altimetry missions (e.g., Cryosat-2, Sentinel-3, Sentinel-6) [73,74]. This advanced algorithm provides two new ocean observations related to vertical wave motion and surface velocity. Including these parameters reconciles differences between conventional (i.e., TOPEX and Jason series) and SAR-Altimetry processing of sea level and waves, ensuring continuity in the ocean topography climate record from altimeters using these different technologies.

STAR, in conjunction with EUMETSAT and the Delft University of Technology, maintains the Radar Altimeter Database System (RADS), which provides multi-mission, consistent sea level anomalies, waves, and ocean surface wind speed products that NOAA operationally employs for ocean modeling and prediction [75]. Daily RADS gridded altimetry, a derived product, comprises a homogenized, high-resolution, gridded, gap-free sea level anomaly product, extending from 2017 to the present, with a major algorithm update slated for release in 2025 and a concomitant extension of the product time series to encompass 2000 onwards. In addition to sea level anomaly and absolute dynamic topography, this product also includes geostrophic surface velocity and eddy kinetic energy (EKE).

The NOAA Ocean Heat Content (OHC) Product Suite provides real-time, daily tropical cyclone heat potential (TCHP) and mixed-layer depth (MLD) estimates for hurricane prediction, using the Navy's ALtimeter Processing System-2 (ALPS2) and the Geopolar Blended 5 km SST [76] as inputs. The Next-Gen Enterprise OHC (in development at STAR), which employs RADS output, as well as Geopolar Blended 5 km SST, ocean surface salinity, and ocean winds, uses a dynamically consistent formulation to reduce error and increase the correlation of TCHP estimates for extremely warm waters [75,77], adding temperature and salinity estimates with 2 m vertical resolution for the upper ocean. The Blended Ocean Surface Currents (BOSC) product combines gridded altimetry with several other data sources to provide a near-real-time blended 1/6-degree gridded output, primarily depending on the RADS to provide accurate global daily surface current estimates. STAR's complementary new Multiparameter Eddy Significance Index (MESI) provides a blended-input estimate of an eddy's physical dynamics and biogeochemical impacts [32].

An innovative STAR algorithm for radar altimeter-based sea ice thickness, currently completing validation and preparations for operational production, exploits the FF-SAR Altimeter Processor [70] to dramatically increase (approximately 50 times) the spatial resolution of along-track altimetry-based sea ice thickness (freeboard) measurements. For more accurate estimates of surface elevation, a robust physical model helps account for ice properties in radar returns from sea ice surfaces, potentially producing more reliable and accurate estimates of sea ice thickness (freeboard), enabled by improved measurement of small sea ice leads (Figure 5) and being unaffected by cloud cover.

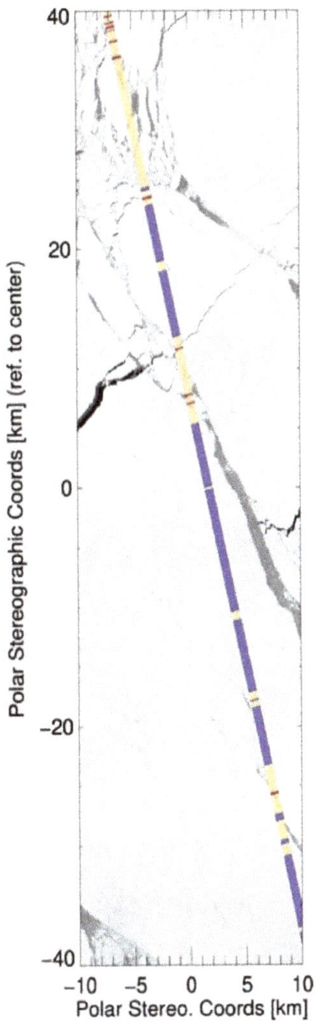

Figure 5. SAR altimeter processor lead detection: blue—floe, yellow—ambiguous, red—lead.

This new product will provide enhanced information for sea ice forecasting, modeling, and ice charting for navigation, maritime operations, and safety at sea. STAR is working with NOAA's National Weather Service (NWS) on implementing operational assimilation of the SAR-Altimeter Processor Sea Ice Thickness product into NOAA's United Forecast System Coupled Model (UFSCM). A new sea ice thickness product line under development aims to produce a sustained long-term time series using CryoSat-2 data, with continuity provided by data from the operational Sentinel-6/Jason-CS and Sentinel-3 missions, and the pending Copernicus Polar Ice and Snow Topography Altimeter (CRISTAL) mission.

Crucial for accurately and reliably estimating sea ice thickness and building upon previous experiences with both airborne and satellite altimetry data, STAR pursues exploiting dual-band satellite altimetry techniques to provide snow depth on arctic sea ice, subtracting IceSat-2 laser-based altimetry observations from radar-based altimetry observations to estimate snow thickness on sea ice. Quantifying and monitoring snow on sea ice provides critical information, constraining precipitation over the Arctic Ocean, thereby improving estimates of ocean–atmosphere heat flux, and enabling more accurate satellite-derived sea ice thickness retrievals through improved knowledge of hydrostatic snow loading. These innovations and associated infrastructure pave the way for exploiting future dual-band altimeter missions, such as CRISTAL, for enhanced sea ice products and their continuity.

Scatterometry

NOAA support for ocean surface wind vector (OSVW) product development and satellite scatterometer calibration/validation activities began in the early 1990s with the ERS-1 scatterometer. Lessons learned from the ERS scatterometer missions, combined with the large areal coverage (broad measurement swath) and consistent quality OSVW retrievals provided by NASA's QuikSCAT mission (launched in 1999), made routine use of scatterometer OSVW data by NWS possible, with the largest impact being direct use of OSVW observations by marine forecasters in their day-to-day jobs, notably for marine wind and wave analysis, warning, and prediction. STAR scientists, working closely with their peers in the NASA QuikSCAT science team and forecasters at the NWS Ocean Prediction Center (OPC) and National Hurricane Center (NHC), improved the near-real-time QuikSCAT OSVW products, thereby supporting NOAA's weather mission. This effort included improving and validating high-wind-speed retrievals in extratropical cyclones through the Ocean Winds flight experiment program described later.

Designed for a 3-year mission life, QuikSCAT lasted 10 years, during which its OSVW products drastically improved marine wind analyses, forecasts, and warnings by significantly extending the available observations beyond sparse buoy and ship reports and NWP model output. Actual observations now provide 12.5 km spatial resolution for an 1800 km wide swath for each QuikSCAT pass. The observed extent of gale, storm, and, sometimes, hurricane-force winds in tropical cyclones notably improved, allowing improved NWS wind warnings and resulting concomitant positive consequences for maritime commerce. In 2006, realizing the significant benefits of satellite scatterometer OSVW products and that NASA and NOAA had no follow-on mission planned for QuikSCAT, NOAA made a concerted effort to document the benefits and requirements for scatterometer OSVW data for operational weather forecasting and warnings [78–82], as well as the economic impact of satellite OSVW data [83,84]. To better address the impacts of satellite ocean observations, further studies of this type are crucially required, capturing not just the straightforward cost–benefit perspective, but also, more broadly, enabling improved societal outcomes. With no US follow-on scatterometer to QuikSCAT, STAR's focus shifted to exploiting observations from the Advanced Scatterometer (ASCAT) on EUMETSAT's Meteorological Operational satellites (MetOp-B/C). STAR, building upon previous experience with QuikSCAT and its flight experiment program, determined that ASCAT measurements had sensitivity to higher winds than currently being exploited [85]. Resulting updates to STAR's ASCAT processing improved detection of hurricane-force winds in extratropical cyclones, helping mitigate the loss of QuikSCAT for the NOAA forecasters. STAR continues extending the information extracted from ASCAT by notably employing different processing techniques to achieve a high-resolution ASCAT coastal wind and ice product [81]. This new technique enables ASCAT measurements within a few kilometers of the coast while also improving ASCAT's ability to measure winds in tropical cyclones (Figure 6). Efforts also focus on developing a near-real-time ice coverage product from this enhanced-resolution ASCAT product, with validation and refinement of this product being a component of STAR's winter flight experiments.

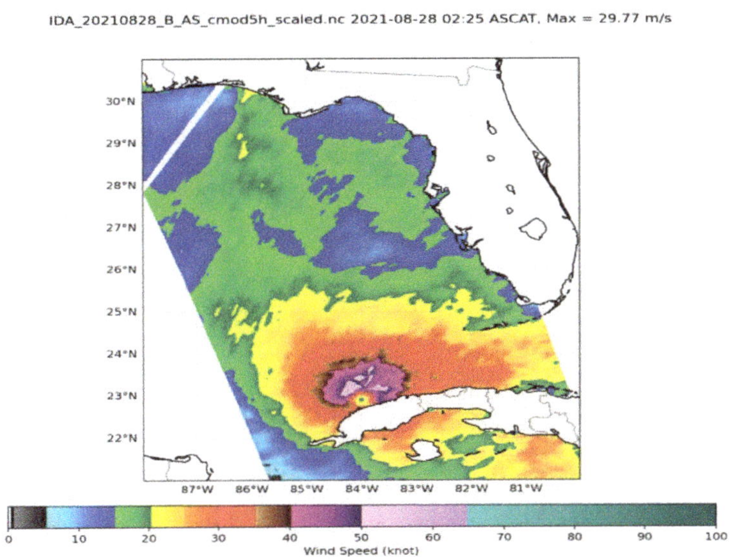

Figure 6. High-resolution ASCAT product utilizing the coastal and tropical cyclone wind speed retrieval improvements for Hurricane Ida on 28 August 2021.

Working closely with the Indian Space Research Organization (ISRO) on calibrating and validating the OSCAT instrument on the Oceansat-2 mission, a scatterometer similar to QuikSCAT in frequency and measurement geometry, STAR developed and produced its own OSCAT OSVW products in near-real time using NOAA's SCATSAT processing system. With both EUMETSAT and ISRO committed to follow-on scatterometer OSVW missions, STAR actively works with both agencies in support of satellite OSVW data for NOAA's weather mission. Currently, STAR OSVW efforts focus on calibrating, validating, and developing products for ISRO's Oceansat-3, as well as working with EUMETSAT in preparing for the launch of ASCAT's successor, the EPS-SG Scatterometer (SCA), in late 2025.

Satellite scatterometer-based products derived from ASCAT measurements help characterize sea ice properties, with spatial resolution enhanced through processing that exploits frequent swath overlap in the polar regions [86], including daily products for ice type, normalized radar cross-section (NRCS), and 10-day NRCS standard deviation. The frequency and resolution of these products provide metrics suitable for sea ice forecasting, input for sea ice thickness estimation, model input, and geophysical process studies.

Synthetic Aperture Radar (SAR)

Exploiting satellite Synthetic Aperture Radar (SAR) observations, STAR provides estimates of boundary layer conditions for the surface of oceans, lakes, rivers, and wetlands. The primary measurements, determined from spaceborne SAR imagery, map the surface microwave radar reflectivity at resolutions from the sub-meter scale to 100 m, depending on the particular SAR satellite and mode. At typical radar frequencies, SAR can image through clouds; consequently, SAR serves as an "all-weather" instrument, independent of time of day because the radar provides its own illumination. STAR currently derives several SAR-based geophysical parameters, enabling high-resolution monitoring of sea surface wind speed, tropical cyclones, sea and lake ice characteristics, flooding, marine oils and surfactants, and waves.

All STAR SAR products transform original SAR source data to NRCS imagery, with the associated backscatter of these NRCS data being proportional to surface roughness on the scale of the radar wavelength (from 3 to 30 cm). NOAA's principal sources for SAR data, C-Band imagery, come through partnerships with ESA (Sentinel 1) and the Canadian Space Agency (RadarSat Constellation Mission (RCM)), as well as commercial procurement for RadarSat-2 data.

For wind speed and direction estimates, STAR employs empirically determined geophysical model functions based on backscatter intensity from SAR NRCS imagery [87], with the same SAR polarization employed for transmission and reception: vertical (V), horizontal (H), co-polarization (VH or HV), or compact polarization (C). SAR fine-resolution estimates for cyclone winds, surface eye locations, and other cyclone characteristics provide tropical forecasters with a new tool [88]. When estimating tropical cyclone force wind speeds, SAR retrievals apply transmission backscatter cross-polarization (VH or HV) [89]. These SAR-derived tropical cyclone products (500 m, 3 km, and 12 km resolution) now routinely contribute to forecast products from the Joint Typhoon Warning Center (JTWC), National Hurricane Center (NHC), and the Central Pacific Hurricane Center (CPHC). Figure 7 depicts the SAR 500 m wind speed estimates for Cyclone Freddy on 16 Feb 2023.

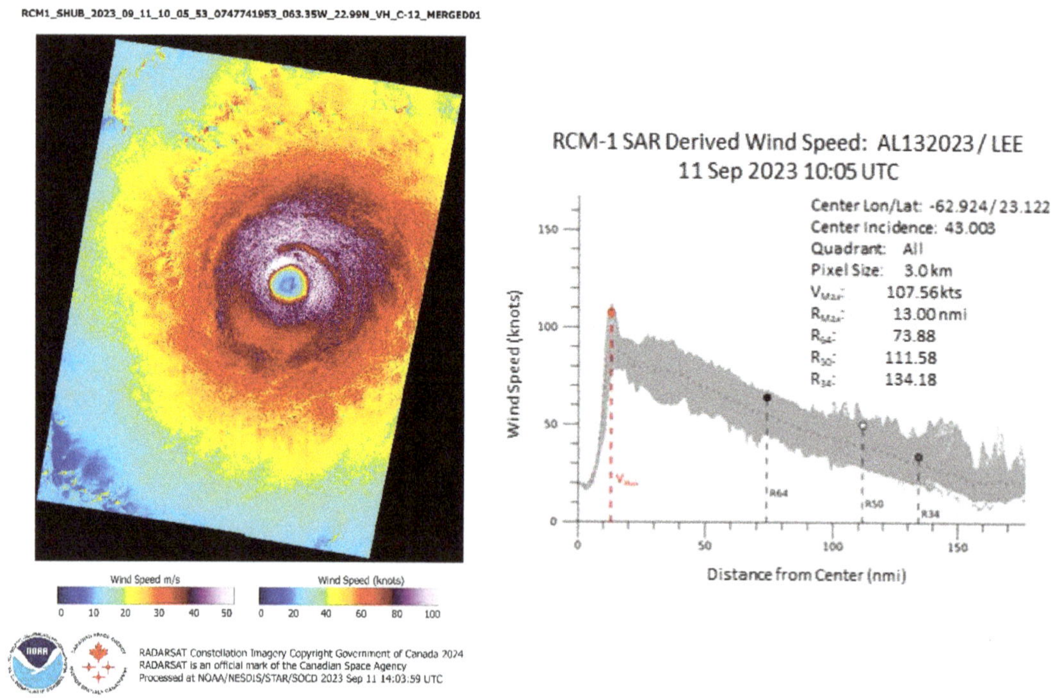

Figure 7. Synthetic Aperture Radar (SAR, Radarsat-2) data for Tropical Cyclone Freddy 11 September 2023 at 10:05 UTC: (**left**) 0.5 km resolution wind speed and (**right**) full storm radial profile, depicting that the 0.5 km processing extracts a maximum velocity (*VMax*) of 136.3 kts.

For most operational ice monitoring agencies, e.g., the US National Ice Center (USNIC) and the Alaskan Sea Ice Program (ASIP), SAR data serve as a primary source, recognizing that SAR data provide an ideal tool for ice observation, particularly due to persistent cloudy conditions, low solar illumination, SAR's high spatial resolution, and SAR's ability to track ice types. STAR provides NRCS, normalized mosaic imagery, ice motion, and ice extent products to USNIC and ASIP for ice charting. The ice motion products, using a customized algorithm, are generated on tailored resolutions and time intervals.

2.3. *Suborbital and In Situ Data Collections Supporting Satellite Observations*

2.3.1. Sea Ice

Validating satellite-based sea ice products with in situ observations presents notoriously difficult challenges due to the remote and harsh polar ocean environments. STAR pursues increased capacity for collecting airborne sea ice coverage measurements by leveraging NOAA's extensive expertise and resources in airborne operations. Using long-range NOAA aircraft, STAR observes remote regions of the arctic, collecting measurements over long transects (approximately 500–1500 km), thereby providing extensive data for validating satellite-based sea ice products over a variety of ice types, effectively bridging the gap between in situ point measurements and large-scale satellite measurements. STAR partnered with the European Space Agency (ESA) and NASA Goddard Space Flight Center (GSFC) during Operation IceBridge to perform long-transect, near-coincident under-flights of Envisat, ICESat, CryoSat-2, Sentinel 3A/B, and ICESat-2 for satellite sea ice product validation [90].

During the winters of 2021, 2022, and 2023, STAR conducted interdisciplinary airborne campaigns over the Beaufort Sea and Bering Sea regions using a WP-3D aircraft equipped with several advanced radar systems, collecting both wind and sea ice surface measurements (e.g., Figure 8), targeting near-coincident and coincident under-flights of multiple satellite altimeters, along with low-altitude surveys of the Sea Ice Dynamic Experiment (SIDEx) ice camp. STAR plans to add additional sea ice measurement instrumentation (e.g., lidar) for future field efforts and pursue using additional NOAA aircraft for product validation and sea ice monitoring, aiming to leverage emerging UAS platforms.

Through coordination with NOAA CoastWatch/PolarWatch and using NOAA's OceanView geophysical data visualization application, STAR has established a system designed to enhance and distribute scientific and monitoring information for the arctic and antarctic regions, aiming to provide access to data collected over a spectrum of spatial and temporal scales, along with active tools for analyzing and comparing multiple datasets. System capabilities include data mapping and visualization, on-the-fly statistical and time series analysis tools, and multivariable comparison tools. The system targets providing reliable polar in situ, airborne, and satellite-based measurements for satellite calibration/validation activities, as well as analysis capabilities for examining regional earth system processes; serving national and international cryospheric operational and science communities; and enabling downstream stakeholders through more useable satellite-based products serving fisheries, commercial efforts, security, navigation, arctic communities, etc.

2.3.2. Extreme Winds

STAR airborne field experiments employ several advanced radar systems, installed on the NOAA P-3 aircraft and routinely operated during flights into tropical cyclones and winter storms, to obtain measurements of extreme conditions. Over the past several years, STAR implemented inflight real-time processing for wind and wave products, enabling collected data to simultaneously inform critical National Hurricane Center (NHC) decision making. Targeting improvements to existing products and future mission instrumentation, these flight experiments, coordinated with satellite passes when possible, have multiple objectives: improve understanding of what microwave remote sensing instruments are really measuring at the extremes; advance the knowledge of processes at the air–sea interface in these extreme environments; test new remote sensing measurement instrumentation and techniques; and validate satellite products.

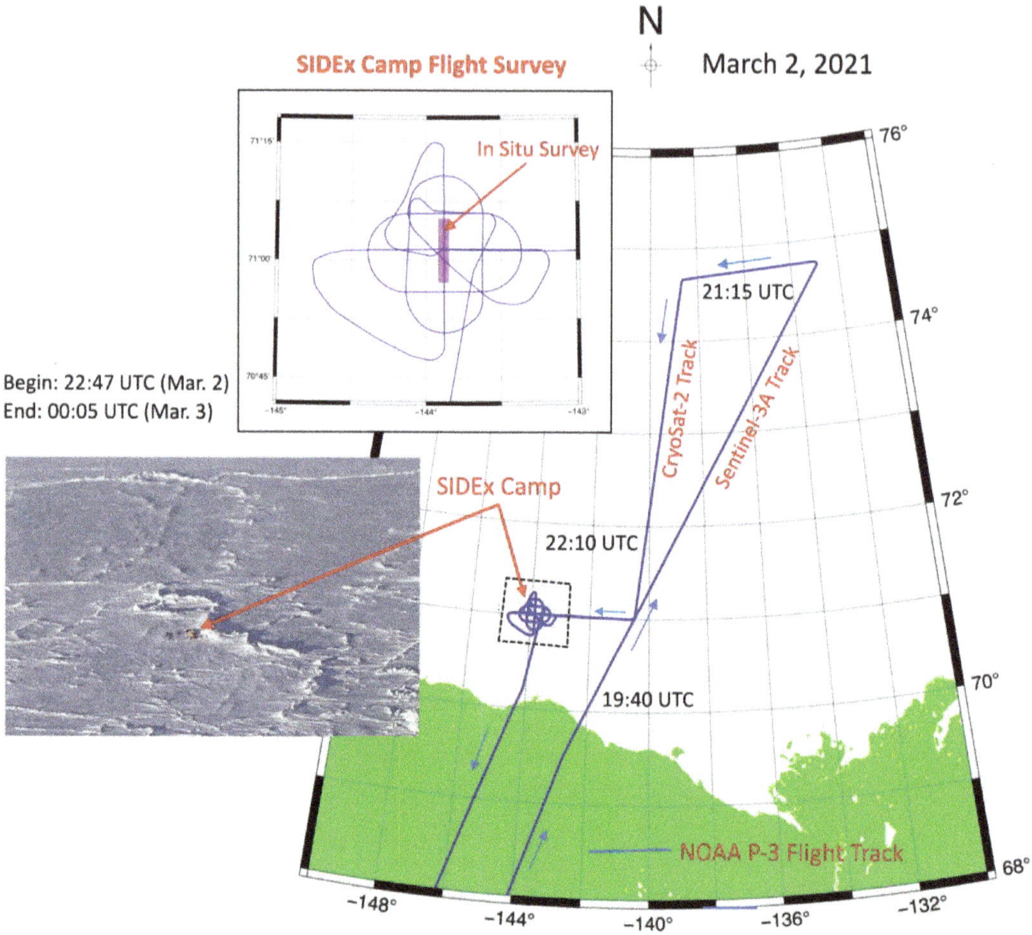

Figure 8. NOAA Ocean Winds and Sea Ice Winter field experiment—2 March 2021. Flight track included near-coincident under-flights of CryoSat-2 and Sentinel-3A satellites and a survey of the SIDEx ice camp.

While the extreme wind/wave conditions found in extratropical and tropical cyclones occupy a small fraction of the ocean surface at any moment, accurate knowledge and monitoring of these conditions are crucial for short-term and longer-term weather and climate applications. NOAA's unique heavy aircraft capability provides the platform with support for current remote sensing products while also enabling investigations targeting improvements in knowledge and future instrumentation. A current STAR priority targets using STAR airborne measurements to understand and address the issues being seen in the airborne Stepped Frequency Microwave Radiometer (SFMR) retrievals in high-wind/rain gradient areas. Hurricane specialists at the NHC question the accuracy of SFMR data in these regions; however, with STAR's suite of measurements, STAR aims to untangle what is happening at the surface and with the retrievals.

3. Models and Assimilation

3.1. General

The assimilation of satellite ocean observations into predictive modeling, namely NOAA's Unified Forecasting System, is the principal use, by volume and parameter type, of STAR's satellite ocean observations, supporting both NWS and NOS operational modeling. Operational assimilation of satellite ocean observation data requires careful calibration and validation to minimize the insertion of excess uncertainty into NOAA's operational forecasts.

Passive retrieval algorithm developments and improvements support the ongoing evolution of NOAA's Unified Forecast System to radiance-based assimilation, where feasible. The STAR-managed Community Surface Emissivity Model (CSEM) and Community Radiative Transfer Model (CRTM) implement STAR's passive retrieval algorithms for operationally translating between observed radiances and modeled geophysical parameters within the NOAA Unified Forecasting System. These passive retrieval algorithms enable the focus on radiance-based observation assimilation supporting the Joint Effort for Data assimilation Integration (JEDI), comprising US agencies and the United Kingdom (UK) Met Office.

3.2. Assimilation of Ocean Data into Models

For near-real-time and subseasonal–seasonal applications, NOAA employs satellite observations to assess the ocean state and its dynamics, as well as coupled ocean, atmosphere, and sea ice processes and fluxes of heat, moisture, and momentum for improved forecasts and outlooks. NOAA's operational global ocean model, the Real-Time Ocean Forecast System, currently assimilates SST, SSS, Absolute Dynamic Topography (ADT, similar to SSH), and sea ice concentration (passive microwave). Satellite ocean color observations inform predictive modeling on the absorption of solar radiance in the near-surface ocean, which has a biophysical feedback to ocean atmosphere heat and moisture fluxes. Developmental work is being carried out to assimilate satellite ocean color (chlorophyll) observations for initiating representation of the biological carbon cycle's photosynthetic base. Satellite sea surface roughness observations inform and constrain modeling of the surface wave state and sea ice extent, informing weather forecasting and subseasonal–seasonal outlooks. STAR recently brought satellite ocean data assimilation to the US NOS for operational regional/coastal ocean modeling, leading development of the US West Coast Operational Forecast System, which currently assimilates SST, ADT, and coastal high-frequency radar observations. Satellite SSS spatial resolution is an assimilation challenge due to finer scales of eastern boundary ocean dynamical features, notably eddies and filaments. STAR currently pursues better assimilation of high-latitude satellite SSS observations through an artificial intelligence approach to counter the impact of colder high-latitude waters on accurate satellite SSS retrievals.

4. Applications and User Support

Through leadership and satellite ocean observations, STAR works to address societal issues and develop relevant policies. Examples include co-leading the UN Ocean Decade Action titled "The Committee on Earth Observation Satellites–Coastal Observations, Applications, Services, and Tools (CEOS COAST)" [91], as well as the Group on Earth Observations' Blue Planet Initiative, which focuses on ocean and coastal observations for policy and decision making, aiming to bridge the gap between ocean and coastal observational data and societal needs for actionable information.

The STAR-led NOAA CoastWatch/OceanWatch [31], PolarWatch [92], and Coral Reef Watch [93] provide broad access to satellite-based ocean products, analyses, tools, and situational awareness applications that address the global oceans, focused regional and coastal ocean domains, and ocean cryosphere regions. The CoastWatch program, organized in four pillars (Regional or Thematic Nodes; Central Operations; Training, Communications and Outreach; and Innovation, Research and Applications), encompasses seven CoastWatch Nodes, each managed by a user-focused NOAA service, notably the National Ocean Service

(NOS), the National Weather Service (NWS), the National Marine Fisheries Service (NMFS), and NOAA Research. All CoastWatch Nodes share the same mission—connect people and applications to global and regional satellite data products for use in understanding, managing, and protecting ocean and coastal resources, and for assessing impacts of change in ecosystems, the environment, and climate. CoastWatch scientists lead research and co-design, co-develop new or enhanced exploitation of satellite ocean data for applications, and disseminate knowledge needed for informed decision making. For training, communications, and outreach, CoastWatch provides training courses, develops learning modules, and leads user engagement workshops. The CoastWatch Learning Portal [94], a collaboration with University of Maryland, hosts training modules for free and provides open access. The CoastWatch Data Portal [95] encompasses a browser-based one-stop-shop for finding, visualizing, and downloading oceanographic data, with over 150 near-real-time and science quality (delayed) products available for mapping. Through the integration of ArcGIS and CoastWatch Utilities Software (v 4.0.0), only a web browser is needed to access, visualize, and obtain the data available in the portal. CoastWatch performed over 35.8 million data transfers in 2022 alone. The OceanWatch Monitor [96] and OceanView [97] provide user-selected current and past analyses and visualizations of global and regional satellite ocean observations, spanning the available parameters.

4.1. Applications

A significant part of the ocean observation value chain entails extracting information from the data to provide actionable knowledge; consequently, the CoastWatch/OceanWatch Program develops and provides user-driven applications. Examples of successfully developed applications that have transitioned to use include fisheries and aquaculture, water quality assessment, ecosystem monitoring, ocean and coastal dynamics, transportation and safety, and climate and weather. Fisheries and aquaculture applications (SST, ocean color, currents, SSH) include nutrient assessments, bycatch minimization, favorable conditions for commercial fisheries, and ocean acidification. Water quality and habitat applications exploit SST, ocean color, optical imagery, and SAR data to assess estuary conditions, harmful algal blooms, ocean acidification, oil spills, etc. Ecosystem monitoring applications employ SST, SSH, ocean color, and SSS data for assessing habitats of temperature-sensitive species, as well as marine heat waves, currents, upwelling, distributions of ecologically important ecosystems, sediment plumes, freshwater influences in estuaries, and sea level rise impacts. Satellite SST, SSH, SSS, currents, and ocean color observations provide a basis for assessing ocean and coastal dynamics, including upwelling and mesoscale dynamic features, contributing to optimizing marine transportation and fishing efforts. Satellite ocean surface winds, sea surface roughness, and SSH data provide the marine transportation and safety sector information on dangerous conditions, surf and swell conditions, ice type and thickness, as well as improve tide and wave height forecasting for greater efficiency and safety. Optical imagery (e.g., from VIIRS) and SAR observations provide coastal inundation assessments for public safety and recovery. Additionally, STAR creates extended consistent satellite ocean observation datasets to address climate-scale questions and issues.

4.2. Polar Regions

Analysis tools and geophysical products developed by STAR enhance the understanding, monitoring, and situational awareness of Earth system environments and processes in the polar regions, notably remote sensing techniques and fit-for-purpose sea ice products supporting sea ice monitoring, numerical modeling, and prediction, advancing knowledge of the changing arctic sea ice cover across time scales. On the broader scale, to consolidate observing requirements and identify observing system gaps for properties of the cryosphere [98], STAR participates in the World Meteorological Organization (WMO) Rolling Review of Requirements (RRR) and Global Climate Observing System (GCOS) efforts.

Climate data records can enable better understanding of the relationships between various components of the climate system. For example, NOAA's satellite climate data records produced new information and insights on multi-decadal changes in arctic sea ice area, thickness, and volume. A recent STAR study [99] introduces a new perspective, based on ice longevity, for determining where ice is persistent and where it is disappearing (Figure 9). That research highlights the presence and persistence of sea ice in an area that directly influences arctic weather and climate, marine transportation, and ecosystems. These findings highlight that the arctic has become less ice-covered in all seasons, with summer and autumn having the most changes, with areal loss of perennial sea ice being the major factor in total sea ice loss in all seasons. Different mechanisms of the changing climate affect the thermodynamic and dynamic sea ice thickness processes.

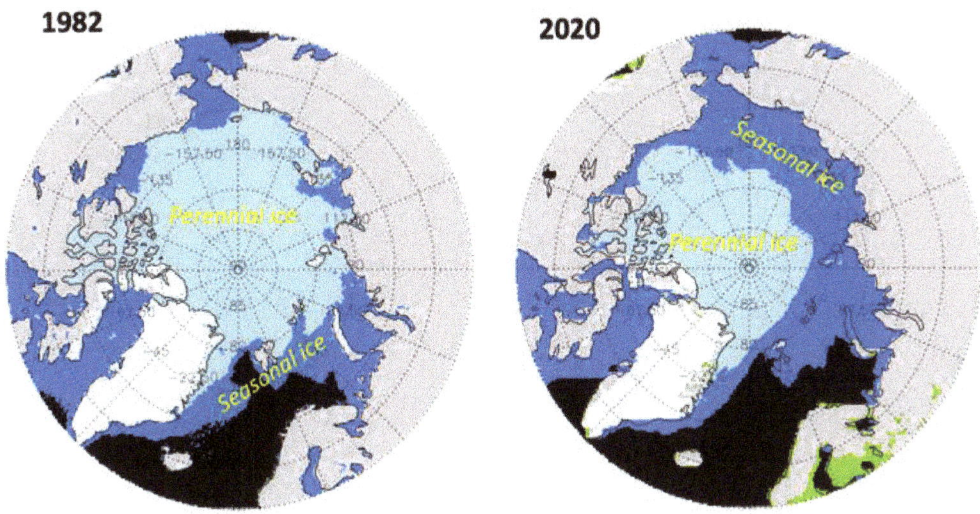

Figure 9. Spatial distribution of Arctic sea ice in 1982 (**left**) and 2020 (**right**) for perennial and seasonal sea ice and snow on land.

Results from passive microwave satellite data (AMSR2) show that Lagrangian dynamics account for almost half of ice thickness growth, depending on the region, providing critical new information for assessing ice processes in coupled models [61,100].

Recent work with satellite-derived cloud and sea ice products highlighted a somewhat surprising relationship between wintertime cloud anomalies and sea ice anomalies the following summer [101,102], with those summertime sea ice anomalies, in turn, driving an apparent positive feedback loop impacting cloud cover anomalies the subsequent winter [103]. The APP-x CDR enabled an assessment of the importance of arctic ice albedo feedback relative to snow albedo feedback [104]. Interestingly, in the arctic, the positive (increasing) trend of solar radiation absorption at the ocean surface due to sea ice decline is more than double the trend resulting from reductions in snow cover over land. Furthermore, in summer, the magnitude of the ice albedo feedback is four times larger than that of the snow albedo feedback; therefore, the dominant radiative feedback for the last few decades appears due to decreasing sea ice cover, not changes in terrestrial snow cover.

4.3. Ecosystem and Biogeochemical Applications

Actively targeting satellite remote sensing approaches for addressing marine ecosystem and biogeochemical data and information needs and challenges, STAR pursues enhancing and applying satellite ocean color measurements of chlorophyll, detritus, colored dissolved organic material (CDOM), and sediment to water quality applications, such as harmful algal bloom assessments and forecasting. The STAR-led CRW, employing SST observations to assess global SST anomalies, trends, hotspots, degree heating weeks, and coral bleaching alert areas, provides free, automated email alerts and critical decision support for coral reef ecosystem management and protection efforts. Additional relevant satellite-derived products address marine heatwaves, coral disease outbreak risk, light stress damage, and monitoring land-based sources of pollution, among other threats to the coral reef environment. Research efforts target projections of bleaching for coral reef areas.

4.3.1. Coral Reefs

The STAR-led NOAA CRW provides satellite and modeled products supporting conservation, restoration, and resilience-based research and management projects that aim to protect and/or restore coral reefs in a rapidly warming world. In times of low or no heat stress, users also apply CRW products to identify appropriate locations to implement conservation and restoration initiatives, for instance, to give transplanted corals or corals grown in situ the best chance at survival.

CRW satellite products, primarily based on SST measurements and updated in near-real time, aim to answer the question, "Is this coral reef currently at risk for bleaching?" These products (Figure 10) include operational daily global 5 km satellite coral bleaching heat stress products (including composites and animations), and daily 5 km satellite Virtual Stations (Regional Virtual Stations currently serving 214 coral reef regions and newly created single-pixel Virtual Stations for multiple US coral reefs). A free, automated Satellite Bleaching Alert email system alerts subscribers, twice per week, when heat stress conditions change on their reefs of interest. CRW also offers, among other tools, a daily global 5 km Marine Heatwave Watch (MHW) product (including new, regional MHW products); historical 5 km satellite Thermal History products; Coral Disease Outbreak Risk products (for Hawaii and the Great Barrier Reef, Australia); and Satellite Ocean Color Monitoring products (for Puerto Rico and Hawaii coral reefs). Additionally, CRW produces a modeled Four-Month Coral Bleaching Heat Stress Outlook (Figure 9) based on NOAA's Climate Forecast System version 2. The Outlook, updated weekly, provides information about potential future heat stress on reefs around the world, answering the question, "Will this reef be at risk for bleaching soon?".

In July 2023, responding to an exigent need, CRW developed single-pixel Virtual Stations [105] for key Florida reefs, consolidating critical CRW heat stress metrics for the individual 5 km × 5 km satellite pixels that overlay the reef sites. Stakeholders can now compare current heat stress levels across sites to assist with prioritization of current efforts by location; communication among stakeholders; subsequent relocation of coral colonies back into the water when the heat stress subsides; and monitoring and restoration activities. Working with local management partners, CRW also developed single-pixel Virtual Stations for key restoration and nursery reef sites in Puerto Rico and the US Virgin Islands. Going forward, CRW will develop additional single-pixel Virtual Stations for all remaining US coral reef areas and for selected key international coral reef sites.

4.3.2. Coastal Zones

Exploiting satellite ocean color observations, NOAA CoastWatch supports risk assessments in coastal zones, including the California-Harmful Algae Risk Mapping (C-HARM) system [106], developed in partnership with the Southern California Coastal Ocean Observing System (SCCOOS). C-HARM HAB products provide probability predictions of high levels of *Pseudo nitzschia* and their harmful neurotoxin, domoic acid. Human and wildlife consumers of shellfish contaminated with domoic acid are at risk of contracting Amnesic Shellfish Poisoning, which can be fatal, if not treated properly. Fisheries, aquaculture, and marine mammal rescue groups, including the West Coast Marine Mammal Stranding Network, need accurate information on a daily basis for making public health decisions and aiding preparations; consequently, C-HARM products serve a vital role in this regard.

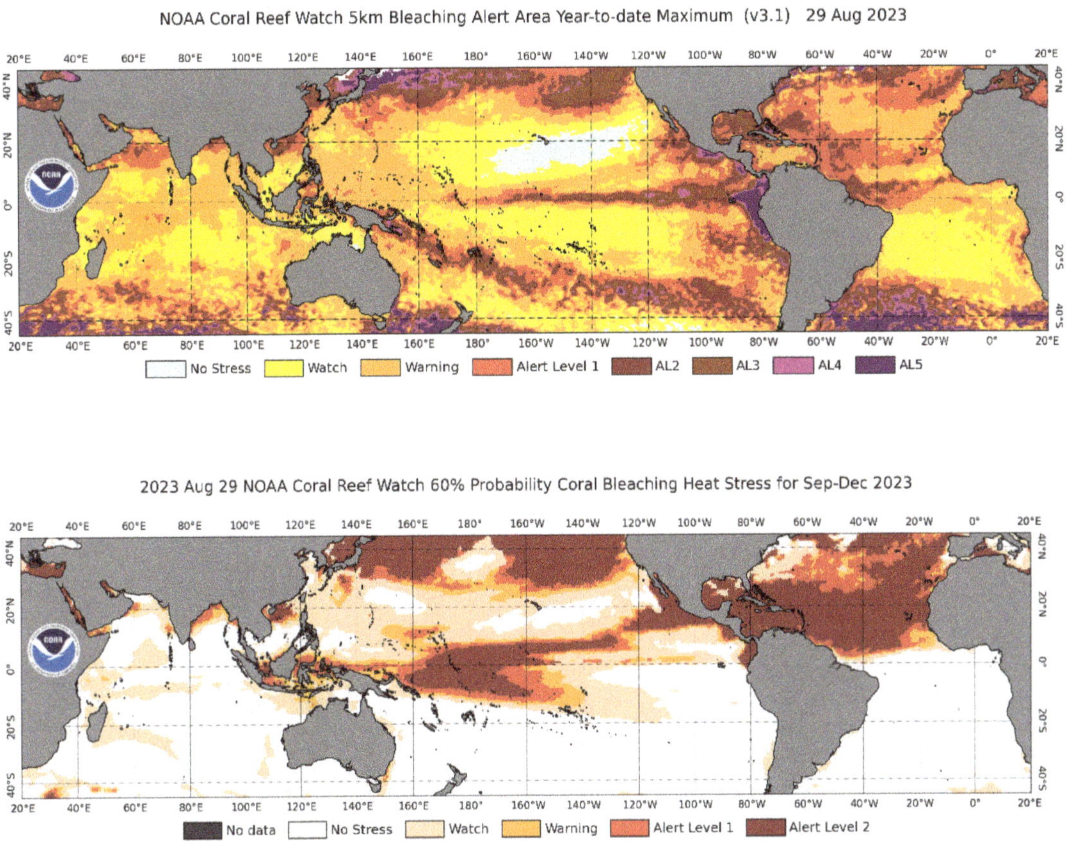

Figure 10. (**Top**) NOAA Coral Reef Watch composite 5 km Satellite Bleaching Alert Area 2023 Year-to-Date Maximum map, depicting the highest bleaching alert levels experienced by tropical coral reefs as of 29 August 2023. In 2023, severe marine heat stress (Bleaching Alert Levels 1 and 2) associated with mass coral bleaching and mortality occurred along Florida, in the Caribbean and Gulf of Mexico, throughout the eastern Tropical Pacific, and in swaths extending from the Sea of Japan to the South China Sea, and from eastern Papua New Guinea to the Cook Islands. (**Bottom**) NOAA Coral Reef Watch modeled Four-Month Coral Bleaching Heat Stress Outlook for 29 August 2023, showing predicted ocean heat stress (and corresponding bleaching alert levels) from September to December 2023.

Timely, accurate, and sustained observations for global coastal zones are essential for addressing existing and emerging societal issues, needs, and concerns, from coastal resiliency, flood forecasting, and disaster recovery to habitat monitoring, eutrophication, water quality, challenges of continued growth in coastal regions, and assessment of climate change impacts. The NOAA co-led CEOS COAST provides new and improved scientific/technical capabilities and continues to build global capacity for a more robust, end-to-end value chain supporting coastal stakeholders and global sustainable development [107,108].

COAST's user-driven efforts encourage broader utilization of Earth observations and other CEOS capabilities for societal benefits within coastal zones (e.g., Blue Economy; Sustainable Development Goals) [109]. User co-design and co-development are key to implementing COAST solutions [109]. High-impact regional pilot projects implemented by CEOS-COAST leverage the CEOS Analysis Ready Data (ARD) framework, employing higher-resolution datasets for the development of shoreline change assessments within Chesapeake Bay [110]. The CEOS-COAST team has additional products and applications in development.

Many global and regional partners seek such scientific and technical (e.g., geospatial and computational) support, affording specific opportunities for CEOS engagement, particularly through the Group on Earth Observations (GEO) Blue Planet and AquaWatch Initiatives, which are actively led and supported by NOAA/STAR and external stakeholders, including the Intergovernmental Oceanographic Commission (IOC)/Global Ocean Observing System (GOOS), United Nations (UN) Environment, the World Meteorological Organization, and the UN Decade of Ocean Science for Sustainable Development (2021–2030). In June 2021, the IOC endorsed CEOS COAST as an official UN Ocean Decade Contribution [91].

4.4. User Engagement and Decision-Making Support

In addition to employing regional nodes for user engagement, NOAA CoastWatch provides training on exploiting satellite ocean observations. Navigating vast amounts of satellite data can be overwhelming; consequently, CoastWatch offers structured courses to introduce users to oceanographic satellite products, the software and tools commonly used to analyze them, and ways to effectively use the data. CoastWatch/OceanWatch resources are free and accessible to all, regardless of class enrollment. An online Learning Portal, launched in 2023 in partnership with the University of Maryland, provides organization for the CoastWatch lectures, guides, and tutorials, receiving positive feedback.

An extensive and diverse user community worldwide regularly employs NOAA CRW's online global decision support system, employing its satellite and modeled products for coral reef management, monitoring, research, and protection of coral reef ecosystems in a rapidly warming world. CRW's broad engagement in research, development, and operations, and its extensive communication and outreach with its users and partners, addresses a critical global need for coral health remote prediction, monitoring, and assessment and allows NOAA to meet user needs and established NOAA missions, goals, and objectives.

Marine resource managers, scientists, decision makers (including political officials), in-water coral reef monitoring networks, and other worldwide coral reef ecosystem stakeholders rely on NOAA's coral reef satellite and modeled products and alerts to predict and monitor in near-real time changes in thermal stress in the coral reef environment; prepare and prioritize resources for events (e.g., mass coral bleaching or disease) that have long-term, ecologically significant impacts on coral and reef health and function; communicate, quickly and broadly, among agencies, the press, and the public, changes in the status of local reefs; implement timely, protective responses and adaptation actions; analyze climate change impacts (e.g., bleaching, disease, and mortality) on coral reefs; and assess when specific reefs are vulnerable or resilient to anthropogenic climate change and its impacts. Using information that the CRW team and its products provide, users have activated local coral bleaching and disease response plans, incident action plans, and associated in-water monitoring networks; rescued native and rare corals; and shaded/cooled key nursery reefs. Stakeholders have also worked to reduce local stressors, such as by closing scuba diving and fishing areas during times of high-bleaching-level heat stress. For example, in May 2016, following CRW's alerts and communications, Thailand's Government closed multiple key reef sites to scuba diving and other activities. In June 2022, the Nha Trang Bay Management Authority closed popular scuba diving areas off of Hon Mun Island, Vietnam. Critical product and data updates are quickly and broadly communicated to the user community and partners worldwide, along with significant contributions to quarterly and monthly climate reports and "Current Conditions Reports" issued by major partners to their in-water monitoring networks, including the Southeast Florida Action Network BleachWatch Program, Mote Marine Laboratory/Florida Keys National Marine Sanctuary BleachWatch Program, Hawaii Eyes of the Reef, the Caribbean Regional Climate Centre/Caribbean Institute for Meteorology and Hydrology, and the Coastal Oceans Research and Development Indian Ocean–East Africa (CORDIO) network, among others. CRW efforts inform many national and international assessments of coral reef conditions, with staff regularly publishing in the scientific literature. User engagement efforts include responding to frequent requests for information about the coral reef products and datasets; consequently, CRW maintains an online tutorial and detailed methodology page about its daily global 5 km satellite products and conducts other outreach and educational activities to enhance understanding and ensure appropriate application of the decision support products and data.

With the developing El Niño of 2023–2024, NOAA CRW directly supports in-water monitoring, conservation, rescue, restoration, and communication efforts, working closely with partners and decision makers in areas impacted by coral reef heat stress. In Florida, where a severe marine heatwave and mass bleaching event continues (as of this writing), CRW continues its extensive communication with in-water partners, guiding the application of CRW products (including newly developed [105] single-pixel satellite Virtual Stations) to guide monitoring, mitigation, and triage efforts, including relocating and protecting thousands of corals in onshore or deeper-water nurseries. Such efforts extend to communicating with global partners that are experiencing heat stress events and documented mass coral bleaching, as well as providing guidance on the strong El Niño as it relates to current/projected heat stress on their reefs. In direct collaboration with the Australian Government and other local stakeholders coordinating information sharing and messaging to global media outlets, as in 2016, 2017, 2020, and 2021, CRW provided advance warning and near-real-time satellite monitoring for the 2022–2023 heat stress and subsequent mass coral bleaching event on the Great Barrier Reef. Leading up to and during severe heat stress and mass bleaching events around the world, CRW frequently provides interviews for US and international press articles, with its products featured across press and social media, notably the daily global 5 km satellite coral bleaching heat stress products and modeled Four-Month Coral Bleaching Heat Stress Outlook.

Through leadership and satellite ocean observations, STAR works to address societal issues and support relevant policies. Examples include co-leading the UN Ocean Decade Action CEOS-COAST, as well as the Group on Earth Observations' Blue Planet Initiative, focusing on ocean and coastal observations for policy and decision making, aiming to bridge the gap between ocean and coastal observations and delivering actionable information for societal needs.

5. Discussion

Fulfilling its crucial role in NOAA, STAR continues to support and advance satellite oceanography from the R&D and operational perspectives, with these inherently coupled domains furthered through extensive internal and external collaborations and partnerships. From the R&D perspective, key foci and activity areas by STAR and the broader Earth observing community over the next decade and beyond follow.

STAR will actively facilitate development and exploitation of new missions in the domains of ocean, coastal, and inland waters. The implementation of geostationary ocean color, as part of the forthcoming GEO-XO satellite system, represents an extremely exciting development that will significantly advance ecosystem monitoring and forecasting for US waters, significantly expanding capabilities for research, applications, and operations. NOAA will also continue to rely on leveraging non-NOAA satellite data streams for its ocean and aquatic observations. Recent and upcoming agency launches, including those of SWOT, PACE, and NISAR, will provide important new experimental and operational data streams for NESDIS to develop and exploit in support of NOAA and non-NOAA stakeholders. Greater utilization of data from commercial platforms likewise will be essential as complementary to the observations from agency platforms.

Developing new and improved products for ocean basins, coastal zones, and inland waters continues to be a high-priority need, particularly for the rapidly changing polar regions (Arctic and Antarctic). In particular, pursuing and exploiting hyperspectral (e.g.,

PACE), higher spatial, temporal, and spectral resolution capabilities of existing and planned satellite sensors is a crucial need and priority; notably, in this regard, new PACE hyperspectral data for ocean color research and applications, as well as commercial data, will be invaluable, especially high-resolution optical and SAR imagery, notably enabling efforts in critical coastal regimes. Required efforts include continued development of fully focused SAR altimetry approaches for various aquatic applications. Identifying, tracking, and extracting dynamic processes, events, and features from multi-sensor and multi-parameter satellite imagery (e.g., MESI [32]) present compelling capabilities supporting diverse users and applications, e.g., plumes, blooms, habitat assessments for fisheries and aquaculture management, marine carbon dioxide removal (MCDR), and coastal blue carbon for climate assessments.

Users need further development of gap-reduced or gap-filled products, including multi-mission merged/fused Level-3 products, as well as gap-filled blended Level-4 analyses, with spatial and temporal feature resolution maximally preserved. Many users indicate a strong preference for these types of products to facilitate exploiting increasing information content while reducing data volumes, number of files, cloud impacts, etc. Additionally, long-identified requirements across diverse NOAA users exist for developing consistent, long-term, multi-mission times series for complementary physical, biological, and biogeochemical ocean parameters and their anomalies. Within NOAA, addressing the wide range of relevant climate research and other longer-term applications requires using STAR enterprise algorithm systems. Ocean color radiometry, ocean surface vector winds, sea surface height, and sea surface temperature time series currently are high priorities within NESDIS that will leverage and integrate both NOAA and non-NOAA data streams.

Significant advances in data science and informatics for ocean research and applications need to complement developments in new and improved data and derived products. The broader community progresses with exciting ongoing transitions into cloud-based solutions, as well as broader exploitation of Artificial Intelligence (AI) and Machine Learning (ML) approaches for data processing, product development, forecasting, etc. STAR research and development activities support and advance these trends, which are expected to provide more equitable access to data and derived information. Emerging ecosystem applications include using AI approaches for identifying and forecasting eutrophication and hypoxia in coastal waters, which will be particularly important for developing nations and transitioning economies as they strive for sustainable development in a rapidly changing environment.

Curation and translation of ocean, coastal, and inland water data products to provide better information and greater knowledge for users proves to be increasingly important. Given petabytes of data and hundreds of products, STAR scientists, as part of broader global community efforts, increasingly play the role of trusted advisor and curator to users, identifying fit-for-purpose products for their particular needs and applications within their domain. Immediate priorities include advancing the use of satellite products in ocean and ecological forecasting, aiming to address underutilization of observation data in marine modeling. This underutilization can be attributed to challenges in modeling complex interrelationships between observed parameters; data latency, accuracy, and uncertainties in what is measured and inferred; and spatial resolution. With respect to spatial resolution challenges, data thinning may be required to align with model resolution, as well as the data may be non-representative when pixel resolution exceeds the resolution of targeted modeled features and processes. Numerical ocean forecasting and prediction have predominantly focused on representing physical states and processes, for which STAR has recently brought new data streams into operational assimilation, notably sea surface salinity and absolute dynamic topography derived from satellite altimetry. Developmental efforts include exploiting satellite chlorophyll-a observations for both biophysical feedback processes and ecosystem modeling, as well as new satellite sea ice observations, notably sea ice thickness from satellite infrared observations. New sea ice observation capabilities from altimetry, synthetic aperture radar, and passive microwave are currently in progress.

A significant focus on exploiting satellite observations of sea ice aims to enable and support NOAA's pending new operational coupled (ocean, atmosphere, and sea ice) model. Integration of data across missions (NOAA and non-NOAA), parameters (e.g., physical, optical, biological, and biogeochemical), domains (e.g., land-sea, ocean-atmosphere), and disciplines (e.g., natural sciences and social sciences) needs to provide a foundation for augmenting the development of new and improved satellite-based indicators that measure both outputs and outcomes, as well as enhanced scientific support and service delivery.

6. Conclusions

The NOAA Center for Satellite Applications and Research (STAR) broadly engages in research, development, and operational support to improve formulation, availability, and utilization of ocean and coastal observations from space; keep NOAA on the leading edge of monitoring and predicting the global environment; and provide, for the benefit of society, critical, timely, and fit-for-purpose information for global stakeholders about changing ocean conditions and associated impacts.

The complexity and criticality of these scientific and technical challenges and efforts and, more so, desired societal impacts, require a global partnership. In advancing the routine and sustained use of satellite oceanography data, NOAA actively coordinates with EUMETSAT and other global and regional partners to develop and implement the collaborative international operational satellite oceanography (OSO) initiative. The OSO initiative, started in 2018, has, to date, held three related international community symposia. The most recent symposium (2023), held in Busan, South Korea, was recognized as an official event of the U.N. Decade of Ocean Science for Sustainable Development. Further details of these events are available in the respective OSO Symposia (OSOS) Reports [111,112]. Working together as a community of EO data providers, the collective ability to positively impact the future is very promising.

Author Contributions: Conceptualization, E.B.; Writing–Original Draft Preparation, E.B., P.S.C., J.L.D.L.C., S.R.H., A.I., J.K., V.L., E.W.L., D.A.B., Y.L., X.L., M.W., J.W. and P.M.D.; Writing–Review and Editing, E.B. and P.M.D. All authors have read and agreed to the published version of the manuscript.

Funding: This NOAA effort was funded by the NOAA Product Development, Readiness, and Application (PDRA) Ocean Re-mote Sensing (ORS) and Jason/Sentinel-6 Programs. The NOAA Coral Reef Conservation Program provided co-author Jacqueline De La Cour's funding support via NOAA grant NA19NES4320002 (Cooperative Institute for Satellite Earth System Studies) at the University of Maryland/ESSIC.

Conflicts of Interest: Author Jianwei Wei was employed by the company Global Science and Technology, Inc. The remaining authors declare that the research was conducted in the absence of any commercial or financial relationships that could be construed as a potential conflict of interest.

Disclaimer: The scientific results and conclusions, as well as any views or opinions expressed herein, are those of the author(s) and do not necessarily reflect those of NOAA or the U.S. Department of Commerce.

Abbreviations

Acronyms (alphabetical order)

ABI	Advanced Baseline Imager
ACSPO	Advanced Clear Sky Processor for Ocean
$a_{dg}(443)$	absorption coefficient of detritus and colored dissolved organic matter at 443 nm
ADT	Absolute Dynamic Topography
AHI	Advanced Himawari Imager
AI	artificial intelligence
AM	aggregated mid-morning data from the MetOp-FG-A/B/C satellites
AMSR2	Advanced Microwave Scanning Radiometer 2

AMSR3	Advanced Microwave Scanning Radiometer 3
$a_{ph}(443)$	light absorption coefficient of phytoplankton at 443 nm
APP-x	Polar Pathfinder
ArcGIS	a family of client, server and online geographic information system software developed and maintained by Esri.
ARMS	ACSPO Regional Monitor for SST
ASIP	Alaskan Sea Ice Program
AVHR	Advanced Very High-Resolution Radiometer
$b_{bp}(443)$	light backscattering coefficient of particles at 443 nm
BOSC	Blended Ocean Surface Currents
Cal/Val	calibration/validation
CDOM	colored dissolved organic matter
CDR	climate data records
CEOS	Committee on Earth Observation Satellites
CEOS-COAST	Committee on Earth Observation Satellite's Coastal Observations, Applications, Services and Tools team
CGMS	Coordination Group for Meteorological Satellites
C-HARM	California-Harmful Algae Risk Mapping
Chl-a	chlorophyll-a
CLW	cloud liquid water
CORDIO	Coastal Oceans Research and Development Indian Ocean–East Africa
CPHC	Central Pacific Hurricane Center
CRISTAL	Copernicus Polar Ice and Snow Topography Altimeter
CRTM	Community Radiative Transfer Model
CRW	Coral Reef Watch
CryoSat-2	a European Space Agency Earth Explorer Mission dedicated to measuring polar sea ice thickness and monitoring changes in ice sheets
CSEM	Community Surface Emissivity Model
CW Utilities	NOAA CoastWatch Utilities data acquisition tool
DINEOF	Data Interpolating Empirical Orthogonal Function
DM	delayed mode
DY	a single daily product
EKE	eddy kinetic Energy
ENSO	El Nino Southern Oscillation
EOF	Empirical Orthogonal Functions
EPS-SG	EUMETSAT Polar System–Second Generation
ERS-1	European Remote-Sensing Satellite-1
ESA	European Space Agency
EUMETCast	a method of disseminating various (mainly satellite based) meteorological data operated by the European Organisation for the Exploitation of Meteorological Satellites
EUMETSAT	European Organisation for the Exploitation of Meteorological Satellites
FEMA	Federal Emergency Management Agency
FRAC	Full Resolution Area Coverage
GAC	Global Area Coverage
GCOM-C	Global Change Observation Mission-Climate
GCOM-W1	Global Change Observation Mission–Water 1
GEO	geostationary
GHRSST	Group for High Resolution Sea Surface Temperature
GOES-R	Geostationary Operational Environmental Satellites R-series
GOSAT-GW	Global Observing SATellite for Greenhouse gases and Water cycle
GPB	Geo-Polar Blended
GSFC	Goddard Space Flight Center
HAB	harmful algal bloom
ICESat-2	Ice, Cloud and land Elevation Satellite 2
IOCCG	International Ocean Colour Coordinating Group
iQuam	NOAAs in situ SST Quality Monitor
ISRO	Indian Space Research Organization
IUU	illegal, unreported, and unregulated fishing
JAXA	Japan Aerospace Exploration Agency
JEDI	Joint Effort for Data assimilation Integration

JPSS	Joint Polar Satellite System
JTWC	Joint Typhoon Warning Center
K_d	diffuse attenuation coefficient
L3C	Level three collated
L3S	Level three super-collated
L3U	level three uncollated
L4	Level-4
LEO	low Earth orbit
Lidar	Light Detection and Ranging
MBT	microwave brightness temperature
MDA	an international space mission partner based in Canada
MESI	Multiparameter Eddy Significance Index
FCI	Flexible Combined Imager on the Meteosat Third Generation Satellite
METImage	Meteorological Imager
METOP	(ASCAT) Meteorological Operational satellites advanced scatterometer
MetOp SG	MetOp Second Generation
MICROS	Monitor of IR Clear-sky Radiances over Ocean for SST
MLD	Mixed Layer Depth
MOBY	Marine Optical BuoY
MODIS	Moderate Resolution Imaging Spectroradiometer
MSG SEVIRI	Meteosat Second Generation Spinning Enhanced Visible Infra-Red Imager
MSL12	Multi-Sensor Level-1 to Level-2
MWI	Microwave Imager
NASA	National Aviation and Space Administration
NESDIS	National Environmental Satellite, Data and Information Service
NHC	National Hurricane Center
NIR	near-infrared
NIR	RBG-red, green, and blue bands
$nL_w(\lambda)$	Normalized water-leaving radiance
NMFS	National Marine Fisheries Service
NOAA	National Oceanic and Atmospheric Administration
NOAA-20	the first of NOAA's latest generation of US JPSS polar-orbiting satellites
NOAA-21	the second of NOAA's latest generation of US JPSS polar-orbiting satellites
NOS	National Ocean Service
NRCS	normalized radar cross section
NRT	near-real-time
NRTAVS	Near Real-Time Altimeter Validation System
NWS	National Weather Service
OCI	ocean color index
OCR	ocean color radiometry
OHC	operational ocean heat content
OLCI	Ocean and Land Colour Instrument
OPC	Ocean Prediction Center
OSCAT	OceanSat 2 Scatterometer
OSTST	Ocean Surface Topography Science Team
OSVW	ocean surface vector winds
P-3	a type of military aircraft
PADI	Professional Association of Diving Instructors
PAR	photosynthetically available radiation
PM	aggregated afternoon data from the JPSS NPP/N20/N21 and Aqua MODIS satellites
PO.DAAC	NASA Physical Oceanography Distributed Active Archive Center
QuikSCAT	Quick Scatterometer
R&D	research and development
RADARSAT RCM	Canadian Space agency's RadarSat Constellation Mission
RADS	Radar Altimeter Database System
RAN	reanalysis
R_{rs}	Remote sensing reflectance

RTOFS	Real-Time Ocean Forecast System
S3A	Sentinel-3A
S3B	Sentinel-3B
SAR	synthetic aperture radar
SCA	Scatterometer
SCCOOS	Southern California Coastal Ocean Observing System
SFMR	Stepped Frequency Microwave Radiometer
SGLI	Second-Generation Global Imager
SIDEx	Sea Ice Dynamic Experiment
SIR	Sargassum Inundation Reports
SMAP	Soil Moisture Active Passive
SMOS	Soil Moisture and Ocean Salinity
SNPP	Suomi National Polar-orbiting Partnership
SPM	suspended particulate matter
SQUAM	SST Quality Monitor
SSH	sea-surface height
SSS	sea-surface salinity
SST	sea-surface temperature
SSW	sea surface wind speed
STAR	Center for Satellite Applications and Research
TC	Tropical cyclone
TCHP	tropical cyclone heat potential
TPW	total precipitable water
UAS	unmanned aircraft system
UFSCM	NOAA's United Forecast System Coupled Model
UK	United Kingdom
UN	United Nations
UN Ocean Decade	United Nations Decade of Ocean Science for Sustainable Development
US	United States
USNIC	US National Ice Center
VIIRS	Visual Infrared Imaging Radiometer Suite
VMax	Vertical Maximum
WP-3D	a type of Lockheed Orion plane used to take measurements in storms
WSC	water surface conditions

References

1. Schmit, T.J.; Griffith, P.; Gunshor, M.M.; Daniels, J.M.; Goodman, S.J.; Lebair, W.J. A Closer Look at the ABI on the GOES-R Series. *Bull. Am. Meteorol. Soc.* **2017**, *98*, 681–698. [CrossRef]
2. Chen, H.; Xiong, X.; Link, D.O.; Sun, C.; Chiang, K. V NOAA-20 Visible Infrared Imaging Radiometer Suite Day–Night Band on-Orbit Calibration and Performance. *J. Appl. Remote Sens.* **2020**, *14*, 34516. [CrossRef]
3. Eldardiry, H.; Hossain, F.; Srinivasan, M.; Tsontos, V.M. Success Stories of Satellite Radar Altimeter Applications. *Bull. Am. Meteorol. Soc.* **2022**, *103*, E33–E53. [CrossRef]
4. AMSRs Advanced Microwave Scanning Radiometer Series. Available online: https://www.eorc.jaxa.jp/AMSR/index_en.html (accessed on 12 April 2024).
5. Entekhabi, D.; Yueh, S.H.; Bindlish, R.; Entin, J.K.; Garcia, M. SMAP Science and Application Results. In Proceedings of the IGARSS 2022-2022 IEEE International Geoscience and Remote Sensing Symposium, Kuala Lumpur, Malaysia, 17–22 July 2022; pp. 4224–4227.
6. Ricciardulli, L.; Manaster, A. Intercalibration of ASCAT Scatterometer Winds from MetOp-A,-B, and-C, for a Stable Climate Data Record. *Remote Sens.* **2021**, *13*, 3678. [CrossRef]
7. Dabboor, M.; Olthof, I.; Mahdianpari, M.; Mohammadimanesh, F.; Shokr, M.; Brisco, B.; Homayouni, S. The RADARSAT Constellation Mission Core Applications: First Results. *Remote Sens.* **2022**, *14*, 301. [CrossRef]
8. Potin, P.; Rosich, B.; Miranda, N.; Grimont, P. Sentinel-1 Mission Status. *Procedia Comput. Sci.* **2016**, *100*, 1297–1304. [CrossRef]
9. Gordon, H.R.; Wang, M. Retrieval of Water-Leaving Radiance and Aerosol Optical Thickness over the Oceans with SeaWiFS: A Preliminary Algorithm. *Appl. Opt.* **1994**, *33*, 443–452. [CrossRef] [PubMed]
10. Wang, M. Remote Sensing of the Ocean Contributions from Ultraviolet to Near-Infrared Using the Shortwave Infrared Bands: Simulations. *Appl. Opt.* **2007**, *46*, 1535–1547. [CrossRef]
11. Wang, M.; Son, S.H. VIIRS-Derived Chlorophyll-a Using the Ocean Color Index Method. *Remote Sens. Environ.* **2016**, *182*, 141–149. [CrossRef]

12. Wang, M.; Jiang, L.; Mikelsons, K.; Liu, X. Satellite-Derived Global Chlorophyll-a Anomaly Products. *Int. J. Appl. Earth Obs. Geoinf.* **2021**, *97*, 102288. [CrossRef]
13. Wang, M.; Son, S.H.; Harding, L.W. Retrieval of Diffuse Attenuation Coefficient in the Chesapeake Bay and Turbid Ocean Regions for Satellite Ocean Color Applications. *J. Geophys. Res. Ocean.* **2009**, *114*, C10011. [CrossRef]
14. Wei, J.; Wang, M.; Jiang, L.; Yu, X.; Mikelsons, K.; Shen, F. Global Estimation of Suspended Particulate Matter from Satellite Ocean Color Imagery. *J. Geophys. Res. Ocean.* **2021**, *126*, e2021JC017303. [CrossRef] [PubMed]
15. Wei, J.; Wang, M.; Mikelsons, K.; Jiang, L.; Kratzer, S.; Lee, Z.; Moore, T.; Sosik, H.M.; Van der Zande, D. Global Satellite Water Classification Data Products over Oceanic, Coastal, and Inland Waters. *Remote Sens. Environ.* **2022**, *282*, 113233. [CrossRef]
16. Goldberg, M.D.; Kilcoyne, H.; Cikanek, H.; Mehta, A. Joint Polar Satellite System: The United States next Generation Civilian Polar-Orbiting Environmental Satellite System. *J. Geophys. Res. Atmos.* **2013**, *118*, 13463–13475. [CrossRef]
17. Donlon, C.; Berruti, B.; Buongiorno, A.; Ferreira, M.H.; Féménias, P.; Frerick, J.; Goryl, P.; Klein, U.; Laur, H.; Mavrocordatos, C.; et al. The Global Monitoring for Environment and Security (GMES) Sentinel-3 Mission. *Remote Sens. Environ.* **2012**, *120*, 37–57. [CrossRef]
18. Hu, C.; Lee, Z.; Franz, B. Chlorophyll a Algorithms for Oligotrophic Oceans: A Novel Approach Based on Three-Band Reflectance Difference. *J. Geophys. Res. Ocean.* **2012**, *117*, C01011. [CrossRef]
19. Clark, D.K.; Gordon, H.R.; Voss, K.J.; Ge, Y.; Broenkow, W.; Treess, C. Validation of Atmospheric Correction over the Oceans. *J. Geophys. Res. Atmos.* **1997**, *102*, 17209–17217. [CrossRef]
20. Wang, M.; Liu, X.; Tan, L.; Jiang, L.; Son, S.; Shi, W.; Rausch, K.; Voss, K. Impacts of VIIRS SDR Performance on Ocean Color Products. *J. Geophys. Res. Atmos.* **2013**, *118*, 10347–10360. [CrossRef]
21. Wang, M.; Liu, X.; Jiang, L.; Son, S. Visible Infrared Imaging Radiometer Suite Ocean Color Products. *VIIRS Ocean Color Algorithm Theoretical Basis Document (ATBD)*, 68 pp., NOAA/NESDIS/STAR, On-Line Version 1.0 NOAA, June 2017. Available online: https://www.star.nesdis.noaa.gov/jpss/documents/ATBD/ATBD_OceanColor_v1.0.pdf (accessed on 26 June 2024).
22. Wang, M.; Jiang, L.; Son, S.; Liu, X.; Voss, K.J. Deriving Consistent Ocean Biological and Biogeochemical Products from Multiple Satellite Ocean Color Sensors. *Opt. Express* **2020**, *28*, 2661. [CrossRef]
23. Platt, T.; Sathyendranath, S.; Caverhill, C.M.; Lewis, M.R. Ocean Primary Production and Available Light: Further Algorithms for Remote Sensing. *Deep Sea Res. Part A Oceanogr. Res. Pap.* **1988**, *35*, 855–879. [CrossRef]
24. Yu, X.; Lee, Z.; Shen, F.; Wang, M.; Wei, J.; Jiang, L.; Shang, Z. An Empirical Algorithm to Seamlessly Retrieve the Concentration of Suspended Particulate Matter from Water Color across Ocean to Turbid River Mouths. *Remote Sens. Environ.* **2019**, *235*, 111491. [CrossRef]
25. Liu, X.; Wang, M. Gap Filling of Missing Data for VIIRS Global Ocean Color Products Using the DINEOF Method. *IEEE Trans. Geosci. Remote Sens.* **2018**, *56*, 4464–4476. [CrossRef]
26. Liu, X.; Wang, M. High Spatial Resolution Gap-Free Global and Regional Ocean Color Products. *IEEE Trans. Geosci. Remote Sens.* **2023**, *61*, 3271465. [CrossRef]
27. Liu, X.; Wang, M. Global Daily Gap-Free Ocean Color Products from Multi-Satellite Measurements. *Int. J. Appl. Earth Obs. Geoinf.* **2022**, *108*, 102714. [CrossRef]
28. Alvera-Azcárate, A.; Barth, A.; Rixen, M.; Beckers, J.M. Reconstruction of Incomplete Oceanographic Data Sets Using Empirical Orthogonal Functions: Application to the Adriatic Sea Surface Temperature. *Ocean Model.* **2005**, *9*, 325–346. [CrossRef]
29. Beckers, J.M.; Rixen, M. EOF Calculations and Data Filling from Incomplete Oceanographic Datasets. *J. Atmos. Ocean. Technol.* **2003**, *20*, 1839–1856. [CrossRef]
30. NOAA STAR NOAA STAR Ocean Color Science Team. Available online: https://www.star.nesdis.noaa.gov/socd/mecb/color/index.php (accessed on 19 February 2024).
31. NOAA CoastWatch. Available online: https://coastwatch.noaa.gov/cwn/index.html (accessed on 19 February 2024).
32. Roman-Stork, H.L.; Byrne, D.A.; Leuliette, E.W. MESI: A Multiparameter Eddy Significance Index. *Earth Sp. Sci.* **2023**, *10*, e2022EA002583. [CrossRef]
33. Walton, C.C.; Pichel, W.G.; Sapper, J.F.; May, D.A. The Development and Operational Application of Nonlinear Algorithms for the Measurement of Sea Surface Temperatures with the NOAA Polar-Orbiting Environmental Satellites. *J. Geophys. Res. Oceans* **1998**, *103*, 27999–28012. [CrossRef]
34. McClain, E.P.; Pichel, W.G.; Walton, C.C. Comparative Performance of AVHRR-Based Multichannel Sea Surface Temperatures. *J. Geophys. Res. Ocean.* **1985**, *90*, 11587–11601. [CrossRef]
35. Wu, X.; Menzel, W.P.; Wade, G.S. Estimation of Sea Surface Temperatures Using GOES-8/9 Radiance Measurements. *Bull. Am. Meteorol. Soc.* **1999**, *80*, 1127–1138. [CrossRef]
36. Maturi, E.; Harris, A.; Merchant, C.; Mittaz, J.; Potash, B.; Meng, W.; Sapper, J. NOAA's Sea Surface Temperature Products from Operational Geostationary Satellites. *Bull. Am. Meteorol. Soc.* **2008**, *89*, 1877–1888. [CrossRef]
37. Petrenko, B.; Ignatov, A.; Kihai, Y.; Heidinger, A. Clear-Sky Mask for the Advanced Clear-Sky Processor for Oceans. *J. Atmos. Ocean. Technol.* **2010**, *27*, 1609–1623. [CrossRef]
38. Petrenko, B.; Ignatov, A.; Kihai, Y.; Stroup, J.; Dash, P. Evaluation and Selection of SST Regression Algorithms for JPSS VIIRS. *J. Geophys. Res.* **2014**, *119*, 4580–4599. [CrossRef]
39. Petrenko, B.; Ignatov, A.; Kihai, Y.; Dash, P. Full Access Sensor-Specific Error Statistics for SST in the Advanced Clear-Sky Processor for Oceans. *J. Atmos. Ocean. Technol.* **2016**, *33*, 345–359. [CrossRef]

40. Ignatov, A.; Gladkova, I.; Ding, Y.; Shahriar, F.; Kihai, Y.; Zhou, X. JPSS VIIRS Level 3 Uncollated Sea Surface Temperature Produc at NOAA. *J. Appl. Remote Sens.* **2017**, *11*, 32405. [CrossRef]
41. Petrenko, B.; Pryamitsyn, V.; Ignatov, A.; Jonasson, O.; Kihai, Y. AVHRR GAC Sea Surface Temperature Reanalysis Version 2 *Remote Sens.* **2022**, *14*, 3165. [CrossRef]
42. Jonasson, O.; Ignatov, A.; Pryamitsyn, V.; Petrenko, B.; Kihai, Y. JPSS VIIRS SST Reanalysis Version 3. *Remote Sens.* **2022**, *14*, 3476 [CrossRef]
43. Pryamitsyn, V.; Petrenko, B.; Ignatov, A.; Kihai, Y. Metop First Generation AVHRR FRAC SST Reanalysis Version 1. *Remote. Sens* **2021**, *13*, 4046. [CrossRef]
44. Jonasson, O.; Ignatov, A.; Petrenko, B.; Pryamitsyn, V.; Kihai, Y. NOAA MODIS SST Reanalysis Version 1. *Remote Sens.* **2023** *15*, 5589. [CrossRef]
45. Dash, P.; Ignatov, A.; Kihai, Y.; Sapper, J. The SST Quality Monitor (SQUAM). *J. Atmos. Ocean. Technol.* **2010**, *27*, 1899–1917 [CrossRef]
46. Liang, X.; Ignatov, A. Monitoring of IR Clear-Sky Radiances over Oceans for SST (MICROS). *J. Atmos. Ocean. Technol.* **2011**, *28* 1228–1242. [CrossRef]
47. Ding, Y.; Ignatov, A.; He, K.; Gladkova, I. ACSPO Regional Monitor for SST (ARMS). In Proceedings of the GHRSTT XIX: Science Team Meeting, Darmstadt, Germany, 4–8 June 2018.
48. Xu, F.; Ignatov, A. In Situ SST Quality Monitor (IQuam). *J. Atmos. Ocean. Technol.* **2014**, *31*, 164–180. [CrossRef]
49. He, K.; Ignatov, A.; Kihai, Y.; Cao, C.; Stroup, J. Sensor Stability for SST (3s): Toward Improved Long-Term Characterization of AVHRR Thermal Bands. *Remote Sens.* **2016**, *8*, 346. [CrossRef]
50. Maturi, E.; Harris, A.; Mittaz, J.; Sapper, J.; Wick, G.; Zhu, X.; Dash, P.; Koner, P. A New High-Resolution Sea Surface Temperature Blended Analysis. *Bull. Am. Meteorol. Soc.* **2017**, *98*, 1015–1026. [CrossRef]
51. Lance, V.P. (NOAA) First Report of the NESDIS SST Working Group August 2021 Advancing NESDIS Sea Surface Temperature to Fulfill NOAA Requirements and; 2021.
52. Key, J.R.; Mahoney, R.; Liu, Y.; Romanov, P.; Tschudi, M.; Appel, I.; Maslanik, J.; Baldwin, D.; Wang, X.; Meade, P. Snow and Ice Products from Suomi NPP VIIRS. *J. Geophys. Res. Atmos.* **2013**, *118*, 12816–12830. [CrossRef]
53. Meier, W.N.; Stewart, J.S.; Liu, Y.; Key, J.; Miller, J.A. Operational Implementation of Sea Ice Concentration Estimates from the AMSR2 Sensor. *IEEE J. Sel. Top. Appl. Earth Obs. Remote Sens.* **2017**, *10*, 3904–3911. [CrossRef]
54. Dworak, R.; Liu, Y.; Key, J.; Meier, W.N. A Blended Sea Ice Concentration Product from Amsr2 and Viirs. *Remote Sens.* **2021** *13*, 2982. [CrossRef]
55. Hoffman, J.P.; Ackerman, S.A.; Liu, Y.; Key, J.R. A 20-Year Climatology of Sea Ice Leads Detected in Infrared Satellite Imagery Using a Convolutional Neural Network. *Remote Sens.* **2022**, *14*, 5763. [CrossRef]
56. Hoffman, J.P.; Ackerman, S.A.; Liu, Y.; Key, J.R.; McConnell, I.L. Application of a Convolutional Neural Network for the Detection of Sea Ice Leads. *Remote Sens.* **2021**, *13*, 4571. [CrossRef]
57. Liu, Y.; Dworak, R.; Key, J. Ice Surface Temperature Retrieval from a Single Satellite Imager Band. *Remote Sens.* **2018**, *10*, 1909 [CrossRef]
58. Key, J.; Wang, X.; Liu, Y.; Dworak, R.; Letterly, A. The AVHRR Polar Pathfinder Climate Data Records. *Remote Sens.* **2016**, *8*, 167 [CrossRef]
59. Liu, Y.; Key, J.R.; Wang, X.; Tschudi, M. Multidecadal Arctic Sea Ice Thickness and Volume Derived from Ice Age. *Cryosphere* **2020**, *14*, 1325–1345. [CrossRef]
60. Alsweiss, S.; Jelenak, Z.; Chang, P.S. Extending the Usability of Radiometer Ocean Surface Wind Measurements to All-Weather Conditions for NOAA Operations: Application to AMSR2. *IEEE Trans. Geosci. Remote Sens.* **2023**, *61*, 5301112. [CrossRef]
61. Anheuser, J.; Liu, Y.; Key, J.R. A Climatology of Thermodynamic vs. Dynamic Arctic Wintertime Sea Ice Thickness Effects during the CryoSat-2 Era. *Cryosphere* **2023**, *17*, 2871–2889. [CrossRef]
62. Xie, P.; Boyer, T.; Bayler, E.; Xue, Y.; Byrne, D.; Reagan, J.; Locarnini, R.; Sun, F.; Joyce, R.; Kumar, A. An in Situ-Satellite Blended Analysis of Global Sea Surface Salinity. *J. Geophys. Res. Ocean.* **2014**, *119*, 6140–6160. [CrossRef]
63. Abdalla, S.; Abdeh Kolahchi, A.; Ablain, M.; Adusumilli, S.; Aich Bhowmick, S.; Alou-Font, E.; Amarouche, L.; Andersen, O.B.; Antich, H.; Aouf, L.; et al. Altimetry for the Future: Building on 25 Years of Progress. *Adv. Sp. Res.* **2021**, *68*, 319–363. [CrossRef]
64. Mitchum, G.T. An Improved Calibration of Satellite Altimetric Heights Using Tide Gauge Sea Levels with Adjustment for Land Motion. *Mar. Geod.* **2000**, *23*, 145–166. [CrossRef]
65. Zhang, B.; Cao, C.; Lillibridge, J.L.; Miller, L. Assessing the Measurement Consistency Between the Jason-2/AMR and SARAL/AltiKa/DFMR Microwave Radiometers Using Simultaneous Nadir Observations. *Mar. Geod.* **2015**, *38*, 143–155. [CrossRef]
66. Leuliette, E.W.; Willis, J.K. Balancing the Sea Level Budget. *Oceanography* **2011**, *24*, 122–129. [CrossRef]
67. Merrifield, M.A.; Thompson, P.R.; Chambers, D.P.; Mitchum, G.T.; Menéndez, M.; Nerem, R.S.; Leuliette, E.W.; Miller, L.; Holgate, S.J.; Marra, J.J.; et al. Sea Level Variability and Change. *Bull. Am. Meteorol. Soc.* **2017**, *93*, S81–S84.
68. Sandwell, D.T.; Goff, J.A.; Gevorgian, J.; Harper, H.; Kim, S.; Yu, Y.; Tozer, B.; Wessel, P.; Smith, W.H.F. Improved Bathymetric Prediction Using Geological Information: SYNBATH. *Earth Sp. Sci.* **2022**, *9*, e2021EA002069. [CrossRef]
69. Marks, K.M.; Smith, W.H.F. Comparison of Stacked Sentinel-3 A&B and AltiKa Repeat Cycle Data. *Earth Sp. Sci.* **2022**, *9*, e2021EA001892. [CrossRef]

70. Egido, A.; Smith, W.H.F. Fully Focused SAR Altimetry: Theory and Applications. *IEEE Trans. Geosci. Remote Sens.* **2017**, *55*, 392–406. [CrossRef]
71. Altiparmaki, O.; Kleinherenbrink, M.; Naeije, M.; Slobbe, C.; Visser, P. SAR Altimetry Data as a New Source for Swell Monitoring. *Geophys. Res. Lett.* **2022**, *49*, e2021GL096224. [CrossRef]
72. Kleinherenbrink, M.; Naeije, M.; Slobbe, C.; Egido, A.; Smith, W.H.F. The Performance of CryoSat-2 Fully-Focussed SAR for Inland Water-Level Estimation. *Remote Sens. Environ.* **2020**, *237*, 111589. [CrossRef]
73. Buchhaupt, C.K.; Egido, A.; Smith, W.H.F.; Fenoglio, L. Conditional Sea Surface Statistics and Their Impact on Geophysical Sea Surface Parameters Retrieved from SAR Altimetry Signals. *Adv. Sp. Res.* **2023**, *71*, 2332–2347. [CrossRef]
74. Buchhaupt, C.K.; Egido, A.; Vandemark, D.; Smith, W.H.F.; Fenoglio, L.; Leuliette, E. Towards the Mitigation of Discrepancies in Sea Surface Parameters Estimated from Low- and High-Resolution Satellite Altimetry. *Remote Sens.* **2023**, *15*, 4206. [CrossRef]
75. Scharroo, R.; Lillibridge, J.; Byrne, D.; Naeije, M. RADS: Consistent Multi-Mission Products RADS: CONSISTENT MULTI-MISSION PRODUCTS. In Proceedings of the Symposium on 20 Years of Progress in Radar Altimetry, Venice, Italy, 20–28 September 2012; European Space Agency European Space Agency Special Publication: Paris, France, 2013; p. 4.
76. Maturi, E.; Shay, L.; Donahue, D.; Byrne, D. NOAA's Operational Satellite Ocean Heat Content Products. *J. Oper. Oceanogr.* **2022**, *17*, 93–102. [CrossRef]
77. Byrne, D.A.; Maturi, E.M.; Burns, J.; Leuliette, E.W. New Developments for NOAA's Operational Upper Ocean Heat Content Product Suite. In Proceedings of the Ocean Surface Topography Science Team (OSTST) Meeting, Chicago, IL, USA, 21–25 October 2019. Available online: https://ostst.aviso.altimetry.fr/fileadmin/user_upload/Poster-APOP2019-Byrne.pdf (accessed on 26 June 2024).
78. Chang, P.; Jelenak, Z. NOAA Operational Satellite Ocean Surface Vector Winds Requirements Workshop Report; 2006. Available online: https://manati.star.nesdis.noaa.gov/SVW_nextgen/SVW_workshop_report_final.pdf (accessed on 26 June 2024).
79. Chang, P.S.; Jelenak, Z.; Sienkiewicz, J.M.; Knabb, R.; Brennan, M.J.; Long, D.G.; Freeberg, M. Operational Use and Impact of Satellite Remotely Sensed Ocean Surface Vector Winds in the Marine Warning and Forecasting Environment. *Oceanography* **2009**, *22*, 194–207. [CrossRef]
80. Jelenak, Z.; Chang, P.S.; Sienkiewicz, J.; Brennan, M.; Knabb, R.; Ward, B.; Edson, R.; Hufford, G.; Scott, C.; Lorens, J.; et al. *Dual Frequency Scatterometer (DFS) on GCOM-W2 Satellite: Instrument Design, Expected Performance and User Impact Addendum to QuikSCAT Follow-On User Impact Study*; NESDIS: Silver Spring, MD, USA, 2009. Available online: https://manati.star.nesdis.noaa.gov/SVW_nextgen/DFS_design_performance_and_user_impact_rev3.pdf (accessed on 26 June 2024).
81. Soisuvarn, S.; Jelenak, Z.; Chang, P.S.; Zhu, Q.; Shoup, C.G. High-Resolution Coastal Winds from the NOAA Near Real-Time ASCAT Processor. *IEEE Trans. Geosci. Remote Sens.* **2023**, *61*, 5500212. [CrossRef]
82. Jelenak, Z.; Chang, P.S.; Sienkiewicz, J.; Knabb, R.; Ward, B.; Edson, R.; Hufford, G.; Lorens, J.; Murphy, V.; Johnson, K.; et al. QuikSCAT Follow—On Mission: User Impact Study Report NOAA Operational Satellite Ocean Surface Vector Winds; 2008. Available online: https://manati.star.nesdis.noaa.gov/SVW_nextgen/QFO_user_impact_study_final.pdf (accessed on 26 June 2024).
83. Kite-Powell, H. The Value of Ocean Surface Wind Information for Maritime Commerce. *Mar. Technol. Soc. J.* **2011**, *45*, 75–84. [CrossRef]
84. Powell, H.K. Economic Benefits from Ocean Surface Vector Wind Observations and Forecasts Cumulative Activity by Distance from Shore. 2009, 1–5. Available online: https://manati.star.nesdis.noaa.gov/SVW_nextgen/ocean_surface_wind_obs_benefits_study_09SEPT2009.pdf (accessed on 26 June 2024).
85. Soisuvarn, S.; Jelenak, Z.; Chang, P.S.; Alsweiss, S.O.; Zhu, Q. CMOD5.H-A High Wind Geophysical Model Function for C-Band Vertically Polarized Satellite Scatterometer Measurements. *IEEE Trans. Geosci. Remote Sens.* **2013**, *51*, 3744–3760. [CrossRef]
86. Long, D.G.; Hardin, P.J.; Whiting, P.T. Resolution Enhancement of Spaceborne Scatterometer Data. *IEEE Trans. Geosci. Remote Sens.* **1993**, *31*, 700–715. [CrossRef]
87. Monaldo, F.; Jackson, C.; Pichel, W. Seasat to Radarsat-2: Research to Operations. *Oceanography* **2013**, *26*, 34–45. [CrossRef]
88. Jackson, C.R.; Ruff, T.W.; Knaff, J.A.; Mouche, A.; Sampson, C.R. Chasing Cyclones from Space. *EOS Earth Space Sci. News* **2021**, *102*. [CrossRef]
89. Mouche, A.; Chapron, B.; Knaff, J.; Zhao, Y.; Zhang, B.; Combot, C. Copolarized and Cross-Polarized SAR Measurements for High-Resolution Description of Major Hurricane Wind Structures: Application to Irma Category 5 Hurricane. *J. Geophys. Res. Ocean.* **2019**, *124*, 3905–3922. [CrossRef]
90. Duncan, K.; Farrell, S.L.; Richter-Menge, J.; Casal, T.; Connor, L.; Forsberg, R.; Haas, C.; Hendricks, S.; Leuliette, E.; McAdoo, D.C.; et al. Dedicated Airborne Experiments to Validate Satellite Altimetry over Arctic Sea Ice: A Review. In Proceedings of the Poster, CryoSat 10th Anniversary Science Conference, Virtual Event, 14–17 June 2021.
91. CEOS COAST UN Ocean Decade Contribution. Available online: https://oceandecade.org/actions/committee-on-earth-observation-satellites-coastal-observations-applications-services-and-tools-ceos-coast/ (accessed on 19 February 2024).
92. NOAA PolarWatch. Available online: https://polarwatch.noaa.gov/ (accessed on 19 February 2024).
93. NOAA Coral Reef Watch. Available online: https://coralreefwatch.noaa.gov/ (accessed on 19 February 2024).
94. NOAA CoastWatch Learning Portal. Available online: https://umd.instructure.com/courses/1336575 (accessed on 19 February 2024).

95. NOAA CoastWatch Data Portal. Available online: https://coastwatch.noaa.gov/cw_html/cwViewer.html (accessed on 19 February 2024).
96. NOAA OceanWatch Monitor. Available online: https://www.star.nesdis.noaa.gov/socd/om/ (accessed on 19 February 2024).
97. NOAA Ocean Viewer. Available online: https://www.star.nesdis.noaa.gov/socd/ov/ (accessed on 19 February 2024).
98. Lavergne, T.; Kern, S.; Aaboe, S.; Derby, L.; Dybkjaer, G.; Garric, G.; Heil, P.; Hendricks, S.; Holfort, J.; Howell, S.; et al. A New Structure for the Sea Ice Essential Climate Variables of the Global Climate Observing System. *Bull. Am. Meteorol. Soc.* **2022**, *103*, E1502–E1521. [CrossRef]
99. Wang, X.; Liu, Y.; Key, J.R.; Dworak, R. A New Perspective on Four Decades of Changes in Arctic Sea Ice from Satellite Observations. *Remote Sens.* **2022**, *14*, 1846. [CrossRef]
100. Anheuser, J.; Liu, Y.; Key, J.R. A Simple Model for Daily Basin-Wide Thermodynamic Sea Ice Thickness Growth Retrieval. *Cryosphere* **2022**, *16*, 4403–4421. [CrossRef]
101. Letterly, A.; Key, J.; Liu, Y. The Influence of Winter Cloud on Summer Sea Ice in the Arctic, 1983–2013. *J. Geophys. Res.* **2016**, *121*, 2178–2187. [CrossRef]
102. Liu, Y.; Key, J.R. Less Winter Cloud Aids Summer 2013 Arctic Sea Ice Return from 2012 Minimum. *Environ. Res. Lett.* **2014**, *9*, 44002. [CrossRef]
103. Liu, Y.; Key, J. Cold Season Cloud Response to Sea Ice Loss in the Arctic. *J. Climate* **2024**. *submitted*.
104. Letterly, A.; Key, J.; Liu, Y. Arctic Climate: Changes in Sea Ice Extent Outweigh Changes in Snow Cover. *Cryosphere* **2018**, *12*, 3373–3382. [CrossRef]
105. NOAA Coral Reef Watch Virtual Stations. Available online: https://coralreefwatch.noaa.gov/product/vs_single_pixel_exp/florida_keys.php (accessed on 19 February 2024).
106. Anderson, C.R.; Kudela, R.M.; Kahru, M.; Chao, Y.; Rosenfeld, L.K.; Bahr, F.L.; Anderson, D.M.; Norris, T.A. Initial Skill Assessment of the California Harmful Algae Risk Mapping (C-HARM) System. *Harmful Algae* **2016**, *59*, 1–18. [CrossRef]
107. Introduction to CEOS COAST. Available online: https://www.star.nesdis.noaa.gov/socd/coast/ (accessed on 19 February 2024).
108. CEOS COAST Ad Hoc Team Implementation Plan. Available online: https://ceos.org/document_management/Ad_Hoc_Teams/COAST/Documents/CEOS-COAST%20Implementation%20Plan%20Final%20Vers_1.2.pdf (accessed on 19 February 2024).
109. Neely, M.B.; Nezlin, N.; Lazzaro, R.; Bishop-Taylor, R.; Saga, S.; DiGiacomo, P.M.; Sharma, R. CEOS COAST: Progress on the Chesapeake Bay Pilot and Future Work. In Proceedings of the OCEANS 2022 Hampton Roads, Virginia Beach, VA, USA, 17–20 October 2022; pp. 1–5.
110. Nezlin, N.P.; Herman, J.D.; Hodge, J.; Sagar, S.; Bishop-Taylor, R.; Zheng, G.; DiGiacomo, P.M. Assessment of Changes of Complex Shoreline from Medium-Resolution Satellite Imagery. *Estuaries Coasts* **2023**, *46*, 1723–1739. [CrossRef]
111. Davies, G.; Digiacomo, P.; Lance, V.; Obligis, E.; Carroll, A.O. *Report of the Second International Operational Satellite Oceanography Symposium*, Virtual Meeting Held 25–27 May 2021, 2022; EUMETSAT, Darmstadt, Germany. Available online: https://cdn.eventsforce.net/files/ef-xnn67yq56ylu/website/18/osos-2_final_report_released_31_march_2022.pdf (accessed on 26 June 2024).
112. Brown, C.; Digiacomo, P.; Foli, B.; Lance, V.; Levy, G.; Montagner, F.; Neely, M.B.; Rayner, R.; Reppucci, A.; Wilson, C. Report of the First International Operational Satellite Oceanography; 2019; NOAA, College Park, MD, USA. Available online: https://www.geoaquawatch.org/wp-content/uploads/2020/06/Final-2019-OSOS-Summary_Final_vers2.1.pdf (accessed on 26 June 2024).

remote sensing

MDPI

Review

The Bright Decade of Ocean Salinity from Space

Roberto Sabia [1,*], Jacqueline Boutin [2], Nicolas Reul [3], Tong Lee [4] and Simon H. Yueh [4]

1 ESA ESRIN, Climate Action, Sustainability and Science Department, 00044 Frascati, Italy
2 LOCEAN IPSL Laboratory, 75252 Paris, France
3 Laboratoire d'Océanographie Physique et Spatiale (LOPS), CNRS, IRD, IUEM, Ifremer, University Brest, 29280 Plouzané, France
4 NASA JPL, California Institute of Technology, Pasadena, CA 91125, USA
* Correspondence: roberto.sabia@esa.int

Abstract: Sea Surface Salinity is a crucial climatic variable due to its twofold role as both a passive and an active tracer of oceanic processes. Despite its relevance, however, it could not be measured from space, mainly because of technological limitations, until 2009. Since then, the generation and assessment of satellite salinity has become a game-changer in physical and biogeochemical oceanography, as well as in climate science. Three satellite sensors with salinity-measuring capabilities (SMOS-Soil Moisture and Ocean Salinity, Aquarius, and SMAP-Soil Moisture Active Passive) have been launched in the previous decade, each characterized by specific measurement concepts and features and ad hoc validation approaches. The increasing usage of spaceborne salinity products has produced a variety of results and applications, which are here summarized under three specific domains: climate, scientific, and operational. Finally, short-to-mid-term perspectives, indicating both the expected improvements in terms of algorithms and also looking at novel mission concepts (that will provide continuation of these measurements in the decade to come) have been described.

Keywords: Sea Surface Salinity; Earth Observation; SMOS; Aquarius; SMAP; radiometry; oceanography; climate

1. Sea Surface Salinity: Background and Ongoing Efforts

Sea Surface Salinity (SSS) is a crucial environmental and climatic variable due to its twofold role as both a passive and an active tracer of oceanic processes. Despite its relevance, however, it has not been measurable from space, mainly because of technological limitations, up to the advent of the first satellite with this capability, in 2009.

The "Oceans from Space" conference series, which reviews progress and challenges in the satellite oceanography field with a decadal pace [1], tracked the development of SSS remote sensing since its inception. At the conference held in 2000, a friendly wager on the future of spaceborne satellites was made among the conference participants [2], foreseeing a bright future for technologies aiming at addressing SSS from space. At the subsequent 2010 edition, reflections on that wager were elaborated by the Principal Investigators of the recently launched Soil Moisture and Ocean Salinity (SMOS) mission and the soon-to-be-launched Aquarius mission, focusing on the challenges and excitement of the first-ever salinity measurements from space [3]. At the 2020 conference (deferred to October 2022 due to the COVID-19 pandemic restrictions), a compendium of a full decade of achievements in the broad oceanographic and climate perspective was presented [4], stressing how decisive the impact of SSS remote sensing actually was and currently is. Feedback was also gathered on residual concerns and limitations, as well as on future developments.

In this paper, we take stock of the broad scientific advancements and related discussions, assessing to what extent salinity from satellites has been a game-changer in physical and biogeochemical oceanography, and in climate science at large. After describing the three satellite sensors with salinity-measuring capabilities deployed so far—SMOS; Aquarius; and Soil Moisture Active Passive (SMAP)—along with their measurement concepts and features; the related validation approaches; and the inventory of products generated; we shall concentrate on the portfolio of scientific results that have been obtained by exploiting these data sources. The results will be opportunely divided into climate, scientific, and operational applications. Lastly, we will focus on the short-to-mid-term perspectives, indicating the expected improvements in terms of algorithms, but also looking at novel mission concepts that will provide continuity and enhancement to these measurements in the next decade. The focus of the present review is intentionally broad but necessarily succinct to cover the ample variety of studies on oceanic processes that were initiated, enhanced, or better characterized by the increasingly accurate and de-biased continuous stream of SSS data from the available variety of platforms and sensors.

1.1. Ocean Salinity Relevance in Oceanography and Water Cycle

The salinity of marine waters has long been recognized as a crucial parameter in the monitoring, diagnosis, and understanding of oceanic processes. In recognition of its relevance, it has been labeled as an Essential Climate Variable (ECV) and Essential Ocean Variable (EOV) by the Global Climate Observing System (GCOS) and Global Ocean Observing System (GOOS) committees [5]. SSS has a prominent role as a passive tracer, being the resulting diagnostic of the interplay between evaporation and precipitation and, in specific areas, of the additional influence of formation/melting of ice and freshwater runoff, together with horizontal/vertical advection. As such, salinity relevance within the water cycle is critical. Besides, salinity also plays an active tracer role, whereas horizontal and vertical variations of salinity might imply a change in seawater density and therefore indicate a possible trigger of the thermohaline circulation.

The advent of salinity-measuring satellites since the end of the 2000s (Figure 1) has dramatically reduced the observing-system capability gaps, given that the existing Argo floats, although essential, were insufficient to resolve salinity features with spatial scales smaller than hundreds of km and time scales of less than a month [6].

Figure 1. Artistic view of the three L-band satellite radiometers measuring sea surface salinity. From left to right: SMOS, Aquarius, and SMAP.

Before the launch of current satellite SSS sensors, and as repeatedly discussed at the early Oceans from Space conferences recalled above, it was anticipated that the synoptic and frequent monitoring of this parameter from space, in spite of the limitations of a coarse spatial resolution and the inherent noise and biases, might prove essential in giving insights into various aspects of oceanography and climate, such as (and not limited to) air-sea interactions, river discharge monitoring, ocean dynamics, mesoscale characterization, ocean modeling, and so forth.

1.2. Satellite SSS Sensors and Measurement Principles/Challenges (SMOS, Aquarius, SMAP)

1.2.1. SMOS

The Soil Moisture and Ocean Salinity (SMOS) is a European Space Agency (ESA) Earth Explorer opportunity mission [7]. It was launched in November 2009, with a nominal lifetime of 3–5 years, currently extended until the end of 2025, with a recent request for continuation up until 2028. It was a proof of concept of a novel Earth Observation (EO) technique demonstration: microwave radiometry by aperture synthesis.

SMOS is the first L-band radiometer in space, achieving global and continuous coverage. It was also the first satellite to infer SSS, as well as Soil Moisture (a land hydrology variable not covered in the present paper), directly and in an absolute fashion. The 2D interferometer was conceived to provide fully polarized multi-angular measurements of the surface, with a native spatial resolution of about 30 km at 3 dB below the satellite and 43 km on average over the whole field of view, sampled over the ISEA 4H9 Discrete Global Grid (DGG) at 15 km resolution [8]. The L-band channel was selected as the best compromise among minimum atmospheric effect, reasonable spatial resolution, and maximum sensitivity to the target variable of ocean salinity. However, being the first L-band mission to be deployed in space, it experienced an unexpectedly high level of Radio Frequency Interference (RFI) in the supposedly protected 1400–1427 MHz band [9], which called for sustained efforts over the years in devising mitigation and filtering techniques, and that was also instrumental in the design of the forthcoming L-band sensors.

The mission design allows global coverage of the Earth in less than 3 days at the equator, with a dusk/dawn orbit to minimize temperature gradients at the ocean-atmosphere interface. The rationale of SSS retrieval exploits the variation of the seawater conductivity properties with salinity, which in turn affects the emissivity and ultimately the Brightness Temperature (TB) measured with microwave radiometers. Specifically, the multangular acquisition of TBs over the so-called dwell-line is translated into a single SSS estimate in each pixel once the contribution of additional auxiliary data that also affect the TB value—namely; Sea Surface Temperature (SST); and Wind Speed (WS) as prime descriptors of sea roughness—are also accounted for [10–12]. Regarding the latter, colocation/uncertainty of the needed auxiliary data of SST and WS are also crucial. Parametrizations of the dielectric constant model [13] and of the emissivity of the sea surface due to roughness and foam have been constantly improving [14], alongside a debiasing module known as Ocean Target Transformation (OTT) [15]. In recent years, an empirical correction to handle the land signal leakage into the sea scenes that was jeopardizing measurements in the coastal ocean has been devised and implemented: The so-called Land-Sea Contamination (LSC) correction [16]. As such, both uncorrected and corrected salinity fields are served. Moreover, additional perturbation sources such as Galactic noise and Sun glint have been addressed with a set of algorithms that are still evolving. To date, three overall reprocessing campaigns have been carried out, and a fourth one is imminent.

ESA delivers both SMOS L1 and L2 products. Level 1 is multi-angular fully polarized TB, and Level 2 products over the ocean are organized by ascending and descending half-orbits SSS. Level 3 SSS global maps are produced operationally in various research and operational centers, as described afterwards. The prescribed mission accuracy is defined at L3, with SSS aimed to be 0.1 pss STDD (Standard Deviation of the Difference) over a 200 × 200 km spatial scale and over a month window [17].

1.2.2. Aquarius

The joint National Aeronautics and Space Administration (NASA) and Comisíon Nacional de Actividades Espaciales (CONAE) Aquarius mission [18] was launched on the Satélite de Aplicaciones Científicas (SAC)-D spacecraft, provided by Argentina's space agency, in June 2011. It carried a real aperture L-band radiometer and a scatterometer as the primary sensors. The main scientific objectives were to measure SSS over the global ice-free oceans with a 150 km spatial resolution for a 7-day revisit and to achieve a measurement accuracy of less than 0.2 pss on a 30-day time scale, taking into account all sensor and geophysical random errors and biases. Due to an electronic failure on the platform, the Aquarius mission ended prematurely in June 2015. Aquarius Level 3 from NASA (version 5) is the reference end-of-mission data release of the Aquarius/SAC-D mission.

The Aquarius instrument had an onboard Radio Frequency Interference (RFI) filtering system, which took stock of the RFI pollution contaminating SMOS measurements [19]. The unique capability of Aquarius was the availability of an L-band scatterometer onboard, which could be used to provide coincident estimates of the ocean's surface roughness, overcoming one of the hurdles of both SMOS and later SMAP, that is, the spatio-temporal collocation of parameters suitable to correct for the roughness effect on the ocean emissivity.

1.2.3. SMAP

The SMAP mission [20], jointly developed by NASA and the Jet Propulsion Laboratory (JPL), was launched in January 2015. SMAP included originally an L-band radar and an L-band radiometer for concomitant active/passive measurements. The radiometer and radar instruments became operational in April 2015. The radar transmitter regrettably ceased operating only a few months after (in July 2015) due to a hardware glitch, and therefore the radiometer remained the only operational instrument collecting science observations [21].

SMAP employs a 6 m rotating mesh antenna, which provides a wide swath coverage (of about 1000 km), obtaining TB observations at an effective resolution close to 43 km [21]. Benefiting from lessons learned by the SMOS and Aquarius missions, the SMAP receivers have the ability to record time–frequency sub-band data in order to detect emission contamination caused by manmade sources of RFI [22].

The SMAP science data products are provided with different granularity. Operationally, the Level 1 products are TB observations from the radiometer arranged in ascending and descending half-orbits. Specifically, these are the Level 1B geolocated TBs and the Level 1C TBs resampled on the Equal-Area Scalable Earth (EASE)-Grid map projections at 9 km and 36 km grid resolutions. The Level 2 products over the ocean are swath-based salinity produced in two institutions, i.e., Remote Sensing Systems (RSS) and JPL.

1.3. *Features of the Current Version of the Satellite Salinity Mission Processors*

Over the last decade, space agencies issued different releases of their salinity products to reflect the continuous research and development on both the algorithmic side and on the correction of the numerous signal perturbations.

Regarding SMOS, several data product versions have been deployed in the official ESA baseline (Figure 2), associated with three overall archive reprocessing campaigns performed to ensure consistency of the whole dataset. These versions aimed at refining the geophysical forward models, at providing improved versions for dielectric constant models, at devising specific corrections to tackle the perturbing effects of sky, sun, and galactic glints, at addressing instrument biases propagating at L2, and at mitigating the effect of RFI.

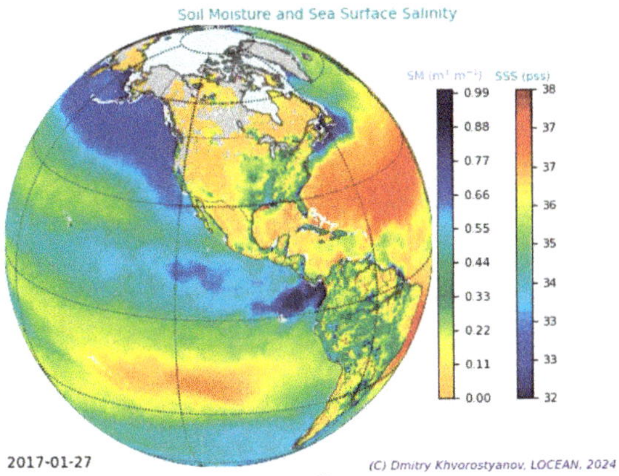

Figure 2. Composite of the two water-cycle-related parameters—soil moisture and ocean salinity—for a sample date in 2017. Over the ocean, visible patterns of high salinity are reflecting the evaporative gyres in the Atlantic and tropical Pacific, while low salinity is associated with the ITCZ (Inter-Tropical Convergence Zone) or high latitudes.

The latest SMOS version that is operational is v700, featuring a novel dielectric constant model [13], an improved computation of a de-biased salinity anomaly (with respect to a self-consistent SMOS climatology), an improved estimation of the auxiliary data errors, and an improved characterization of the L2 SSS uncertainty. Upcoming versions, which will soon undergo a 4th SMOS mission reprocessing, are now dealing with an improved correction to mitigate the influence of ice contamination and ad hoc corrections for latitudinal and seasonal biases experienced in the measurements. An alternative retrieval scheme based on a novel algorithm, the so-called Debiased non-Bayesian-DnB [23], has been implemented and is subject to a dedicated round-robin comparison exercise to assess its merits and limitations as compared to the nominal inversion scheme.

1.4. Mission Supporting Field Campaigns

A variety of experimental field campaigns supporting pre-launch simulations, modelling advancements, and error characterization took place in the decade preceding the satellite's launch, such as WISE, FROG, CAROL, and COSMOS (to name just a few), whose major outcomes are summarized in [24]. In the last decade, two major comprehensive field experiments funded by NASA took place, the so-called Salinity Processes in the Upper-ocean Regional Study (SPURS). Namely, SPURS-1 [25] and SPURS-2 [26] were two process studies aiming at characterizing dynamics in two different regimes, one dominated by evaporation and the other one dominated by precipitation. The results were compiled in [27,28] and helped shed light on various processes dominated or mediated by SSS.

More recently, an intense field campaign activity, funded by NASA, the Salinity and Stratification at the Sea Ice Edge (SASSIE) experiment, was performed in 2022 to enhance the understanding of polar-ocean and sea-ice dynamics. SASSIE has been a comprehensive experiment with consistent deployment of instrumentation and collection of a wealth of in-situ data. A strong emphasis was given to instrumentations able to capture the skin surface salinity, being therefore comparable to satellite measurements and advancing in the characterization of the vertical mismatch problem (Figure 3). Most notably, SASSIE was aiming at investigating how the salinity patterns could be helpful in the prediction of the extent and timing of subsequent sea-ice formation in the Arctic region. A campaign dataset [29] and a dedicated website are currently available, while a specific experiment was recently carried out to build a machine-learning algorithm aimed at inferring sea-ice formation timing out of salinity inputs. Preliminary results indicate that, when including L4 merged salinity products, the estimation of sea-ice formation/retreat is remarkably improving, paving the way to new applications of salinity as a prognostic variable.

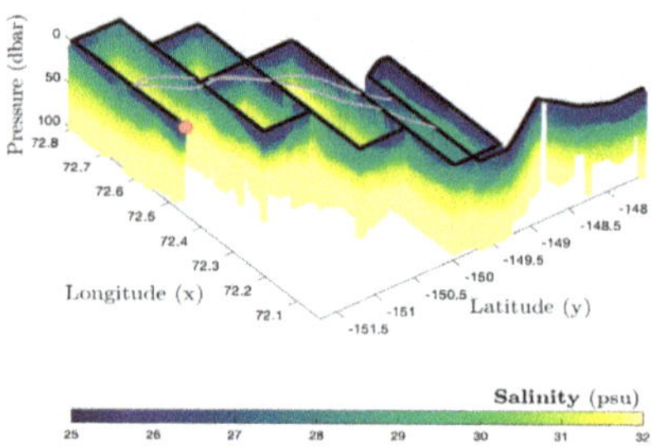

Figure 3. Three-dimensional plot showing salinity evolution over latitude, longitude, and depth for a transect in the SASSIE campaign (credits: J. Schanze, Earth and Space Research—ESR).

1.5. Pi-MEP Salinity—Validation Strategies and Metrics

Addressing a growing demand for coordinated validation and performance assessment, the Pilot Mission Exploitation Platform (Pi-MEP) for salinity was established in 2017 [30]. It includes a vast collection of in situ data coming from moorings, drifters, profilers, thermosalinographs (TSG), etc., as well as models, climatologies, and additional auxiliary data to properly verify, validate, and quantify performance metrics regarding satellite salinity products. It provides related visualization tools and sizeable match-up database reports for both the global ocean and a variety of ocean basins with associated plots and metrics (Figure 4). Specifically, in the Platform, a Match-up DataBases (MDB) Validation Report is provided for any single triplet consisting of one of the 20+ considered regions, per each satellite SSS product and each in-situ SSS considered database. A set of user-selectable criteria for the collocation of ground-based and satellite data are also available.

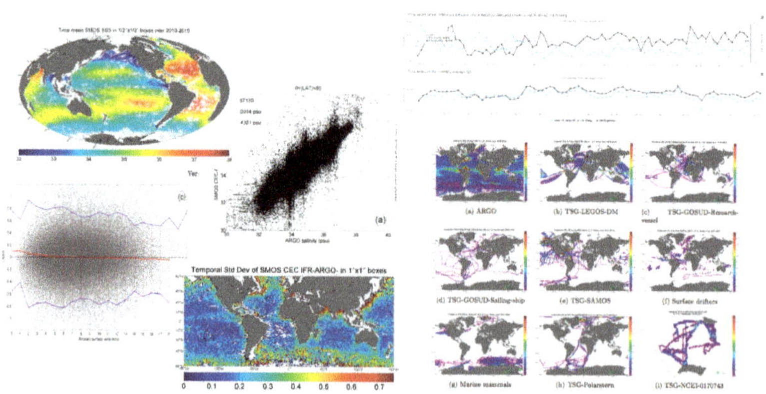

Figure 4. Pictorial chart assembling some of the various metrics reported in the PI-MEP platform, ranging from temporal mean and dispersion of the salinity fields to scatter plots vs. in-situ data. Variation of salinity data are portrayed with respect to a variety of geophysical parameters (e.g., SST and WS), along with the numerous in-situ data considered in the dedicated match-up reports.

Numerous validation efforts estimated performance metrics for satellite salinity, such as, for instance, ref. [28]. The PI-MEP platform allows us to explore a wide range of geophysical and geographical conditions for an ample set of satellite and in-situ data, obtaining therefore a wide range of validation metrics. As a reference, multi-mission L4 product metrics are described in [31], where it is stated that over the global ocean, when considering all monthly matchups with Argo SSS, a robust STDD of 0.16 pss is obtained.

As of 2019, PI-MEP became an interagency ESA-NASA effort addressing validation and coordinated research and development on specific topics such as enhanced match-up criteria, triple-collocation analyses, and representation error characterization [32].

Over the last years, the Pi-MEP platform was not only serving the scope of versioning the various releases of SMOS L2OS processors and of ESA Climate Change Initiative (CCI) salinity (described afterwards), but also became the benchmark for assessing the entity of representation errors in satellite salinity measurements. Representation errors are the intrinsic sampling differences that emerge when comparing box-average satellite data with punctual in-space and in-time in-situ measurements. It may appear as sub-footprint variability or vertical stratification [33] for the spatial component and temporal aliasing for the time component. Within the Platform, these errors are being assessed through a variety of techniques, including high-resolution model sampling and Triple-Collocation analyses. These errors need to be properly quantified to distinguish the actual satellite inaccuracies from the inherent differences due to the sampling strategies and criteria.

Lastly, very recently, efforts for performing spectral analyses along TSG lines to estimate the effective resolution of all satellite products have been undertaken.

1.6. Additional Operational Production Chains

Only regarding the ocean products—since similar efforts are devoted to Soil Moisture, the hydrology counterpart measured by these L-band sensors—there is ample variety of products being generated globally. In Europe, besides the official ESA processing chain generating Level-2 (L2) Ocean Salinity products, two other distribution centers have been in charge of releasing Level-3 (L3) and Level-4 (L4) enhanced salinity products.

The Centre Aval de Traitement des Données SMOS (CATDS), out of France, is currently distributing a wealth of L2, L3, and L4 products, both operational and experimental, taking into account different algorithmic strategies. More specifically, the current CATDS portfolio embodies L2 (daily) and L3 (weekly-to-monthly) products, with a specific set of corrections on land-sea contamination and seasonal latitudinal biases, rain freshening correction, with different grid sampling and optimal interpolation techniques. More details can be found in [34], while the schematic in Figure 5 illustrates the typology and features of the various clusters of products served at CATDS.

The Barcelona Expert Centre (BEC), out of Spain, has also been distributing for a long time an L2-to-L4 data stream by using a different inversion scheme, the so-called Debiased non-Bayesian (DnB) technique, which uses a different selection, statistical treatment, and filtering approach of the TBs and resulting SSS [35]. On top of these algorithmic strategies, the BEC processing chain also embodies different methodologies at L1, such as the Nodal Sampling technique to obtain a reliable set of TBs before the salinity retrieval procedure. Besides producing a global L3 and L4 product, BEC is also focusing on dedicated quasi-operational datasets on various basins (Arctic, Southern Ocean, Baltic, Black Sea, Mediterranean) with tailored processing adapted to the regional features and constraints (e.g., customized debiasing, temporal correction, land and ice contamination correction, etc.). More details can be found in Section 2.3.3 and in [36].

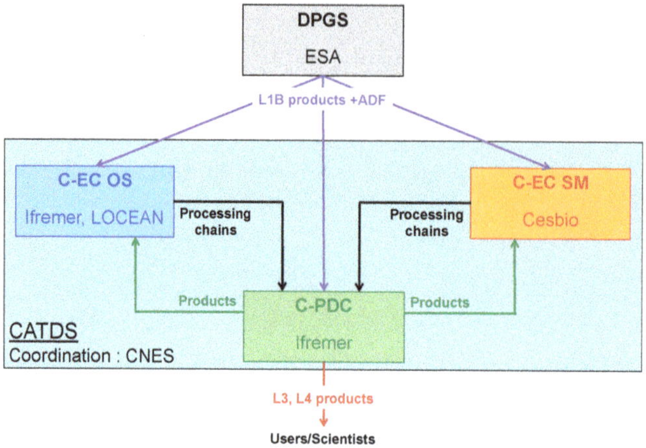

Figure 5. Schematic flowchart of the various CATDS processing lines.

1.7. Climate Data Records

Within the umbrella of the ESA Climate Change Initiative—CCI, the Salinity_cci project started in 2015 with the aim of merging SMOS, SMAP, and Aquarius datasets and generating a consistent, coherent, long-term Climate Data Record (CDR), paving the way for climatic studies using salinity data from space. A specific approach adopted in merging the various satellite salinity sources is to retain the high variability sampled by the satellites, and specifically a temporal Optimal Interpolation (OI) approach is chosen to avoid any spatial smoothing of satellite SSS.

The v3 of the data record embedded a novel dielectric constant model and the usage of a more stable source of auxiliary data, coming from the ECMWF Reanalysis v5 (ERA-5), for SST and wind speed. The current version, v4 (shown as a sample for July 2022 in Figure 6), was extending further the dataset duration and improving performance in challenging ocean basins/regimes. CCI salinity fields are well-suited for monitoring weekly to interannual signals at spatial scales ranging from 50 km to the basin scale. A full summary of the algorithms adopted, the inherent validation (also using Pi-MEP salinity), and the derived scientific studies can be found in [31]. A dedicated effort for the production of a better high-latitude product is underway, and for the imminent v5, a stricter assessment of RFI is foreseen along with a better characterization of latitudinal correction.

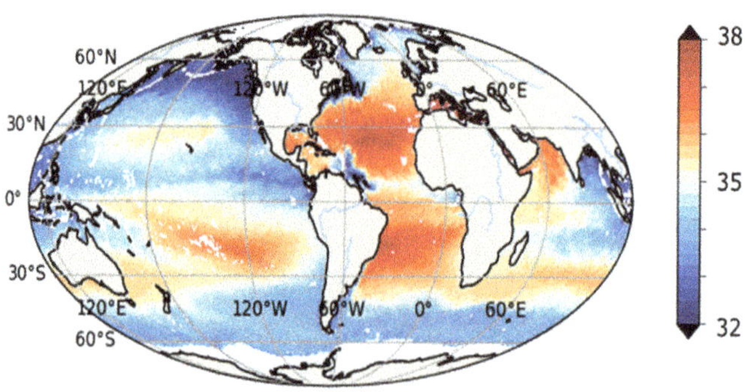

Figure 6. CCI Salinity v4.0 L4 product corresponding to July 2022, out of the merging of SMOS/Aquarius/SMAP data.

Moreover, recent efforts are aimed at extending backward the Salinity CDR by using the ratio between C-band and X-band data from AMSR-E (and Windsat, in the near future) in specific conditions of warm waters and high salinity gradient, i.e., in major tropical river plumes. This is expected to be served as a research product in the CCI salinity CDR.

On the US side, a composite product referred to as OI-SSS is currently produced by Earth and Space Research (ESR), which currently embodies all three satellites in a specific Optimal Interpolation scheme. Namely, the two-month overlap (April–June 2015) between Aquarius and SMAP was used to ensure consistency in the data record. In-situ salinity from Argo floats and moored buoys is used to derive a large-scale bias correction for the entire OI-SSS dataset [37].

2. Research Advances in the Last Decade

The overwhelming collection of results gathered in the last decade, from the launch of the first salinity satellite to date, can be listed and summarized through several angles and clustered in different and even arbitrary ways. Unavoidably, some of the studies could be classified differently or would be overlapping two or more of these domain clusters. Here below, a non-exhaustive mapping of the various efforts is conveniently divided into three main categories, also following the areas identified in the latest SMOS extension review by an independent scientific panel, as

- Climate Applications,
- Science Applications, and
- Operational Applications.

The ocean salinity community is gathering on a fairly regular basis (about every 2 years) in a dedicated "Ocean Salinity Conference" to discuss and take stock of the various scientific, technological, and policy advancements in this domain. Besides, the fifth edition of the decadal "Oceans from Space" conference was also a crucial benchmark to gather the various achievements in several domains, ranging from ocean water cycle linkages to climate prediction. In this paper, the angle chosen is a different splitting of the activities with some updated results, the emphasis on specific efforts such as Pi-MEP Salinity, and updated discussions/feedback gathered at these thematic conferences.

For more in-depth discussion, a number of review papers [38–40] have already been published to give a detailed view of the various results obtained. An original schematic of the various applications addressed from space is given in the word cloud below (Figure 7).

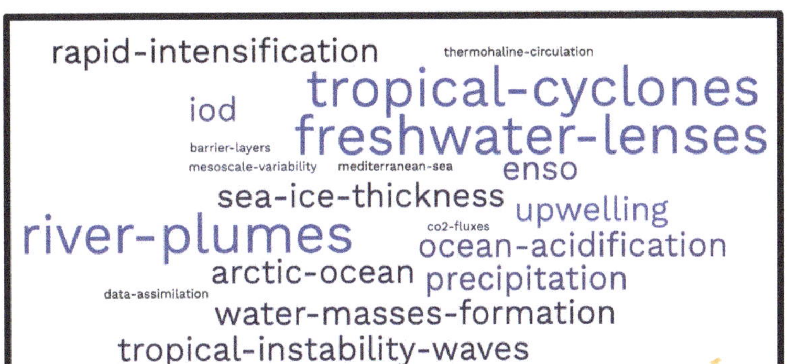

Figure 7. Schematic of the various applications addressed from space. The bigger the size of the word, the more prominent is the attention dedicated to that specific topic. This chart is conceived by the authors and is obtained by using an online word-cloud generator.

2.1. Climate Applications

Over the last decade, SSS fingerprints have been progressively used in climate-related studies. An important aspect that has been elucidated is the relevance of satellite SSS in describing El Niño Southern Oscillations (ENSO) phases, such as the case of extension and contraction of the Pacific freshwater pool in accordance with El Niño and La Niña phases by using SMOS [41] or the ENSO-related displacement of salinity fronts by using Aquarius [42].

Salinity fingerprints related to other climate indices have been studied in the Indian Ocean Dipole (IOD) [43,44] and in relation to the North Atlantic Oscillation (NAO) [45]. Another phenomenon that has been monitored also through the use of spaceborne salinity is the evolution of Tropical Instability Waves (TIW), their speed, and their phasing with ENSO events, with both Aquarius [46] and CCI-Salinity data [47]. SMAP and SMOS SSS have also elucidated that monsoonal rain and runoff in the Indonesian Seas can substantially weaken the upper-layer transport of the Indonesian Throughflow, with significant impact on the transport of heat and other oceanic properties between the Pacific and Indian Oceans [48].

Additionally, the production of a salinity CDR, such as CCI-SSS, is also favoring updated studies on performance metrics. For instance, in [49] it is underlined how, using a high-resolution model reanalysis, most of the largest spatial variability of the satellite-minus-Argo salinity is observed in regions with large sampling mismatches. The authors underline how in dynamic areas (e.g., the Gulf Stream), the sampling mismatch between CCI-SSS and Argo can account for up to about 30% of the Root Mean Square Difference (RMSD).

Regarding climate trends, in [50] it was assessed how salinity trend estimates are influenced by decadal and longer-term salinity variability, while others [51] assessed multi-sensor satellite products in the Mediterranean Sea. Even if the time series is relatively short, a clear interannual trend is found on the Eastern side of the basin, leading to a marked salinification.

2.2. Science Applications

2.2.1. The Freshwater Domain

Major emphasis in the last decade has been devoted to the role of salinity as a tracer for freshwater fluxes. A plethora of studies and applications emerged thanks to the provision of consistent and accurate datasets of salinity from space, as partially described afterwards. A first aspect that has been studied has been the monitoring of major river basins and the assessment of the evolution of their freshwater plumes [52]. This is also one of the process studies consistently monitored in the Pi-MEP platform, which portrays the seasonal/interannual variability in the discharge and advection of freshwater plumes into the ocean. Through salinity, the freshwater plumes—and their direct effect on ocean circulation and gas exchanges—have been better characterized; thanks to the more frequent sampling induced by satellites [53].

Another set of studies [54–56] aimed at elucidating the occurrence of freshwater lenses on the ocean surface by combining salinity datasets and rain rate information (Figure 8). This had a twofold goal: On the one hand, to remove the impact of rain-induced freshening on the validation, and on the other hand, to use salinity as a proxy for studying the fate and dissipation of rain events. Modeling studies also helped in disentangling the rain-induced freshening from, e.g., advection of freshwater from other sources or locations [57].

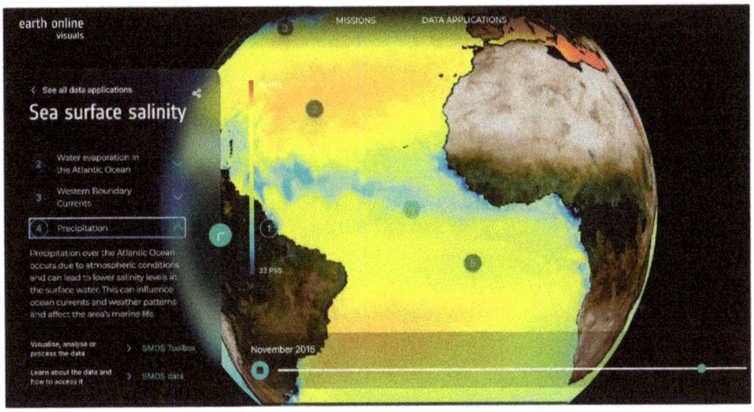

Figure 8. Excerpt of a salinity map elucidating various regimes in specific regions. Here, the meandering low-salinity signal associated with precipitation in the ITCZ is clearly depicted. Relevant regions are: (1) Amazon river plume; (2) Evaporation gyre; (3) Gulf Stream; (4) ITCZ; (5) Upwelling region.

Low surface salinity due to sea ice melting has been investigated by [58]; they found that meltwater lenses may persist more than a month and reach a surface salinity 5 pss fresher than surrounding waters. These studies were also benefiting from dedicated in-situ measurements collected in the uppermost skin layer by instrumentations such as the Salinity Snake, the Sea Surface Scanner (S3), and several drifters [59–61]. SSS as measured from satellites has also been merged with in situ data from profilers to obtain a 3D representation of the spatial and temporal evolution of freshwater pools, as described by [62]. Within the same framework of freshwater fluxes, but with emphasis on the evaporation regimes (which was also the focus of the SPURS-1 field experiment), the location and displacements of the salinity maxima in the various evaporation-dominated ocean gyres have been studied [63].

More generally, the balance of Evaporation–Precipitation (E-P) and its resemblance to salinity patterns has been widely studied by incorporating remotely sensed SSS [64], with special emphasis also on the areas that do not comply with the rain gauge approximation [65] and for which salinity represents the aggregate interplay of a wider set of phenomena (e.g., advection). In an even broader sense, salinity has been used to gather information on water cycle rates, gathering information on Pattern Amplification (PA) under the "rich gets richer" paradigm (causing, for instance, an enhancement of the inter-basin salinity difference) [66].

Additionally, both the pre-conditioning of and the a posteriori impacts of salinity on the passage of Tropical Cyclones (TC) and Extratropical Cyclones (ETC) have been widely assessed in the last ten years. Firstly, it was noted [67] that brightness temperatures, as measured by the available L-band satellite sensors, were sensitive to wind speed even in its severe regimes, unlike other sensors that experienced saturation and perturbation of additional noise sources (e.g., rain). This allowed the derivation of specific Geophysical Model Functions (GMF) that enabled retrieval of very high wind speed in hurricane conditions—recently complemented by Synthetic Aperture Radar (SAR) measurements to gain spatial resolution. Severe winds and Wind Radii products have been available since August 2021 in Near Real Time (within 4–6 h from acquisition) from the Institut Français de Recherche pour l'Exploitation de la Mer (IFREMER) and ESA. In relation to this, low salinity fields denoting stratified barrier layers were shown to be related to the so-called Rapid Intensification of a number of hurricanes in certain geophysical and geographical conditions. This has been shown to be due to the blocking effect of the barrier layer that hampered mixing with colder waters, which conventionally decreases the hurricane's intensity. Lastly, satellite SSS also revealed haline hurricane wakes [68].

2.2.2. The Buoyancy Domain

In this Section, the various studies that relate to the implications of density changes due to salinity variability are described. Several studies looked at the phenomenon of stratification. Importantly, it has been pointed out [69] that, thanks to satellite estimations that are sensing the uppermost layer of the ocean, it is possible to describe an increased stratification of the surface layer if compared to the sub-surface layer conventionally sensed by the in situ measurements (Figure 9). As such, satellite-derived SSS measurements evidence an intensification of the water cycle (the freshest waters become even fresher and vice versa). A related set of studies paid attention to the barrier/slippery layers through investigation of the salinity surface patterns, for instance [70].

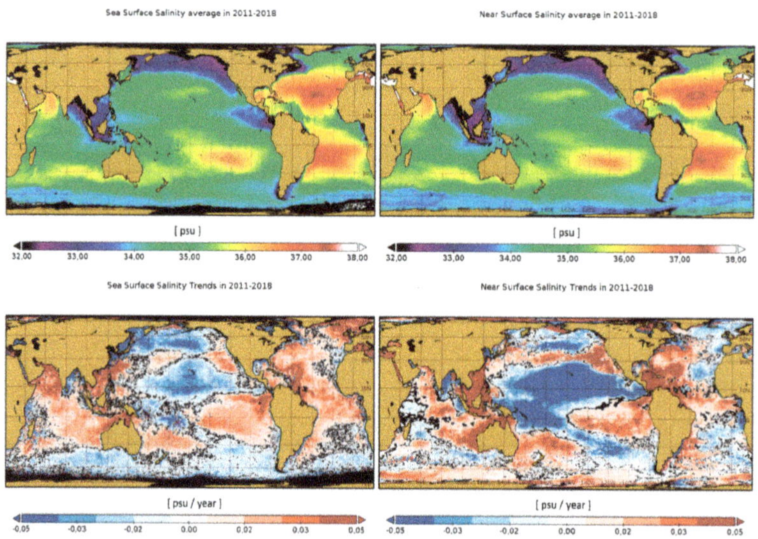

Figure 9. Salinity maps and computed trends for the surface and near-surface layers, evidencing an increased stratification of the uppermost layer sensed from space. Reprinted/adapted with permission from Ref. [69]. Copyright year 2022, copyright owner is Olmedo, E.

The role of salinity in the context of seawater density has been inspected as well [71], paying attention specifically to density-compensation regimes and density ratio. This is especially relevant in the high latitudes, where density is almost completely controlled by salinity. Furthermore, and still within the buoyancy domain, surface Temperature/Salinity (T/S) diagrams have been estimated from space [72], and these have then been used as a diagnostic means to identify density fluxes and water mass formation areas and their geographic extent [73]. This latter approach has then been enhanced to include only satellite-based inputs in the derivation of Water Masses (WM) features of Transformation and Formation [74] and infer estimates of their variability and evolution.

Lastly, another fundamental oceanic process that benefited from the availability of satellite SSS is the characterization of upwelling. For instance, exchanges of salt between the deep ocean and the surface provided additional constraints in the description of upwelling events [75,76].

2.2.3. The Bio-Geo-Chemistry Domain

A relatively novel application framework has been the biogeochemical domain. While applications and usage of satellite salinity in the physical oceanography context were natural, its perusal with the wider biogeochemistry realm was not to be taken for granted. First studies, a few years after the launch of satellite sensors, aimed at estimating pCO_2 also through salinity parameterizations [77]. In parallel, and for the following years, the relationship between salinity and Coloured Dissolved Organic Matter (CDOM) at major river mouths has been used to estimate salinity in the proximity of the coasts that are subject to freshwater runoff dispersal [78]. Later on, the strong relationship between salinity and alkalinity (the buffer capacity of the ocean to neutralize acids) has been exploited [79], eventually estimating the full marine carbonate system, including the surface ocean pH [80]. Eventually, these estimates have also been used for the monitoring of CO_2 fluxes and their current crucial implications as per carbon sequestration and inventory, widening even more the range of applications that emerged by the consistent use of satellite SSS.

2.3. Operational Applications

2.3.1. Data Assimilation

Although in the first years remotely sensed salinity estimates from space suffered from some inertia in their applications within the wider operational community, the increasingly accurate and debiased fields produced from SMOS, Aquarius, and SMAP fostered their use in several Data Assimilation (DA) experiments to constrain ocean state estimation and to initialize ocean and climate predictions. Most notably, regarding SMOS, a first experiment was undertaken by Mercator Ocean in the late 2010s, which showed how the introduction of SSS in their ocean model would be neutral-to-beneficial in the description of the various ocean state variables [81]. This was followed by a similar exercise at the UK Met Office [82]. Both efforts are now being repeated with a more updated merged version of salinity fields coming from CCI. The University of Hamburg, Germany, was also involved in DA and comparison exercises focusing on the seasonal to interannual variability of salinity as underlined by satellites and model intercomparison [83]. On the US side, a set of experiments have been carried out at both NASA and the National Oceanic and Atmospheric Administration (NOAA) with the aim of assessing the potential of satellite SSS to improve El Niño Southern Oscillations (ENSO) predictions [84,85].

In terms of operational data provision, the Copernicus Marine Service, or in full the Copernicus Marine Environment Monitoring Service (CMEMS), delivers a multi-observation global gap-free L4 analysis of Sea Surface Salinity (SSS) that has been obtained through a multivariate optimal interpolation algorithm that combines SMOS satellite estimates and in situ salinity measurements with satellite SST information [86]. Recently, a purely satellite-based product derived from CATDS is also being served.

As mentioned in the previous Section, the R&D efforts in estimating WS parameters in gale-force regimes became operational (Figure 10), and the output products (especially wind radii) are in turn ingested by a variety of Tropical Cyclone Forecast Centers (e.g., US Naval Research Laboratory, NOAA, Meteo-France) to generate warnings and issue evacuation alerts over hazardous areas.

2.3.2. Prognostic

A very interesting focus was portrayed by colleagues at Woods Hole Oceanographic Institution (WHOI) that studied the potential role of salinity as a prognostic variable to identify potential landfall over continental areas. Specifically, a study [87] assessed characteristics of the salinity fields and temporal lags with an atmospheric model to elucidate the transport of evaporated moisture and subsequent intense precipitation over the US Midwest. This has also been studied in the context of Neural Network (NN) schemes with additional ocean variables to forecast rain events [88].

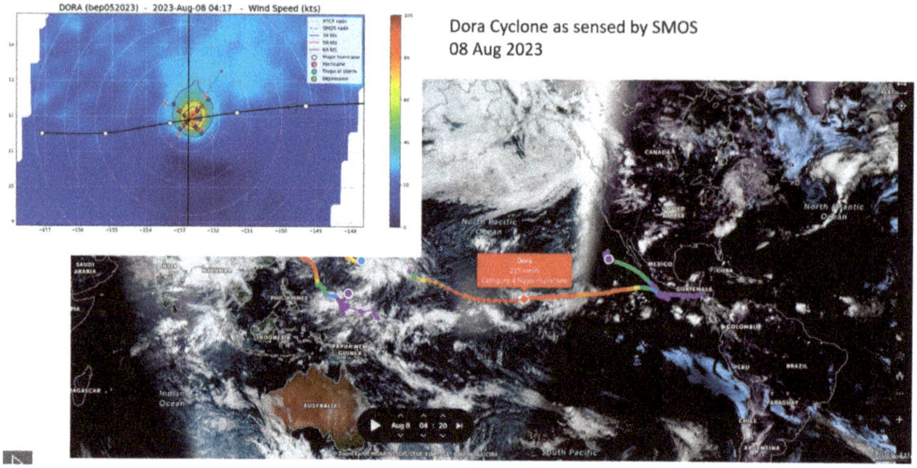

Figure 10. Trajectory, intensity and the evolution of the Cyclone Dora, as sensed operationally by SMOS in August 2023 (Credits: N. Reul, IFREMER).

2.3.3. Dedicated Regional Datasets

Estimations of SSS in regional and/or semi-enclosed basins, or at high latitudes, have always been challenging due to an interplay of their geophysical and geographical connotations and constraints. Specifically, salinity retrieval at high latitudes characterized by low SST and usually high winds has always been difficult due to the decreased sensitivity of TB to SSS in cold waters, enhanced by the difficulty of characterizing the roughness effect at high winds. Additionally, both the relatively coarse spatial resolution that limits the vicinity to coasts and also the presence of land that introduces a disturbance signal leakage hamper reliable retrievals in semi-enclosed basins. Recently, novel retrieval methodologies and data filtering ensured promising results and the generation of quasi-operational bespoke datasets at high latitudes (Arctic, Southern Ocean) [89,90] and also in the Mediterranean regions [91], as well as the Baltic Sea and Black Sea [92,93]. The dedicated Arctic Salinity product has also been ingested in the Copernicus Arctic Ice-Ocean reanalysis TOPAZ4 system [94]. Recently, CATDS also produced a dedicated set of regional products in 8 different regions [95].

3. Missions in the Upcoming Decade

3.1. Short-Term Developments of Ongoing Missions

Although Aquarius was lost in 2015, as of 2024, both SMOS and SMAP are still operational, exceeding by far their nominal expected lifetime. As such, both missions are still subject to ongoing improvements on several aspects, ranging from the algorithm developments to error characterization and from auxiliary data handling to the minimization of related perturbation sources. A large variety of L3 and L4 products stemming from additional processing of these satellite data streams, including merging among these sources and also with in-situ data, are available and collated in the Pi-MEP salinity platform as described above.

In spite of the remarkable advancements described above on a broad variety of oceanographic and climatic topics, discussions on limitations, shortcomings, current concerns, and possible ways forward are still debated and brainstormed in general meetings such as "Oceans from Space". Satellite SSS continuity in itself remains the major objective and desire, given that the generated CDRs necessarily build on long and stable datasets in order to produce relevant climate analyses. For instance, in the near future, attempts will be made to merge the various satellite TBs at L1, homogenizing thus the usage of dielectric constant models and auxiliary data. This is even more relevant, since upcoming missions focused on or including L-band will be available some time in the future (see below), and therefore, ensuring no gaps will be crucial.

Another aspect of the quest for continuity is the preservation of the set-aside passive L-band that is severely threatened by competing interests in the area of mobile telecommunications; under the motto "use it or lose it", the community is advocating for the maintenance of this band and indeed for enforcing regulations to diminish and switch off harmful RFI sources. On the geophysical aspects, conventional concerns in the community are still the improvement of accuracy and spatial resolution, especially for polar oceans and coastal regions, which are evidently paramount areas of interest per their relevance to climate and human activities.

Aiming at addressing scientific gaps and further improving the current estimates, several technological developments and satellite designs have been proposed over the years, with various feasibility studies funded by several space Agencies. At the recent ESA call for Earth Explorer-12 (EE-12), three proposals included L-band sensors with very different designs to estimate salinity from space at various spatial and temporal resolutions. One of them has been retained for further assessment in the initial Phase-0 (CryoRad) [96], which will use an innovative low-frequency wideband radiometer that includes SSS estimations among its primary objectives (Figure 11). The two other proposals, namely the Fine Resolution Explorer for Salinity, Carbon, and Hydrology (FReSCH) and the Sea-Air-Ice-Land Interactions (SAILIN), are under discussion in the community to further explore the innovative technological and development elements that were proposed (although not selected), both aiming at using innovative interferometric concepts to reach a desirable 10 km spatial resolution.

Figure 11. Preliminary field experiments to assess the capability of the wideband radiometer proposed for the EE-12 candidate CryoRad mission.

3.2. The CIMR Mission: Rationale and Multifrequency Capabilities

Aside from the R&D line of the ESA Earth Explorers, in the framework of Copernicus Expansion and within the context of the High Priority Candidate Missions (HPCM), the Copernicus Imaging Microwave Radiometer (CIMR) has been selected for Phase B/C/D after successfully passing a Preliminary Design Review and towards a full Critical Design Review to be held in 2026.

The Copernicus Imaging Microwave Radiometer (CIMR) aims at addressing user requirements as specified by the EC Polar Expert Group (PEG) in its related set of reports. CIMR is a wide-swath imager with multi-frequency and multi-spectral capability geared towards operational monitoring. Major technical features are a conical scanning (Forward scan and Aft scan) with a Sun-synchronous dawn-dusk orbit, with a sub-daily revisit and full Arctic coverage (Figure 12). The multi-frequency capability is represented by the selection of 6 channels at L-, C-, X-, K-, and Ka-band, with L-band intended to give continuity to SMOS and SMAP measurements. All channels have an onboard RFI processor to identify interference and remove it from the measurement.

Figure 12. Artistic view of the CIMR satellite.

As noted, CIMR will include an L-band channel, which will allow estimates of ocean salinity with high revisit time and with collocated measurements of SST and WS, which will aid accurate salinity retrieval in the challenging ocean regions at high latitudes. Primary geophysical variables over the ocean are Sea-Ice Concentration (SIC), thin Sea-Ice Thickness (SIT), SST, SSS, ice sheet, Sea Level change, and Wind Speed [97]. Two satellites are meant to be flown sequentially with an overlap. The expected launch of the first satellite is for 2029.

4. Conclusions

After a couple of decades of technology development and applications for funding, the panorama of salinity-enabled oceanographic and climatic studies radically changed with the advent of SMOS in 2009, which provided the first-ever satellite measurement of SSS. SMOS was followed by Aquarius in 2011 and SMAP in 2015 to create a stream of continuous and increasingly accurate estimates of SSS from space.

SMOS, being the first-ever measurement with a disruptive novel technology (synthetic aperture radiometry), was inherently prone to technical and scientific challenges. With the experience acquired over a decade and via the cross-fertilization of competences among the various sensor communities, many of these shortcomings have been addressed or drastically reduced (RFI, Land-Sea Contamination, and external noise source handling).

Both the current versions of the still-operational sensors (SMOS v700 and SMAP v5) represent a solid and stable dataset to enable science and applications. Novel consolidated platforms (Pi-MEP Salinity) ensure enhanced validation and stimulate oceanographic process studies by embedding a broad set of salinity products from space, altogether with model outputs and a remarkable variety of in-situ ground data, with corresponding match-up reports. Sustained focus over the last years was on the generation of merged salinity products, which establish specific Climate Data Records, most notably CCI-Salinity.

By progressively tackling all the processing issues and improving satellite accuracies, a wide range of oceanographic applications using spaceborne SSS developed (air-sea interactions, ocean circulation and modeling, climate indexes monitoring, marine biogeochemistry, Numerical Weather Prediction, etc.), and they are further enlarging with the release of the latest salinity reprocessing. The planned CIMR mission will perpetuate community expertise in L-band radiometry science, development, and operations, with a distinct focus on high latitudes.

Author Contributions: Writing—original draft preparation, R.S.; writing—review and editing, J.B., N.R. and T.L.; supervision and validation, S.H.Y. All authors have read and agreed to the published version of the manuscript.

Funding: This research received no external funding.

Data Availability Statement: No new data were created or analyzed in this study. Data sharing is not applicable to this article.

Acknowledgments: The authors acknowledge two anonymous reviewers for their constructive remarks that improved the clarity and structure of the paper. The following list of scientists, engineers, and managers involved in the oceanic branch of the Ocean Salinity missions, albeit non-exhaustive, is compiled here below to acknowledge their invaluable contributions to the missions success: J. Font, J. Tenerelli, A. Turiel, J.-L. Vergely, E. Olmedo, E. Lindstrom, N. Vinogradova, G. Lagerloef, J. Schanze, F. Bingham, D. Levine, Y. Kerr, P. Spurgeon, M. Arias, F. D'Amico, R. Catany, E. Jeansou, Y. Rey-Ricord, R. Oliva, S. Guimbard, C. Gabarrò, J. Martinez, V. Gonzalez, C. Gonzalez, A. Garcia, M. Talone, D. Khvorostyanov, A. Parracho, S. Tarot, D. Stammer, M. Senna-Martins, S. Brown, S. Misra, E. Dinnat, A. Camps, F. Bonjean, A. Martin, N. Kolodziejczyk, G. Reverdin, F. Rouffi, J. Jouanno, C. Maes, X. Yin, A. Hasson, A. Supply, F. Gaillard, C. Thouvenin-Masson, N. Martin, G. Alory, A. Gordon, R. Schmitt, T. Meissner, L. Yu, S. Fournier, E. Hackert, L. Li, O. Melnichenko, E. Bayler; K. Druschka, J. Anderson, W.L., Asher, B. Subrahmanyam, B. Ward, W. Tang, S. Grodsky, B. Chapron, B. Buongiorno Nardelli, L. Bertino, A. Soloviev, K. Scipal, S. Mecklenburg, M. Drusch, R. Crapolicchio, R. Diez, M. Martin-Neira, A. de la Fuente, J. Fauste, D. Fernandez, P. Cipollini, C. Donlon, S. Delwart, M. Drinkwater, A. Hahne.

Conflicts of Interest: The authors declare no conflicts of interest.

References

1. Barale, V. Half-a-Century of Oceans from Space: Features & Futures. *Remote Sens.* **2023**, *15*, 4064. [CrossRef]
2. Barale, V.; Gower, J.F.R.; Alberotanza, L. (Eds.) *Oceans from Space, Venice 2000, Abstracts*; EUR 19661 EN; Publication Office of the European Union: Luxemburg, 2010; p. 282. Available online: https://publications.jrc.ec.europa.eu/repository/handle/JRC20666 (accessed on 19 June 2025).

3. Lagerloef, G.; Font, J. SMOS and Aquarius/SAC-D Missions: The Era of Space Borne Salinity Measurements is About to Begin. In *Oceanography from Space, Revisited*; Barale, V., Gower, J.F.R., Alberotanza, L., Eds.; Springer: Dordrecht, The Netherlands; Heidelberg, Germany; London, UK; New York, NY, USA, 2010; pp. 35–58. [CrossRef]
4. Barale, V.; Gower, J.F.R.; Alberotanza, L. (Eds.) *Proceedings "Oceans from Space" V, Venice 2022*; NSA GROUP: Roma, Italy, 2022; p. 252. [CrossRef]
5. Durack, P.J. Ocean salinity and the global water cycle. *Oceanography* **2015**, *28*, 20–31. [CrossRef]
6. Roemmich, D.; Gilson, J. The 2004–2008 mean and annual cycle of temperature, salinity, and steric height in the global ocean from the Argo Program. *Prog. Oceanogr.* **2009**, *82*, 81–100. [CrossRef]
7. Kerr, Y.H.; Waldteufel, P.; Wigneron, J.P.; Delwart, S.; Cabot, F.; Boutin, J.; Escorihuela, M.J.; Font, J.; Reul, N.; Gruhier, C.; et al. The SMOS Mission: New Tool for Monitoring Key Elements of the Global Water Cycle. *Proc. IEEE* **2010**, *98*, 666–687. [CrossRef]
8. Suess, M.; Matos, P.; Gutierrez, A.; Zundo, M.; Martín-Neira, M. Processing of SMOS level 1C data onto a discrete global grid. In Proceedings of the 2004 IEEE International Geoscience and Remote Sensing Symposium (IGARSS 2004), Anchorage, AK, USA, 20–24 September 2004; Volume 3, pp. 1914–1917.
9. Daganzo-Eusebio, E.; Oliva, R.; Kerr, Y.H.; Nieto, S.; Richaume, P.; Mecklenburg, S.M. SMOS Radiometer in the 1400–1427-MHz Passive Band: Impact of the RFI Environment and Approach to Its Mitigation and Cancellation. *IEEE Trans. Geosci. Remote Sens.* **2013**, *51*, 4999–5007. [CrossRef]
10. Boutin, J.; Martin, N.; Yin, X.; Font, J.; Reul, N.; Spurgeon, P. First Assessment of SMOS Data Over Open Ocean: Part II—Sea Surface Salinity. *IEEE Trans. Geosci. Remote Sens.* **2012**, *50*, 1662–1675. [CrossRef]
11. Yueh, S.H.; Chaubell, J. Sea surface salinity and wind retrieval using combined passive and active L-band microwave observations. *IEEE Trans. Geosci. Remote Sens.* **2011**, *50*, 1022–1032. [CrossRef]
12. Reul, N.; Fournier, S.; Boutin, J.; Hernandez, O.; Maes, C.; Chapron, B.; Alory, G.; Quilfen, Y.; Tenerelli, J.; Morisset, S.; et al. Sea Surface Salinity Observations from Space with the SMOS Satellite: A New Means to Monitor the Marine Branch of the Water Cycle. *Surv. Geophys.* **2014**, *35*, 681–722. [CrossRef]
13. Boutin, J.; Vergely, J.-L.; Dinnat, E.P.; Waldteufel, P.; D'Amico, F.; Reul, N.; Supply, A.; Thouvenin-Masson, C. Correcting sea surface temperature spurious effects in salinity retrieved from spaceborne L-band radiometer measurements. *IEEE Trans. Geosci. Remote Sens.* **2021**, *59*, 7256–7269. [CrossRef]
14. Yin, X.; Boutin, J.; Dinnat, E.; Song, Q.; Martin, A. Roughness and foam signature on SMOS-MIRAS brightness temperatures: A semi-theoretical approach. *Remote Sens. Environ.* **2016**, *180*, 221–233. [CrossRef]
15. Spurgeon, P.; Font, J.; Boutin, J.; Reul, N.; Tenerelli, J.; Vergely, J.L.; Gabarro, C.; Yin, X.; Lavender, S.; Chuprin, A.; et al. Ocean salinity retrieval approaches for the SMOS satellite. In Proceedings of the ESA Living Planet Symposium, Bergen, Norway, 28 June–2 July 2010.
16. Martín-Neira, M.; Oliva, R.; Corbella, I.; Torres, F.; Duffo, N.; Durán, I.; Kainulainen, J.; Closa, J.; Zurita, A.; Cabot, F.; et al. Lessons learnt from SMOS after 7 years in orbit. In Proceedings of the 2017 IEEE International Geoscience and Remote Sensing Symposium (IGARSS), Fort Worth, TX, USA, 23–28 July 2017; pp. 255–258.
17. Font, J.; Camps, A.; Borges, A.; Martín-Neira, M.; Boutin, J.; Reul, N.; Kerr, Y.H.; Hahne, A.; Mecklenburg, S. SMOS: The challenging sea surface salinity measurement from space. *Proc. IEEE* **2009**, *98*, 649–665. [CrossRef]
18. Lagerloef, G.; Colomb, F.R.; Le Vine, D.; Wentz, F.; Yueh, S.; Ruf, C.; Lilly, J.; Gunn, J.; Chao, Y.; de Charon, A.; et al. The Aquarius/SAC-D mission—Designed to meet the salinity remote sensing challenge. *Oceanogr. Mag.* **2008**, *21*, 68–81. [CrossRef]
19. Le Vine, D.M.; Matthaeis, P.d. Aquarius active/passive RFI environment at L-band. *IEEE Geosci. Remote Sens. Lett.* **2014**, *11*, 1747–1751. [CrossRef]
20. Entekhabi, D.; Njoku, E.G.; O'Neill, P.E.; Kellogg, K.H.; Crow, W.T.; Edelstein, W.N.; Entin, J.K.; Goodman, S.D.; Jackson, T.J.; Johnson, J.; et al. The soil moisture active passive (SMAP) mission. *Proc. IEEE* **2010**, *98*, 704–716. [CrossRef]
21. Piepmeier, J.R.; Focardi, P.; Horgan, K.A.; Knuble, J.; Ehsan, N.; Lucey, J.; Brambora, C.; Brown, P.R.; Hoffman, P.J.; French, R.T.; et al. SMAP L-Band Microwave Radiometer: Instrument Design and First Year on Orbit. *IEEE Trans. Geosci. Remote Sens.* **2017**, *55*, 1954–1966. [CrossRef]
22. Mohammed, P.N.; Aksoy, M.; Piepmeier, J.R.; Johnson, J.T.; Bringer, A. SMAP L-band microwave radiometer: RFI mitigation prelaunch analysis and first year on-orbit observations. *IEEE Trans. Geosci. Remote Sens.* **2016**, *54*, 6035–6047. [CrossRef]
23. Olmedo, E.; Martínez, J.; Turiel, A.; Ballabrera-Poy, J.; Portabella, M. Debiased non-Bayesian retrieval: A novel approach to SMOS sea surface salinity. *Remote Sens. Environ.* **2017**, *193*, 103–126. [CrossRef]
24. Mecklenburg, S.; Drusch, M.; Kaleschke, L.; Rodriguez-Fernandez, N.; Reul, N.; Kerr, Y.; Font, J.; Martin-Neira, M.; Oliva, R.; Daganzo-Eusebio, E.; et al. ESA's Soil Moisture and Ocean Salinity mission: From science to operational applications. *Remote Sens. Environ.* **2016**, *180*, 3–18. [CrossRef]

25. Lindstrom, E.J.; Bryan, F.; Schmitt, R. SPURS: Salinity processes in the upper-ocean regional study—The North Atlantic experiment. *Oceanography* **2015**, *28*, 14–19. [CrossRef]
26. Lindstrom, E.J.; Edson, J.B.; Schanze, J.J.; Shcherbin, A.Y. SPURS-2: Salinity processes in the upper-ocean regional study 2. The eastern equatorial Pacific experiment. *Oceanography* **2019**, *32*, 15–19. [CrossRef]
27. Gordon, A.L.; Giulivi, C.F.; Busecke, J.; Bingham, F.M. Differences among subtropical surface salinity patterns. *Oceanography* **2015**, *28*, 32–39. [CrossRef]
28. Hernandez, O.; Boutin, J.; Kolodziejczyk, N.; Reverdin, G.; Martin, N.; Gaillard, F.; Reul, N.; Vergely, J.L. SMOS salinity in the subtropical North Atlantic salinity maximum: 1. Comparison with Aquarius and in situ salinity. *J. Geophys. Res. Ocean.* **2014**, *119*, 8878–8896. [CrossRef]
29. Drushka, K.; Westbrook, E.; Bingham, F.M.; Gaube, P.; Dickinson, S.; Fournier, S.; Menezes, V.; Misra, S.; Pérez Valentín, J.; Rainville, E.J.; et al. Salinity and Stratification at the Sea Ice Edge (SASSIE): An oceanographic field campaign in the Beaufort Sea. *Earth Syst. Sci. Data* **2024**, *16*, 4209–4242. [CrossRef]
30. Guimbard, S.; Reul, N.; Sabia, R.; Herlédan, S.; Khoury Hanna, Z.E.; Piollé, J.F.; Paul, F.; Lee, T.; Schanze, J.J.; Bingham FMLe Vine, D. The salinity pilot-mission exploitation platform (Pi-mep): A hub for validation and exploitation of satellite sea surface salinity data. *Remote Sens.* **2021**, *13*, 4600. [CrossRef]
31. Boutin, J.; Reul, N.; Koehler, J.; Martin, A.; Catany, R.; Guimbard, S.; Rouffi, F.; Vergely, J.L.; Arias, M.; Chakroun, M.; et al. Satellite-based sea surface salinity designed for ocean and climate studies. *J. Geophys. Res. Ocean.* **2021**, *126*, e2021JC017676. [CrossRef]
32. Guimbard, S.; Reul, N.; Díez-García, R.; Herlédan, S.; El Khoury Hanna, Z.; Lee, T.; Schanze, J.; Bingham, F.; Scipal, K. Advancing Sea Surface Salinity R&D: The Pi-MEP Initiative for Satellite Salinity Data Validation and Exploitation. In Proceedings of the EGU General Assembly Conference Abstracts, Vienna, Austria, 14–19 April 2024; p. 12907.
33. Boutin, J.; Chao, Y.; Asher, W.E.; Delcroix, T.; Drucker, R.; Drushka, K.; Kolodziejczyk, N.; Lee, T.; Reul, N.; Reverdin, G.; et al. Satellite and In Situ Salinity: Understanding Near-Surface Stratification and Subfootprint Variability. *Bull. Am. Meteorol. Soc.* **2016**, *97*, 1391–1407. [CrossRef]
34. Tarot, S.; Boutin, J.; Kerr, Y.; Vergely, J.L.; Mialon, A.; Vandermarcq, O. CATDS: SMOS L3/L4 salinity products generation and dissemination. In Proceedings of the 7th Ocean Salinity Conference 2024, Noordwijk, The Netherlands, 13–16 May 2024.
35. Hoareau, N.; Turiel, A.; Portabella, M.; Ballabrera-Poy JVogelzang, J. Singularity power spectra: A method to assess geophysical consistency of gridded products—Application to sea-surface salinity remote sensing maps. *IEEE Trans. Geosci. Remote Sens.* **2018**, *56*, 5525–5536. [CrossRef]
36. Olmedo, E.; González-Haro, C.; Hoareau, N.; Umbert, M.; González-Gambau, V.; Martínez, J.; Gabarró, C.; Turiel, A. Nine years of SMOS sea surface salinity global maps at the Barcelona Expert Center. *Earth Syst. Sci. Data Discuss.* **2020**, *2020*, 1–49. [CrossRef]
37. Melnichenko, O.; Hacker, P.; Potemra, J.; Meissner, T.; Wentz, F. A New Multi-Mission Sea Surface Salinity Optimum Interpolation (OISSS) Analysis for Ocean Research and Applications. In Proceedings of the EGU General Assembly Conference Abstracts, Vienna, Austria, 23–28 April 2023; p. EGU-3755.
38. Vinogradova, N.; Lee, T.; Boutin, J.; Drushka, K.; Fournier, S.; Sabia, R.; Stammer, D.; Bayler, E.; Reul, N.; Gordon, A.; et al. Satellite Salinity Observing System: Recent Discoveries and the Way Forward. *Front. Mar. Sci.* **2019**, *6*, 243. [CrossRef]
39. Reul, N.; Grodsky, S.A.; Arias, M.; Boutin, J.; Catany, R.; Chapron, B.; D'amico, F.; Dinnat, E.; Donlon, C.; Fore, A.; et al. Sea surface salinity estimates from spaceborne L-band radiometers: An overview of the first decade of observation (2010–2019). *Remote Sens. Environ.* **2020**, *242*, 111769.
40. Boutin, J.; Yueh, S.; Bindlish, R.; Chan, S.; Entekhabi, D.; Kerr, Y.; Kolodziejczyk, N.; Lee, T.; Reul, N.; Zribi, M. Soil Moisture and Sea Surface Salinity Derived from Satellite-Borne Sensors. *Surv. Geophys.* **2023**, *44*, 1449–1487. [CrossRef]
41. Hasson, A.; Delcroix, T.; Boutin, J.; Dussin, R.; Ballabrera-Poy, J. Analyzing the 2010–2011 La Niña signature in the tropical Pacific sea surface salinity using in situ data, SMOS observations, and a numerical simulation. *J. Geophys. Res. Ocean.* **2014**, *119*, 3855–3867. [CrossRef]
42. Qu, T.; Yu, J.Y. ENSO indices from sea surface salinity observed by Aquarius and Argo. *J. Oceanogr.* **2014**, *70*, 367–375. [CrossRef]
43. Akhil, V.P.; Vialard, J.; Lengaigne, M.; Keerthi, M.G.; Boutin, J.; Vergely, J.L.; Papa, F. Bay of Bengal Sea surface salinity variability using a decade of improved SMOS re-processing. *Remote Sens. Environ.* **2020**, *248*, 111964. [CrossRef]
44. Du, Y.; Zhang, Y. Satellite and argo observed surface salinity variations in the tropical indian ocean and their association with the Indian Ocean dipole mode. *J. Clim.* **2015**, *28*, 695–713. [CrossRef]
45. Da, N.D.; Foltz, G.R. Interannual variability and multiyear trends of sea surface salinity in the Amazon-Orinoco plume region from satellite observations and an ocean reanalysis. *J. Geophys. Res. Ocean.* **2022**, *127*, e2021JC018366. [CrossRef]
46. Lee, T.; Lagerloef, G.; Gierach, M.M.; Kao, H.-Y.; Yueh, S.; Dohan, K. Aquarius reveals salinity structure of tropical instability waves. *Geophys. Res. Lett.* **2012**, *39*, L12610. [CrossRef]

47. Olivier, L.; Reverdin, G.; Hasson, A.; Boutin, J. Tropical Instability Waves in the Atlantic Ocean: Investigating the Relative Role of Sea Surface Salinity and Temperature From 2010 to 2018. *J. Geophys. Res. Ocean.* **2020**, *125*, e2020JC016641. [CrossRef]
48. Lee, T.; Fournier, S.; Gordon, A.L.; Sprintall, J. Maritime Continent water cycle regulates low-latitude chokepoint of global ocean circulation. *Nat. Commun.* **2019**, *10*, 2103. [CrossRef]
49. Thouvenin-Masson, C.; Boutin, J.; Vergely, J.-L.; Reverdin, G.; Martin, A.C.H.; Guimbard, S.; Reul, N.; Sabia, R.; Catany, R.; Fanton-d'Andon, O.H. Satellite and In Situ Sampling Mismatches: Consequences for the Estimation of Satellite Sea Surface Salinity Uncertainties. *Remote Sens.* **2022**, *14*, 1878. [CrossRef]
50. Stammer, D.; Martins, M.S.; Köhler, J.; Köhl, A. How well do we know ocean salinity and its changes? *Prog. Oceanogr.* **2021**, *190*, 102478. [CrossRef]
51. Sammartino, M.; Aronica, S.; Santoleri, R.; Buongiorno Nardelli, B. Retrieving Mediterranean Sea surface salinity distribution and interannual trends from multi-sensor satellite and in situ data. *Remote Sens.* **2022**, *14*, 2502. [CrossRef]
52. Fournier, S.; Lee, T.; Gierach, M.M. Seasonal and interannual variations of sea surface salinity associated with the Mississippi River plume observed by SMOS and Aquarius. *Remote Sens. Environ.* **2016**, *180*, 431–439. [CrossRef]
53. Grodsky, S.A.; Reverdin, G.; Carton, J.A.; Coles, V.J. Year-to-Year Salinity Changes in the Amazon Plume: Contrasting 2011 and 2012 Aquarius/SAC-D and SMOS Satellite Data. *Remote Sens. Environ.* **2014**, *140*, 14–22. [CrossRef]
54. Boutin, J.; Martin, N.; Reverdin, G.; Morisset, S.; Yin, X.; Centurioni, L.; Reul, N. Sea surface salinity under rain cells: SMOS satellite and in situ drifters observations, Journal of Geophysical Research. *Oceans* **2014**, *119*, 5533–5545.
55. Supply, A.; Boutin, J.; Vergely, J.-L.; Martin, N.; Hasson, A.; Reverdin, G.; Mallet, C.; Viltard, N. Precipitation estimates from SMOS sea-surface salinity. *Q. J. R. Meteorol. Soc.* **2018**, *144* (Suppl. S1), 103–119. [CrossRef]
56. Drushka, K.; Asher, W.; Jessup, A.; Thompson, E.; Iyer, S.; Clark, D. Capturing fresh layers with the surface salinity profile. *Oceanography* **2019**, *32*, 76–85. [CrossRef]
57. Santos-Garcia, A.; Jacob, M.M.; Jones, W.L.; Asher, W.E.; Hejazin, Y.; Ebrahimi, H.; Rabolli, M. Investigation of rain effects on Aquarius Sea Surface Salinity measurements. *J. Geophys. Res. Ocean.* **2014**, *119*, 7605–7624. [CrossRef]
58. Supply, A.; Boutin, J.; Reverdin, G.; Vergely, J.-L.; Bellenger, H. Variability of Satellite Sea Surface Salinity Under Rainfall. In *Satellite Precipitation Measurement: Volume 2*; Levizzani, V., Kidd, C., Kirschbaum, D.B., Kummerow, C.D., Nakamura, K., Turk, F.J., Eds.; Springer International Publishing: Cham, Switzerland, 2020.
59. Anderson, J.; Riser, S. Near-surface variability of temperature and salinity in the near-tropical ocean: Observations from profiling floats. *J. Geophys. Res. Ocean.* **2014**, *119*, 2169–9275. [CrossRef]
60. Reverdin, G.; Morisset, S.; Boutin, J.; Martin, N.; Sena-Martins, M.; Gaillard, F.; Blouch, P.; Rolland, J.; Font, J.; Salvador, J.; et al. Validation of salinity data from surface drifters. *J. Atmos. Ocean. Technol.* **2014**, *31*, 967–983. [CrossRef]
61. Ribas-Ribas, M.; Hamizah Mustaffa, N.I.; Rahlff, J.; Stolle, C.; Wurl, O. Sea surface scanner (s 3): A catamaran for high-resolution measurements of biogeochemical properties of the sea surface microlayer. *J. Atmos. Ocean. Technol.* **2017**, *34*, 1433–1448. [CrossRef]
62. Guimbard, S.; Reul, N.; Chapron, B.; Umbert, M.; Maes, C. Seasonal and interannual variability of the Eastern Tropical Pacific Fresh Pool. *J. Geophys. Res. Ocean.* **2017**, *122*, 1749–1771. [CrossRef]
63. Bingham, F.M.; Busecke, J.; Gordon, A.L.; Giulivi, C.F.; Li, Z. The North Atlantic subtropical surface salinity maximum as observed by Aquarius. *J. Geophys. Res. Ocean.* **2014**, *119*, 7741–7755. [CrossRef]
64. Schanze, J.J.; Schmitt, R.W.; Yu, L.L. The global oceanic freshwater cycle: A state-of-the-art quantification. *J. Mar. Res.* **2010**, *68*, 569–595. [CrossRef]
65. Yu, L. A global relationship between the ocean water cycle and near-surface salinity. *J. Geophys. Res.* **2011**, *116*, C10025. [CrossRef]
66. Skliris, N.; Zika, J.D.; Nurser, G.; Josey, S.A.; Marsh, R. Global water cycle amplifying at less than the Clausius-Clapeyron rate. *Sci. Rep.* **2016**, *6*, 752. [CrossRef] [PubMed]
67. Reul, N.; Chapron, B.; Zabolotskikh, E.; Donlon, C.; Quilfen, Y.; Guimbard, S.; Piolle, J.F. A revised L-band radio-brightness sensitivity to extreme winds under Tropical Cyclones: The five year SMOS-storm database. *Remote Sens. Environ.* **2016**, *180*, 274–291. [CrossRef]
68. Grodsky, S.A.; Reul, N.; Lagerloef, G.; Reverdin, G.; Carton, J.A.; Chapron, B.; Quilfen, Y.; Kudryavtsev, V.N.; Kao, H.-Y. Haline hurricane wake in the Amazon/Orinoco plume: AQUARIUS/SACD and SMOS observations. *Geophys. Res. Lett.* **2012**, *39*, L20603. [CrossRef]
69. Olmedo, E.; Turiel, A.; González-Gambau, V.; Gonzalez-Haro, C.; Garcia-Espriu, A.; Gabarrò, C.; Portabella, M.; Corbella, I.; Martin-Neira, M.; Arias, M.; et al. Increasing stratification as observed by satellite sea surface salinity measurements. *Sci. Rep.* **2022**, *12*, 6279. [CrossRef]
70. Felton, C.S.; Subrahmanyam, B.; Murty, V.S.N.; Shriver, J.F. Estimation of the barrier layer thickness in the Indian Ocean using Aquarius Salinity. *J. Geophys. Res. Ocean.* **2014**, *119*, 4200–4213. [CrossRef]

71. Kolodziejczyk, N.; Reverdin, G.; Boutin, J.; Hernandez, O. Observation of the surface horizontal thermohaline variability at mesoscale to submesoscale in the north-eastern subtropical Atlantic Ocean. *J. Geophys. Res. Ocean.* **2015**, *120*, 2588–2600. [CrossRef]
72. Sabia, R.; Klockmann, M.; Fernández-Prieto, D.; Donlon, C. A first estimation of SMOS-based ocean surface T-S diagrams. *J. Geophys. Res. Ocean.* **2014**, *119*, 7357–7371. [CrossRef]
73. Piracha, A.; Sabia, R.; Klockmann, M.; Castaldo, L.; Fernandez, D. Satellite-driven estimates of water mass formation and their spatio-temporal evolution. *Front. Mar. Sci.* **2019**, *6*, 589. [CrossRef]
74. Piracha, A.; Olmedo, E.; Turiel, A.; Portabella, M.; González-Haro, C. Using satellite observations of ocean variables to improve estimates of water mass (trans) formation. *Front. Mar. Sci.* **2023**, *10*, 1020153. [CrossRef]
75. Alory, G.; Maes, C.; Delcroix, T.; Reul, N.; Illig, S. Seasonal dynamics of sea surface salinity off Panama: The far Eastern Pacific Fresh Pool. *J. Geophys. Res.* **2012**, *117*, C04028. [CrossRef]
76. Awo, F.M.; Rouault, M.; Ostrowski, M.; Tomety, F.S.; Da-Allada, C.Y.; Jouanno, J. Seasonal cycle of sea surface salinity in the Angola upwelling system. *J. Geophys. Res. Ocean.* **2022**, *127*, e2022JC018518. [CrossRef]
77. Brown, C.W.; Boutin, J.; Merlivat, L. New insights of pCO2 variability in the tropical eastern Pacific Ocean using SMOS SSS. *Biogeosci. Discuss.* **2015**, *12*, 4595–4625. [CrossRef]
78. Umbert, M.; Gabarro, C.; Olmedo, E.; Gonçalves-Araujo, R.; Guimbard, S.; Martinez, J. Using remotely sensed sea surface salinity and colored detrital matter to characterize freshened surface layers in the kara and laptev seas during the ice-free season. *Remote Sens.* **2021**, *13*, 3828. [CrossRef]
79. Fine, R.A.; Willey, D.A.; Millero, F.J. Global variability and changes in ocean total alkalinity from Aquarius satellite data. *Geophys. Res. Lett.* **2017**, *44*, 261–267. [CrossRef]
80. Gregor, L.; Gruber, N. OceanSODA-ETHZ: A global gridded data set of the surface ocean carbonate system for seasonal to decadal studies of ocean acidification. *Earth Syst. Sci. Data* **2021**, *13*, 777–808. [CrossRef]
81. Tranchant, B.; Remy, E.; Greiner, E.; Legalloudec, O. Data assimilation of Soil Moisture and Ocean Salinity (SMOS) observations into the Mercator Ocean operational system: Focus on the El Niño 2015 event. *Ocean Sci.* **2019**, *15*, 543. [CrossRef]
82. Martin, M.; King, R.R.; While, J.; Aguiar, A. Assimilating satellite sea surface salinity data from SMOS, Aquarius and SMAP into a global ocean forecasting system. *Q. J. R. Meteorol. Soc.* **2018**, *145*, 705–726. [CrossRef]
83. Kohl, A.; Martins, M.S.; Stammer, D. Impact of assimilating surface salinity from SMOS on ocean circulation estimates. *J. Geophys. Res. Ocean.* **2014**, *119*, 5449–5464. [CrossRef]
84. Hackert, E.; Kovach, R.M.; Molod, A.; Vernieres, G.; Borovikov, A.; Marshak, J.; Chang, Y. Satellite Sea Surface Salinity Observations Impact on El Niño/Southern Oscillation Predictions: Case Studies from the NASA GEOS Seasonal Forecast System. *J. Geophys. Res. Ocean.* **2020**, *125*, e2019JC015788. [CrossRef]
85. Bayler, E.; Chang, P.S.; De La Cour, J.L.; Helfrich, S.R.; Ignatov, A.; Key, J.; Lance, V.; Leuliette, E.W.; Byrne, D.A.; Liu, Y.; et al. Satellite Oceanography in NOAA: Research, Development, Applications, and Services Enabling Societal Benefits from Operational and Experimental Missions. *Remote Sens.* **2024**, *16*, 2656. [CrossRef]
86. Nardelli, B.; Droghei, R.; Santoleri, R. Multi-dimensional interpolation of SMOS sea surface salinity with surface temperature and in situ salinity data. *Remote Sens. Environ.* **2016**, *180*, 392–402. [CrossRef]
87. Li, L.; Schmitt, R.W.; Ummenhofer, C.C.; Karnauskas, K.B. Implications of North Atlantic Sea surface salinity for summer precipitation over the U.S. Midwest: Mechanisms and predictive value. *J. Clim.* **2016**, *29*, 3143–3159. [CrossRef]
88. Liu, T.; Schmitt, R.W.; Li, L. Global search for autumn-lead sea surface salinity predictors of winter precipitation in southwestern United States. *Geophys. Res. Lett.* **2018**, *45*, 8445–8454. [CrossRef]
89. Olmedo, E.; Gabarró, C.; González-Gambau, V.; Martínez, J.; Ballabrera-Poy, J.; Turiel, A.; Portabella, M.; Fournier, S.; Tong, L. Seven years of SMOS sea surface salinity at high latitudes: Variability in Arctic and sub-Arctic regions. *Remote Sens.* **2018**, *10*, 1772. [CrossRef]
90. González Gambau, V.; Silvano, A.; Olmedo, E.; González-Haro, C.; García Espriu, A.; Turiel, A.; Gabarró, C.; Catany, R.; Naveira-Garabato, A.; Allen, B.; et al. SO-FRESH project: First satellite regional Sea Surface Salinity maps for further understanding of the Southern Ocean dynamics. In Proceedings of the EC-ESA Joint Earth System Science Initiative Workshop, Frascati, Rome, Italy, 22–24 November 2023. Available online: https://hdl.handle.net/10261/369261 (accessed on 19 June 2025).
91. Grodsky, S.; Reul, N.; Bentamy, A.; Vandemark, D.; Guimbard, S. Eastern Mediterranean salinification observed in satellite salinity from SMAP mission. *J. Mar. Syst.* **2019**, *198*, 103190. [CrossRef]
92. Gonzalez-Gambau, V.; Olmedo, E.; Haro, C.G.; Turiel, A.; Garcia, A.; Gabarro, C.; Martinez, J.; Alenius, P.; Tuomi, L.; Roiha, P.; et al. First regional SMOS Sea Surface Salinity products over the Baltic Sea: Quality assessment and oceanographic added-value (No. EGU21-15254). In Proceedings of the Copernicus Meetings, Vienna, Austria, 19–30 April 2021.

93. Olmedo, E.; González-Gambau, V.; Turiel, A.; González-Haro, C.; García-Espriu, A.; Gregoire, M.; Álvera-Azcárate, A.; Buga, L. Rio, M.H. New SMOS SSS maps in the framework of the Earth Observation data For Science and Innovation in the Black Sea *Earth Syst. Sci. Data Discuss.* **2021**, *2021*, 1–40.
94. Xie, J.; Raj, R.P.; Bertino, L.; Martínez, J.; Gabarró, C.; Catany, R. Assimilation of sea surface salinities from SMOS in an Arctic coupled ocean and sea ice reanalysis. *Ocean Sci.* **2023**, *19*, 269–287. [CrossRef]
95. Boutin, J.; Vergely, J.-L.; Khvorostyanov, D. De-Biased SMOS SSS L3 V9 Maps Generated by LOCEAN/ACRI-ST Expertise Center SEANOE. 2024. Available online: https://www.seanoe.org/data/00417/52804/ (accessed on 19 June 2025).
96. Johnson, J.T.; Jezek, K.C.; Macelloni, G.; Brogioni, M.; Tsang, L.; Dinnat, E.P.; Walker, J.P.; Ye, N.; Misra, S.; Piepmeier, J.R.; et al Microwave Radiometry at Frequencies From 500 to 1400 MHz: An Emerging Technology for Earth Observations. *IEEE J. Sel. Top Appl. Earth Obs. Remote Sens.* **2021**, *14*, 4894–4914. [CrossRef]
97. Jiménez, C.; Tenerelli, J.; Prigent, C.; Kilic, L.; Lavergne, T.; Skarpalezos, S.; Høyer, J.L.; Reul, N.; Donlon, C. Ocean and Sea Ice Retrievals from an End-to-End Simulation of the Copernicus Imaging Microwave Radiometer (CIMR) 1.4–36.5 GHz Measurements. *J. Geophys. Res. Ocean.* **2021**, *126*, e2021JC017610. [CrossRef]

remote sensing

MDPI

Review

A Review of Ocean Color Algorithms to Detect *Trichodesmium* Oceanic Blooms and Quantify Chlorophyll Concentration in Shallow Coral Lagoons of South Pacific Archipelagos

Cécile Dupouy [1,2,*], Andra Whiteside [1,2], Jing Tan [3], Guillaume Wattelez [4], Hiroshi Murakami [5], Rémi Andréoli [6], Jérôme Lefèvre [7], Rüdiger Röttgers [8], Awnesh Singh [2] and Robert Frouin [3]

1 Aix-Marseille Université, Université de Toulon, IRD, CNRS/INSU, Mediterranean Institute of Oceanography, MIO UM 110, 13288 Marseille, France; andrawhiteside@gmail.com
2 Pacific Centre for Environment and Sustainable Development (PaCE-SD), The University of the South Pacific, Laucala Campus, Suva, Fiji; awnesh.singh@usp.ac.fj
3 Scripps Institution of Oceanography, University of California San Diego, 8810 Shellback Way, La Jolla, CA 92093, USA; jit079@ucsd.edu (J.T.); rfrouin@ucsd.edu (R.F.)
4 Interdisciplinary Laboratory for Research in Education, EA 7483, University of New Caledonia, Avenue James Cook, Nouméa 98800, New Caledonia; guillaume.wattelez@univ.nc
5 Japan Aerospace Exploration Agency, Tsukuba 305-8505, Ibaraki, Japan; murakami.hiroshi.eo@jaxa.jp
6 Bluecham SAS, Nouméa 98848, New Caledonia; remi.andreoli@bluecham.net
7 ENTROPIE, Centre IRD de Nouméa, BP A5, Nouméa 98848, New Caledonia; jerome.lefevre@ird.fr
8 Institute of Carbon Cycles, Helmholtz-Zentrum HEREON, Max-Planck-Str. 1, Building 70, D-21502 Geesthacht, Germany; rroettgers@hereon.de
* Correspondence: cecile.dupouy@ird.fr

Abstract: The oceanic waters of the Southwest Tropical Pacific occupy a vast region including multiple Pacific Island Countries. The state of these waters is determinant for fisheries and the blue economy. Ocean color remote sensing is the main tool to survey the variability and long-term evolution of these large areas that are important for economic development but are affected by climate change. Unlike vast oligotrophic gyres, tropical waters are characterized by numerous archipelagos and islands, with deep and shallow lagoons subjected to the large impacts of the land. Strikingly large dendritic phytoplankton (*Trichodesmium*) blooms with high levels of chlorophyll, developing within archipelagos, as well as coastal enrichments from various origins may be observed. Algorithms to detect the presence of *Trichodesmium* have been developed or adapted, as well as algorithms to estimate the chlorophyll concentration ([Chl-a]). Adapting existing [Chl-a] algorithms does not always yield high, i.e., sufficient, accuracy. A review of published regional bio-optical algorithms developed taking into account the specific phytoplankton composition and minimizing the adverse impacts of particles and the seabed bottom on [Chl-a] determination is presented, as well the bio-optical database that allowed their development. The interest of such algorithms for a variety of applications and scientific accomplishments is highlighted, with a view to further addressing the main biology and biogeochemistry questions, e.g., to determine the true impact of diazotrophs and assess lagoon [Chl-a] variability with the highest confidence. This work anticipates the use of future coarse and high-spatial-resolution and multi- and hyper-spectral satellite imagery in the Pacific.

Keywords: chlorophyll-a concentration; ocean color; *Trichodesmium* blooms; bathymetry; water column attenuation correction; MODIS; MERIS; Sentinel 2; remote sensing; reflectance; algorithms; inversion method; seabed mapping; clustering; support vector machine; New Caledonia; Fiji Islands; Southwestern Tropical Pacific

1. Introduction

The ecological status of the seas of the archipelagos of the Southwest Tropical Pacific (SWTP) is vital for the economy of Pacific Island Countries [1,2]. The Pacific waters at 15°S in the tropical band are dominated by oligotrophic conditions with a generalized deep chlorophyll-a concentration [Chl-a] maximum, nitrate (NO_3^-)-depleted surface waters, and surface [Chl-a] < 0.1 mg.m^{-3} ([3,4]). It is an N-limited, oligotrophic area characterized by a permanent deep thermocline and nutricline (up to 130 m) [5–7]. This defines the tropical waters as Case 1, whose inherent optical properties (IOPs) are dominated by phytoplankton (e.g., most open ocean waters). However, summer dentritic blooms develop around the archipelagos (Figure 1) with higher [Chl-a] values (Solomon Islands, New Caledonia, Vanuatu, and Fiji–Tonga), while never appearing in winter. A hypothesis in [8] is that *Trichodesmium*, a major colonial cyanobacterial nitrogen fixer, can form large blooms in NO_3^--depleted tropical oceans, as often observed from satellites [9–13], supported by cruise observations [14–19]. Empirical algorithms have been developed for these blooms based on the specific water reflectance of *Trichodesmium* [20,21]. Applications have been successful with SeaWiFS [22], though the resulting distribution of *Trichodesmium* in the SWTP differed from what had been expected [23]. The first attempt to discriminate surface slicks at a large scale in the open ocean of the SWTP was made using SeaWiFS reflectance at 30 km resolution, based on anomalies in the visible spectrum [24]. Another successful application in the SWTP was using MERIS and the maximum chlorophyll index (MCI) [25]. This last application succeeded in discriminating *Trichodesmium* in summer, where it was expected (around New Caledonia and Vanuatu: Figure 1 left and right). As reviewed by [26], in the SWTP open ocean, the condition that the near-infrared (NIR) signal is above the threshold required for discrimination is rarely observed. Inside the Great Barrier Reef, application of the MCI was successful due to the high concentration of *Trichodesmium* and the associated strong signal [27].

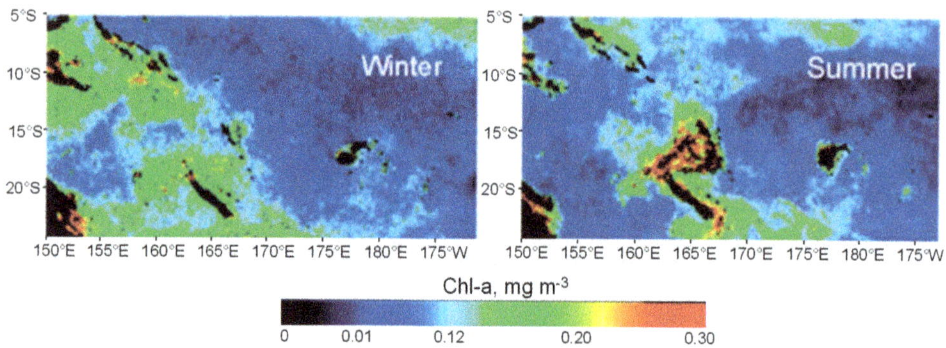

Figure 1. Seasonal variability of phytoplankton in Southwestern Tropical Pacific, shown for winter (**left**) and summer (**right**), as 5-year averages from SeaWiFS data. Extracted as in [13].

In enclosed lagoon ecosystems (e.g., coral reefs, seagrasses), like the one of New Caledonia or Fiji, optical complexity is linked to the influence of colored dissolved organic matter (CDOM) and particulate inorganic material, which implies the use of specific bio-optical algorithms [28,29]. In such Case 2 environments, the ocean color is not only influenced by other light-absorbing and light-scattering components than phytoplankton, but also by bathymetry or the "bottom effect". In the Great Barrier Reef, the bottom effect can be huge and preponderant, and developing algorithms to reduce its impact on the retrieval of water properties from satellite imagery has been the main objective in many studies [30–34]. Similar to the Great Barrier Reef are the lagoons of New Caledonia, the second-longest continuous coral reefs in the world, and the Fijian lagoons that surround the two main islands. They both show large summer chlorophyll-a enrichments, possibly caused by *Trichodesmium* and/or large runoffs (Figure 2a,b). There, brown surface areas correspond to shallow waters, where algorithms have to be adapted (New Caledonia lagoon and south of Fiji) before applying regional algorithms for [Chl-a] or turbidity (Figure 2a,b).

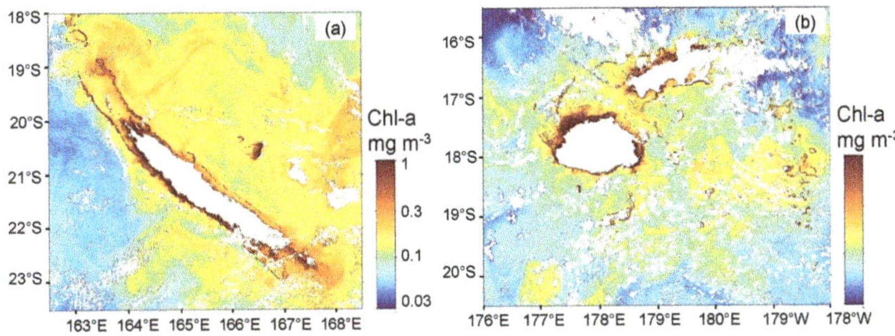

Figure 2. (**a**) A Second-Generation Global Imager (SGLI), JAXA image of 6 March 2022 mapping a large dendritic bloom east of the Loyalty Islands (New Caledonia archipelago). (**b**) An SGLI image of 17 February 2022 showing high chlorophyll concentrations around the main islands of Fiji.

Discriminating *Trichodesmium* as well as mapping bio-optical components from space in such shallow settings as the New Caledonia or Fiji lagoons remains challenging. In order to adapt or develop regional or local algorithms, it appears necessary for us to complete international databases with regional and local data in Case 1 and Case 2 waters, sufficiently representative of the variability of the optical environments encountered in the SWTP.

The objective of this paper is to review the work accomplished in order to answer these questions. The first part will present the in situ data set of inherent optical properties (IOPs) and remote sensing reflectance (R_{rs}) that was used to develop [Chl-a] algorithms for Case 1 and Case 2 waters and the satellite data selected for applications. The second part will summarize ocean color methods and algorithms in the SWTP region to assess lagoon [Chl-a], turbidity, or seabed color. The third part will highlight their effectiveness and limitations and will give an overview of the interest of such results for the understanding of the biogeochemistry of waters in the tropics and to open perspectives.

2. Data and Methods

2.1. Data

2.1.1. In Situ Reflectance and IOPs Collected in the SWTP

In order to validate our algorithms, we collected a comprehensive data set of in situ bio-optical data representative of optical variability in Case 1 and Case 2 waters in various regions and over more than 20 years. This data basis had also the advantage of being the result of validated methodologies and obtained by the same operators and instruments during that period. This set was gathered from 2000 to 2023 from New Caledonia to Fiji during oceanic and lagoon cruises and at the MOISE OSU PYTHEAS observing station. Remote sensing reflectance, R_{rs}, and IOPs were measured during different cruises in the SWTP and are listed in Table 1. Note that the 2000–2010 data set was already presented in [35,36].

Hyperspectral R_{rs} was measured by a TriOS radiometer system consisting of radiance and irradiance sensors with a spectral range of 320 nm to 950 nm, a spectral resolution of about 10 nm (sampled by every 3.3 nm), and a 7-degree field-of-view for the radiance sensor. The method developed by [37] was used to determine R_{rs}, i.e., the radiance sensor was mounted on a small raft to measure water-leaving radiance just below the surface, and R_{rs} was calculated by normalizing the water-leaving radiance with the downward solar irradiance measured on the ship [38]. [Chl-a] was measured by fluorimetry and only at a few cruises by HPLC (NASA laboratory). Turbidity was estimated from CTD Seabird profiles or from dry weights. All IOP measurements followed NASA protocols described in [35,36,39–42].

Table 1. Optical data gathered in the Southwestern Tropical Pacific from different cruises (see Reference/DOI).

Cruise	Year	Situation	a_p, a_{dg} (m^{-1})	b_b-H6 (m^{-1})	R_{rs} (sr^{-1})	Chl-a (mg m^{-3})	DW (g m^{-3})	Reference/DOI
FLUPAC	1994	Equatorial Upwelling	x			x	x	[43]
ZONAL FLUX	1996	Equatorial Upwelling	x			x	x	[44,45]
EBENE	1996	Equatorial Upwelling	x			x	x	[46]
TRICHONESIA 1	1998	NC Vanuatu Fiji Tonga	x	x	x	x	x	[9,16]
TRICHONESIA 2	1999	North Australia (Arafura Sea)	x	x	x	x	x	Dupouy, unpub. Res.
DIAPALIS (1 to 9)	2001–2004	NC Loyalty Isl.	x	x	x	x	x	[19,36]; 10.18142/85
BISSECOTE	2006	NC West. lagoon	x	x	x	x	x	[38]; 10.17600/6100010
ECHOLAG	2007	NC West. lagoon	x	x	x	x	x	[47]; 10.17600/7100010
VALHYBIO	2008	NC West. lagoon	x	x	x	x	x	[42]; 10.17600/8100020
Lagoon surveys	2001–2013	NC West. lagoon	x	x	x	x	x	[48]; 10.17600/3100080
SPOT (1 to 14)	2013–2018	Vanuatu trench (6000 m)	x	x	x	x	x	10.18142/237
COMEVA 1,2	2015–2016	Vanuatu	x	x	x	x	x	10.17600/16003500; 10.17600/16004100
OUTPACE	2015	NC Vanuatu Fiji Tonga Niue	x	x	x	x	x	[39]; 10.17600/15000900
MOANAMATY 1,2	2018	French Polynesia			x	x	x	10.17600/18000580; 10.17600/18000887
CALIOPE 1	2011	NC East. lagoon	x	x	x	x	x	[42]; 10.17600/14000900
CALIOPE 2	2014	NC East. lagoon	x	x	x	x	x	[49]; 10.17600/14003900
CALIOPE 3	2016	NC East. lagoon	x	x	x	x	x	[50,51]; 10.17600/16003400
MOISE station	2012–2018	NC West. lagoon	x	x	x	x	x	[52]
BULA (1 to 5)	2000–2004	Laucala Bay Fiji	x	x	x	x		[53]; 10.18142/71
Laucala Bay (1 to 4)	2015–2017	Laucala Bay Fiji	x	x		x	x	[54]
SOKOWASA	2022	Viti Levu-Kadavu	x	x	x	x	x	10.17600/18002025

2.1.2. Satellite Data Used for Our Studies

Ocean color sensors have been used for *Trichodesmium* algorithms (SeaWiFS: 2000–2004, MODIS on Aqua and OLCI on Sentinel 3). For coastal lagoon waters, algorithms were developed from in situ reflectance and then tested on Level 1b imagery at a 500 m resolution issued from standard MODIS processing [55]. We had the opportunity to use AVNIR-2 images with a 30 m spatial resolution and four bands in the southwest tropical lagoon of New Caledonia [56]. For these data, atmospheric and water reflectance were corrected iteratively through the retrieval of IOPs and after additional correction of the sea floor reflectance. The seabed color inversion was obtained from one of the clearest images of the MERIS sensor acquired over New Caledonia. The development of the support vector machine (SVM) algorithm for turbidity was achieved from the clearest images of MODIS providing the maximum number of match-ups. The development of the seabed color inversion in the small shallow lagoon of Voh-Kone-Pouembout, north of New Caledonia, was achieved from the clearest image of Sentinel 2 (THEIA). The application of C2RCC was performed on the Laucala Bay lagoon with Sentinel 2 data (Copernicus).

2.2. Algorithms

2.2.1. *Trichodesmium* Detection

To address the limitation whereby the *Trichodesmium* concentrations in the open ocean of the SWTP are never high and the NIR signal as measured in [22,35] is rarely above the threshold required, existing algorithms had to be refined. The algorithms all failed to detect only *Trichodesmium* slicks as they also captured all pixels around unmasked clouds. An adaptation of the McKinna algorithm allowed us to eliminate such ring pixels [26].

2.2.2. Estimation of [Chl-a] and Turbidity

In the deeper waters of the east coast of New Caledonia (CALIOPE cruises, Table 1), IOPs, i.e., the absorption coefficients of phytoplankton and dissolved substances + detritus (a_{ph} and a_{dg}, respectively), and the particulate backscattering coefficient (b_{bp}) were estimated from hyperspectral R_{rs} data by applying linear matrix inversion [57]. Local characteristics of the IOP spectra at 45 stations were used for the candidate spectra. The [Chl-a] and IOP inversion algorithm was adapted to MODIS data and applied to Level 1b imagery at a 500 m resolution [57]. The use of a constant bottom reflectance helped in retrieving IOPs after minimizing the influence of the bathymetry. For the estimation of [Chl-a] and turbidity, we used quasi-analytical (GSM, QAA, GIOP, LMI) algorithms from satellite reflectance [49]. The retrieval of IOPs and [Chl-a] was also investigated for AVNIR-2 images with a 30 m spatial resolution and four bands [56].

Algorithms based on statistical analysis (of the support vector regression (SVR) (or support vector machine (SVM) type) were proposed to estimate [Chl-a] in the optically complex waters of the New Caledonian lagoon from MODIS-derived R_{rs} [58]. The algorithm was developed via supervised learning on match-ups of [Chl-a] measurements crossed with MODIS reflectance available from cruises conducted between 2002 and 2010. The best performance was obtained by combining two models, selected according to the ratio of R_{rs} in spectral bands centered at 488 and 555 nm used to discriminate low vs. high [Chl-a]: a log-linear model for low [Chl-a] (AFLC) developed based on support vector regression (SVR) analysis and a classic model (OC3) for high [Chl-a]. This approach outperformed the classical OC3 approach, especially in shallow waters, with the root mean squared error being about 30% lower.

For turbidity in deep waters (>20 m), a specific algorithm [38] was developed for New Caledonia waters, based on the best fits between in situ turbidity data (in FTU) and reflectance from MODIS imagery. Finally, the Case 2 Regional CoastColour (C2RCC) atmospheric correction algorithm for suspended particulate matter (in $g.m^{-3}$) on Sentinel 2 imagery was applied to the deep Eastern coastal lagoon of Hienghène.

For turbidity in the oligotrophic and shallower waters of tropical lagoons, the bottom reflection of downwelling light usually hampers the use of classical optical algorithms. In order to address this issue, an SVR model was developed and tested on a large training sample of in situ turbidity values representative of the annual variability in the Voh-Koné-Pouembout lagoon (lagoon surveys; Table 1) and on the coincident water reflectance from MODIS. The SVR was trained with reflectance and two other explanatory parameters—bathymetry and bottom color. Our approach converged with a 3-parameter model that included $R_{rs}(555)$, $R_{rs}(645)$, and $R_{rs}(667)$ as optical parameters, with bathymetry and bottom color added as explanatory variables [59].

2.2.3. Extraction of Bathymetry and Seabed Color in Shallow Lagoon Waters

As previously highlighted, both bathymetry and seabed reflectance greatly influence remotely sensed biogeochemical parameter retrievals. Many attempts were made to retrieve the bottom reflectance influence [30,33,34]. In the optically complex lagoon of New Caledonia, a test was performed from MERIS multispectral satellite images to test the influence of [Chl-a] on the depth estimation [60,61]. The variation in [Chl-a] on the image chosen, correlated with the variation in diffuse attenuation, explained the error of the depth estimation map.

In order to estimate the bottom color, an unsupervised clustering approach was applied to a clear Sentinel 2 image of the small Voh-Koné-Pouembout lagoon in New Caledonia with a complex bathymetry and bottom environments [62]. Data processing included Lyzenga's correction for the estimation of the water column reflectance, optical spectra standardization for the attenuation of water absorption effects, and clustering using the unsupervised k-means method. This methodological approach was applied to the 497, 560, 664, and 704 nm optical bands [62].

3. Results

3.1. Variability of the Water Reflectance in the SWTP

The variability of R_{rs}, with values ranging from 0.003 to 0.015 sr^{-1} at 400 nm and reaching 0.021 sr^{-1} at 550 nm, was measured in waters deeper than 20 m around New Caledonia (the VALHYBIO cruise, Table 1; Figure 3a). The variability of the IOPs, i.e., the absorption coefficients of phytoplankton (a_p), dissolved substances (a_g, m^{-1} or a_{cdom}, m^{-1}), and the particulate backscattering coefficient issued from Hydroscat-6 in situ (b_b-H6, m^{-1}), was also measured during the same cruise (Figure 3b–d; Table 1). Ocean color variability is caused by [Chl-a], turbidity, CDOM absorption [36,50,51], and the presence of the seabed (old reefs, grey bottom, or sands) as this was found in the southwestern lagoon, by comparison with open ocean waters at a sufficient distance from the barrier reefs of New Caledonia.

Specific cases of reflectance spectra near a small islet, with a peak in the green (500–600 nm) channel, were measured above sand and coral reefs of different colors, at different depths and times, with variable water attenuation due to phytoplankton and/or other (inorganic) particles and/or CDOM (Figure 3e). The R_{rs}, with values ranging from 0.005 sr^{-1} to 0.021 sr^{-1} at 550 nm without an evident distinction possible between corals and sands, illustrates that the remote sensing of [Chl-a], turbidity, the CDOM absorption coefficient, and, generally, IOPs above coral reefs or sandy bottoms is difficult as bottom reflectance is the major component of satellite reflectance (and the atmospheric correction schemes have to be adapted).

Spectra measured with a drifting TriOS radiometer above *Trichodesmium* slicks (Figure 3f) indicate that when the colony concentration is sufficiently high to be visible to the naked eye, a red and NIR signal can be detected [20,22], similar to the spectra of colonies above a filter in the laboratory [26]. The reflectance of *Trichodesmium* slicks between 550 and 600 nm is not easily distinguishable from the blooms of other phytoplankton species. The surface signal in the red and NIR (600–800 nm) is increasing with the thickness of the surface slick (Figure 3f). Unfortunately, this effect is never observed in radiometric vertical profiles because colonies are not concentrated enough, except in the upper 15 m, where the measurement is impacted by large light variations. Additional absorption in the UV

(330 nm) from mycosporine-like amino acids (MAAs), a UV-protective pigment contained in the colony sheet, can be observed [20,26,39] but is not easily measurable by radiometers due to the generally high absorption of the water in this range. Such dense surface slicks are easily observable in closed seas and lagoons (New Caledonia, Great Barrier Reef), but remain exceptional in the open ocean, as they are rapidly dispersed by currents and winds and detectable by satellites only during extremely calm meteorological conditions.

In Laucala Bay of Suva, Fiji, R_{rs} is strongly influenced by the bottom and highly turbid waters issued from the large Rewa river after strong rains, and therefore reflectance is higher by an order of magnitude [38].

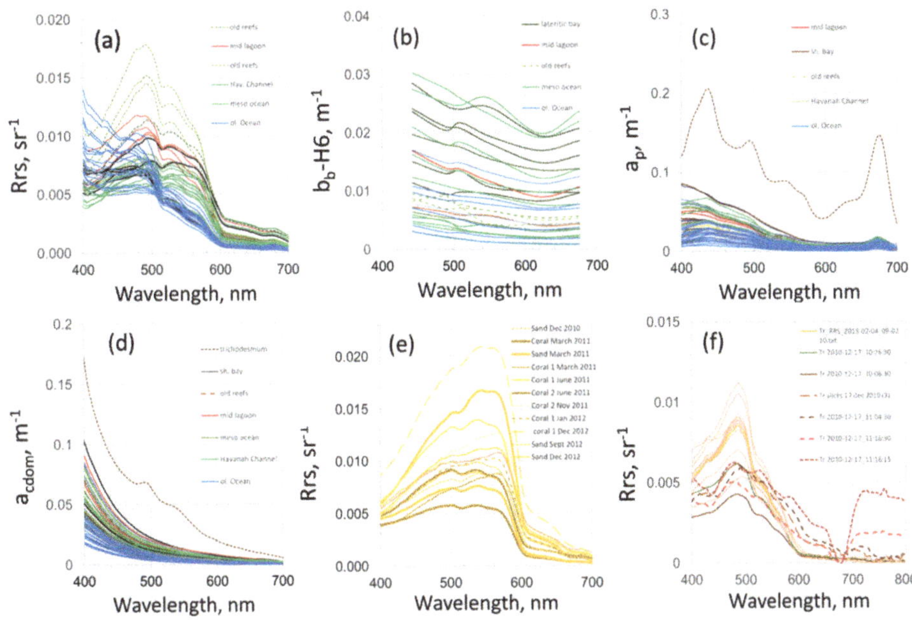

Figure 3. TriOS and IOP radiometric measurements of ocean color spectra collected in the Southwestern Tropical Pacific Ocean showing the variability of ocean color signatures in inshore (influenced by bathymetry, seabed reflectance, [Chl-a], and CDOM absorption and particle backscattering) in comparison to offshore waters. (**a–d**) for waters > 20 m in New Caledonia during the VALHYBIO cruise (unpublished data), with (**a**) R_{rs} (sr^{-1}), (**b**) bb-H6 (m^{-1}) from Hydroscat-6, (**c**) absorption a_p (m^{-1}) from PSICAM measurements, (**d**) a_{cdom} (m^{-1}) from PSICAM measurements. (**e**) TriOS spectra of R_{rs} obtained over white sand and coral reefs of different in situ colors and at different dates. (**f**) TriOS spectra of R_{rs} (shown until 800 nm) taken on *Trichodesmium* slicks in the lagoon (in February 2013; light pink and showing variability during 30 min at low concentrations without the red and NIR signal), and on thick slicks at different times (in December 2010; dashed dark brown) indicating the strong red and NIR signal between 650 and 800 nm.

3.2. *Trichodesmium* Blooms in the SWTP

By using an adapted algorithm allowing the elimination of unmasked pixels corresponding to cloud rings, the empirical algorithms based on the ratios of NIR and red channel reflectance [63] successfully detected *Trichodesmium* slicks that had been observed during the 2015 OUTPACE campaign. A new, striking example of successful discrimination on MODIS and Sentinel 3 images from 6 November 2020 in open waters at the east of New Caledonia and at the south of Fiji is shown (Figure 4a–d). Even if R_{rs}678 nm is not so high in the slicks in Figure 3f, it may become slightly higher due to the large backscatter of large colonies and therefore be detectable at this channel. R_{rs}678 nm can be used to delineate thick accumulations at the surface (Figure 4a–d). Indeed, this corresponded to a period when slicks were concentrated due to exceptionally calm weather and where discrimination was successful thanks to a clear sly.

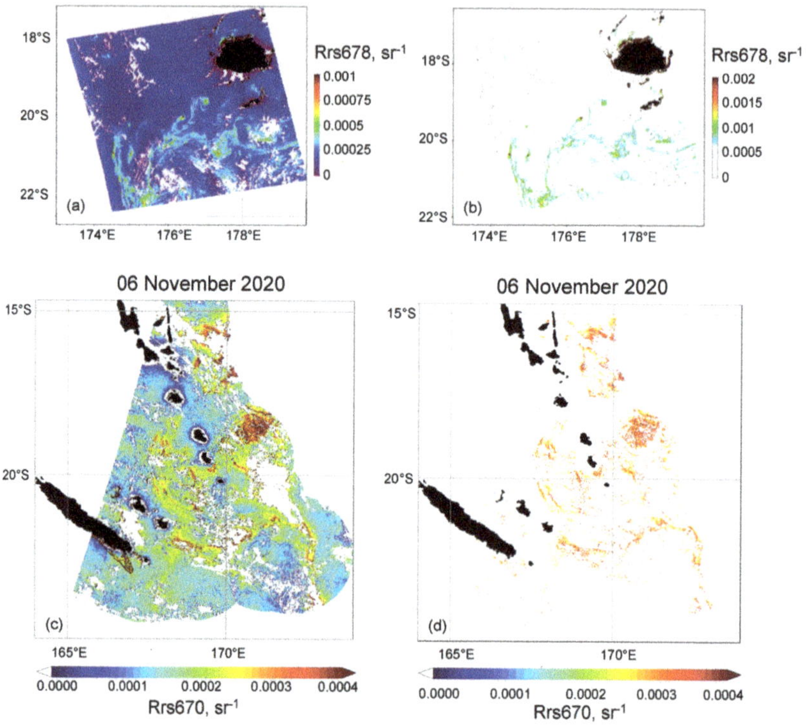

Figure 4. Example of detection of surface slicks with the R_{rs}678 nm channel indicating high concentrations of *Trichodesmium* on 6 November 2020. (**a**) R_{rs}678 nm image from MODIS around Fiji; (**b**) pixels discriminated from a high value of R_{rs}678 nm (>a threshold of 0.0007 sr^{-1}); (**c**) Rrs670 nm image OLCI Sentinel 3 of New Caledonia and Vanuatu; (**d**) pixels discriminated from a high value of R_{rs}670 nm (>a threshold of 0.0003 sr^{-1}). For this last image, sea truth observations of a dense slick across hundreds of miles on 8 November 2020 coincides with the highest R_{rs}678 nm signal at the east of New Caledonia [64].

3.3. *[Chl-a], IOPs, and Turbidity in New Caledonia Waters*

The use of a constant sea floor reflectance helped in retrieving IOPs after minimizing the influence of the bathymetry (Figure 5a,b). The effect of such a correction of the sea floor reduces the impact of the complex bathymetry and various seabed colors on the IOP's distributions (which appear then to be more homogeneous than with classical OC3, where all shallow waters saturate the optical signal). Numerical results show that the correction allows a decrease in the uncertainty in the scatter diagrams (Figure 6). A similar amelioration of the retrieval of inherent optical properties (IOPs) and [Chl-a] distributions from AVNIR-2 images in the southwest tropical lagoon of New Caledonia [56] is achieved after a similar correction of the sea floor reflectance (Figure 7a–e).

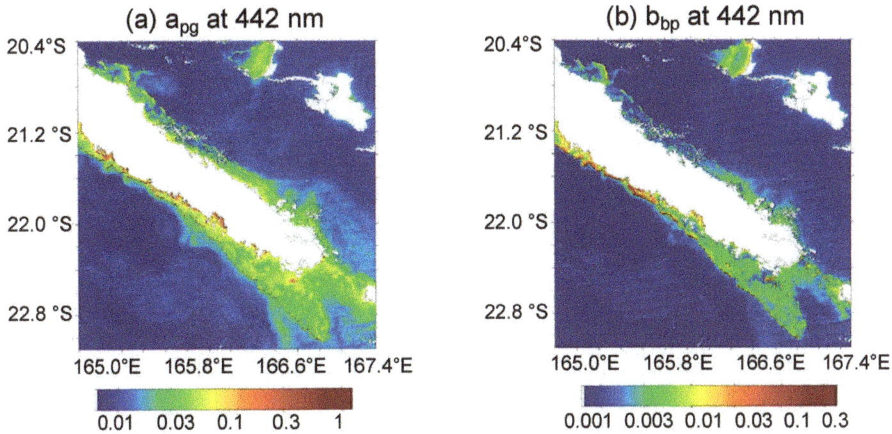

Figure 5. Examples of maps of IOPs at 442 nm on 8 October 2011 estimated from MODIS 500 m data (with the bottom correction). (**a**) a_{pg} (m^{-1}). (**b**) b_{bp} (m^{-1}). (Adapted from [57]).

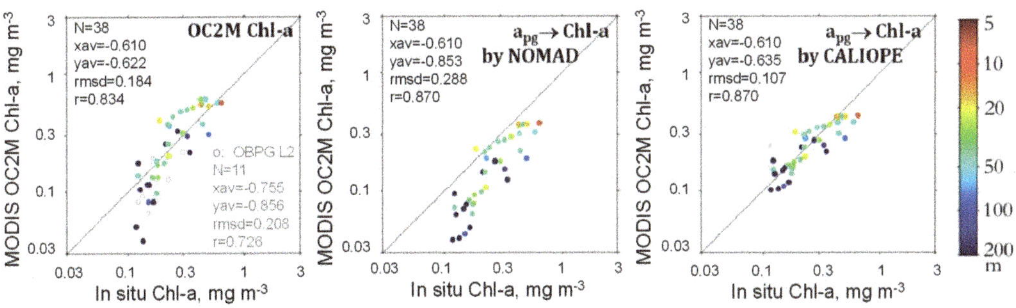

Figure 6. Scatter diagrams between in situ Chl-a and MODIS standard Level 2 Chl-a, or issued from a_{pg} from NOMAD or CALIOPE databases. MODIS OC2M algorithm was applied to our R_{rs} estimates from MODIS 500 m L1B data with the bottom reflectance correction. (Adapted from [57]).

The results of our statistical algorithm applied on the MODIS-derived R_{rs} provided more accurate assessments of [Chl-a] within the whole [Chl-a] range encountered from shallow coastal waters and nearby reefs to deeper waters of the open ocean (Figure 8a–d).

The results of our statistical algorithms for turbidity also performed well. For deep waters (>20 m), the specific algorithm for New Caledonia allowed the mapping of extensive plumes of turbid waters after a tropical storm in December 2011, with turbidity values in the order of those measured during the CALIOPE 1 cruise (Table 1) (Figure 9a). Similarly, the mapping of a large-turbidity plume after a strong rain event in the Hienghène lagoon was validated by data collected during the CALIOPE 3 cruise in 2016 (Table 1) (Figure 9b).

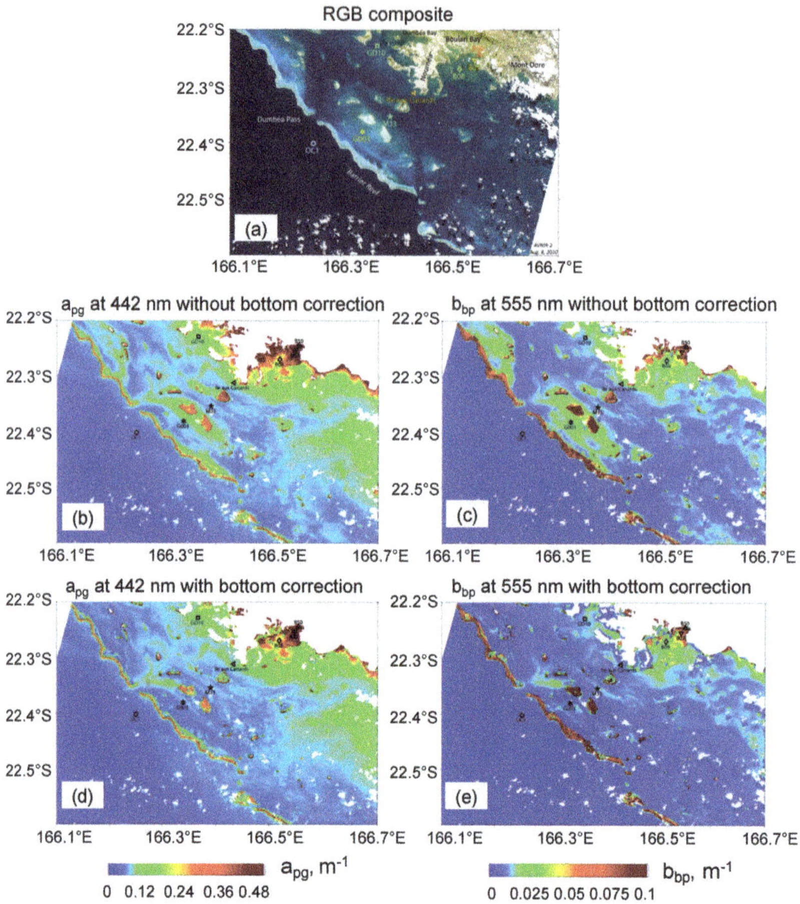

Figure 7. (**a**) AVNIR2 color composite on 3 September 2009 of the southwestern lagoon of New Caledonia area used for validation of different inversion algorithms presented in this paper. (**b**,**c**) Estimations of a_{pg} at 442 nm and b_{bp} at 555 nm, respectively, without bottom reflection correction (upper row). (**d**,**e**) Same as (**b**,**c**) but with bottom correction (lower row). Bottom reflection was considered as constant in this study, and bathymetry was known at each pixel. (Adapted from [56]).

For shallower waters <10 m, such as those of the north lagoon of Voh Koné Pouembout, the best candidate was a three-parameter model that included $R_{rs}(555)$, $R_{rs}(645)$, and $R_{rs}(667)$ as optical parameters and two other explanatory parameters, i.e., bathymetry and bottom color. Figure 10a–d illustrate the drastic reduction in the range of the retrieved turbidity values (0 to 1 NTU, instead of 1 to 30 NTU). This approach significantly improved the model's capacity for retrieving the in situ turbidity range from MODIS images, as compared with algorithms dedicated to deep oligotrophic or turbid waters, which were shown to be totally inadequate ([38]) (Figure 10a–d).

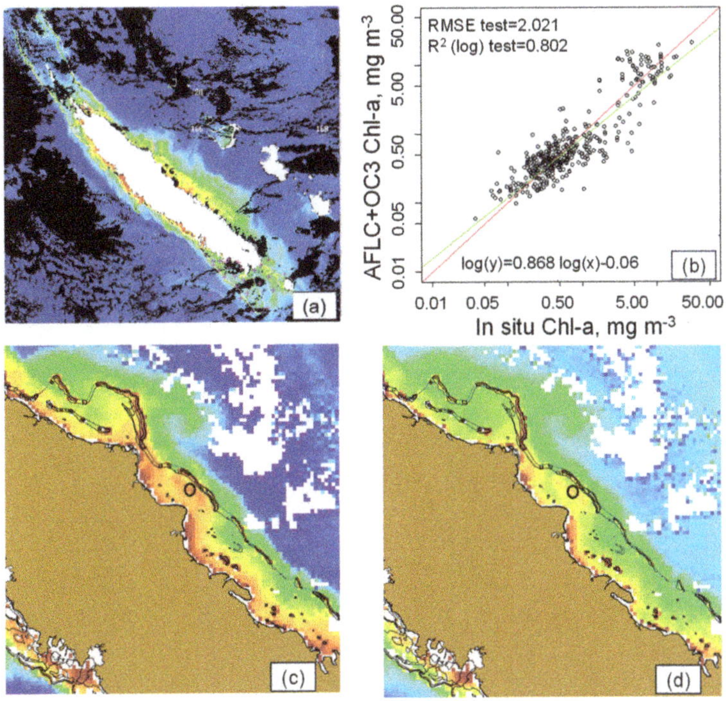

Figure 8. (**a**) MODIS [Chl-a] image of the whole lagoon of New Caledonia. (**b**) Regression between SVM-issued [Chl-a] and in situ [Chl-a]. (**c**,**d**) Maps of Chl-a from OC4 compared with the SVM result showing differences in coastal parts of the lagoon. (Adapted from [58]).

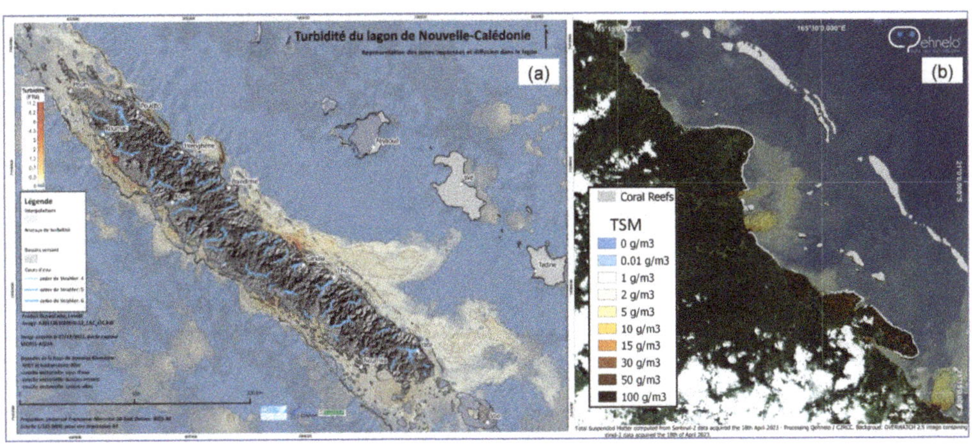

Figure 9. (**a**) Turbidity as calculated from the algorithm from [36] in New Caledonia adapted to deep waters (>20 m) and applied to a MODIS image off the east coast after the tropical storm in December 2011. (**b**) Total suspended matter (TSM) concentration generated from Sentinel 2 immediately after a heavy rain event in the northeastern part of the Caledonia lagoon. Result of CR2CC processing (see details in the text). Coral barriers of the eastern north lagoon are indicated.

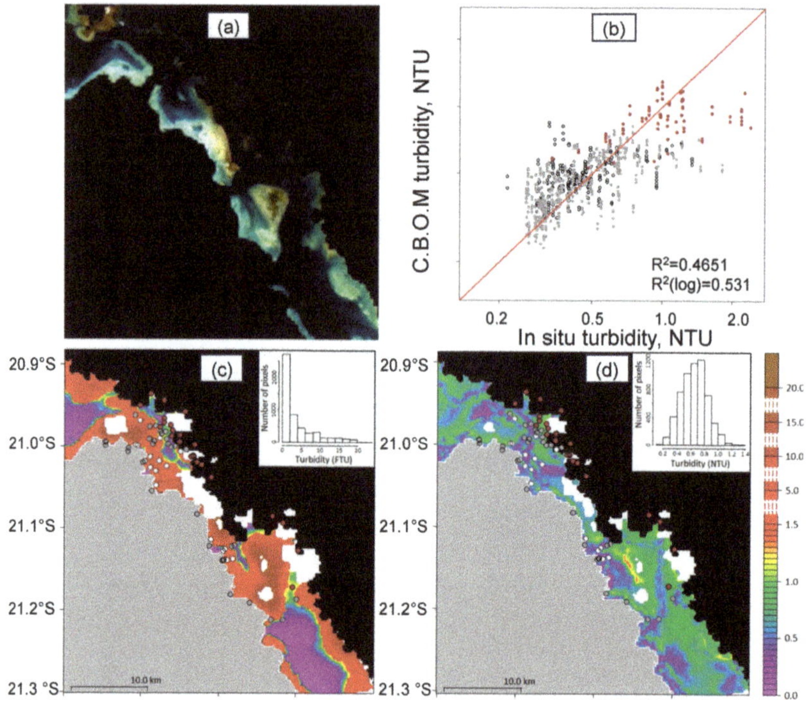

Figure 10. Turbidity estimated in the Voh-Kone-Pouembout (VKP) lagoon area on 21 April 2014 (redrawn from literature cited in the text). (**a**) Sentinel 2 image of the VKP lagoon showing a mosaic of different seabed colors and bathymetry. (**b**) Log-linear regression between in situ turbidity (in NTU) and MODIS remote sensing turbidity retrieved with the SVR model. The red line is the 1:1 line. Only the high turbidity values (corresponding to stations with a brown bottom color) are not well retrieved by the SVR model. Point colors correspond to ○: white bottom; ●: grey bottom; ●: brown bottom. (**c**,**d**) Turbidity maps (in FTU) retrieved from the MODIS image with the [38] New Caledonia turbidity algorithm compared with one from the SVR algorithm (using bathymetry and reflectance combination (see text for details)). (Adapted from [59]).

3.4. Bathymetry and Bottom Classification of the New Caledonia Lagoon

The estimation of the bathymetry from the MERIS image remained valid under a low (<1 mg.m^{-3}) and constant [Chl-a] (<10% of variation). This allowed us to limit the variation in the root-mean-square error (RMSE) on the estimation of the bathymetry (15.5%) (Figure 11a). The non-linear effect of the light attenuation of the water column was corrected to obtain the absolute reflectance of the seabed. The automated supervised classification applied to the MERIS image is shown in Figure 11b,c. The two different mapped features of the seafloor (Figure 11b,c) provided the best overall accuracy (79%), as compared to in situ known seabed colors [61].

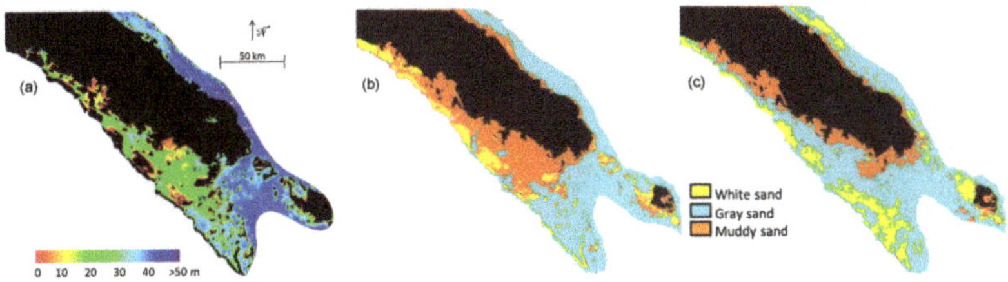

Figure 11. (**a**) The bathymetry of the southern New Caledonia lagoon resulting from Lyzenga's methodology using a MERIS image acquired on 7 December 2008. (**b**) Classification of the seafloor type using the reflectance and SAM distance. (**c**) Classification of the seafloor when the reflectance was corrected for the water attenuation and using SAM distance (see text for details). (Adapted from [60,61]).

On the shallow Voh Koné Pouembout lagoon, the methodological approach applied to the 497, 560, 664, and 704 nm optical bands of a clear Sentinel 2 image (Figure 12a) allowed different bottom types (see [62]). When applied to non-standardized data, our unsupervised classification retrieved three seafloor bottom types (named clusters 1–3 in Figure 12), whereas five bottom types (named clusters 1–5 in Figure 12) could be retrieved using standardized data (standardized R_{rs} whose average is 0 and standard deviation is 1). For each of these two trials, the computed membership values explained more than 75% of the inertia in each Sentinel 2 wavelength band used for the clustering. The accuracy of the method was slightly improved when applied to data corrected for attenuation (Figure 12b,c).

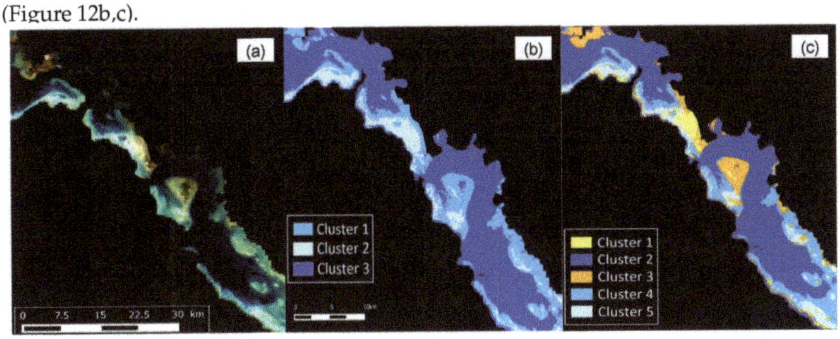

Figure 12. Seafloor classification results of a case study in the Voh-Koné-Pouembout lagoon (New Caledonia) using a clear Sentinel 2 image of 31 July 2017. (**a**) Seafloor spectral signal after applying Lyzenga's correction. Results of an unsupervised classification and clustering of the seabed color in shallow oligotrophic coastal waters. (**b**) Without standardization of Sentinel 2 reflectance spectra in the near-infrared. (**c**) With standardization of Sentinel 2 reflectance spectra in the near-infrared. (Adapted from [62]).

4. Discussion

4.1. Performance and Limitations of Our Algorithms for the SWTP

4.1.1. *Trichodesmium* Algorithms

Many gaps are found in the identification of *Trichodesmium* from satellite remote sensing. Algorithms developed for the discrimination of surface blooms [20,22,24,26] are successful when atmospheric corrections are used appropriately above surface mats of *Trichodesmium* (i.e., keeping the NIR reflectance) and when *Trichodesmium* is dominant [39]. As observed during many cruises, the total [Chl-a] in the upper surface layer reflects not only *Trichodesmium* colonies but also the phytoplankton associated with it (symbiotic diatoms [14,16] and/or non-diazotroph species such as picoplankton [15,19,39]). The ability to positively discriminate and quantify low background concentrations of *Trichodesmium* (e.g., <3200 trichomes L^{-1}), dispersed within the water column, has been described as a difficult task [26]. Moreover, slick formation is strongly governed by highly unstable, dynamic, fine-scale physical structures [65]. Discriminating *Trichodesmium* in such cases was not successful, even in summer in the SWTP, as thick slicks form preferentially in warmer and stable waters. One example was during the Diapalis 07 cruise in February 2003, where a dendritic bloom with high [Chl-a] seen by MODIS corresponded to observed slicks and high *Trichodesmium* concentrations (>3000 trichomes/L [19]) but where the algorithm did not detect the specific spectral signature (Figure 13). Another gap in assessing the spatiotemporal distribution of *Trichodesmium* [19] is the rapid temporal succession between this main diazotroph species and other phytoplankton fueling it or fueled by it, which is yet to be analyzed. In order to validate *Trichodesmium* detection algorithms at a large scale, the global data basis should include their main pigment (phycoerythrin). The phycoerythrin pigment is not detectable by HPLC and almost never measured during cruises. The global data basis should also contain information on size-fractionated pigments [Chl-a] (>10 μm). To conclude, the spectral and spatial resolutions of current ocean-color sensors are a limiting factor for quantitative *Trichodesmium* remote sensing at a large scale. Future efforts with ocean color will concern the estimation of the contribution of *Trichodesmium* to SWTP carbon and nitrogen fixation by diazotrophs, and the understanding of their seasonal and interannual variability in the SWTP.

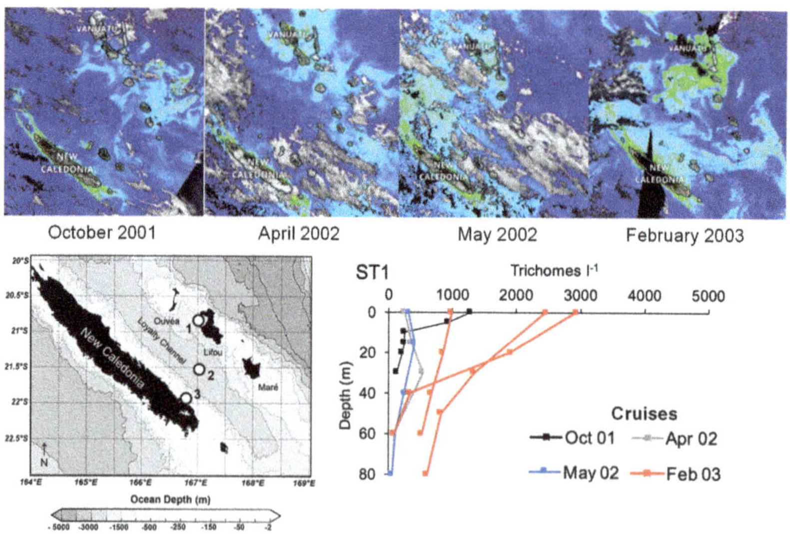

Figure 13. Series of MODIS images acquired during Diapalis cruises in October 2001, April 2002, May 2002, and February 2003. Vertical profiles of *Trichodesmium* colonies counted off Lifou Island during each cruise, showing *Trichodesmium* colony numbers [19] being exceptional in February 2003 compared to October 2001 or April and May 2002. Slicks were numerous but not easily detected by any algorithm, for reasons explained in the text.

4.1.2. Coastal Algorithms

For [Chl-a], regionally tuned algorithms may potentially improve the retrievals, but better parameterization schemes that consider the spatiotemporal variability of the specific IOPs are still needed. Supervised learning methods provided a regional optimal estimation of IOPs and [Chl-a], with better accuracy (<20% for [Chl-a]) and an improvement by a factor of 10 for turbidity in the shallowest lagoons of New Caledonia. Local optimal estimations of IOPs have been successful for low- and high-resolution imagery for the target areas. Our results pointed out that not only the bathymetry but also the seabed color influence had to be determined in order to be injected as an additional variable in the SVRs. The limitation of the incomplete knowledge of the seabed influence to improve the remote sensing of biogeochemical indicators such as [Chl-a] and turbidity in fragile coastal environments was corrected by different methods. Inversion procedures and unsupervised clustering have provided the seabed color at different scales and for some test areas of the New Caledonia lagoon. The estimation of the seabed reflectance from unsupervised classification in the whole lagoon area from high-resolution data will allow us to definitively address the problem and account for the bottom effect (bathymetry + seabed color) in the next SVRs. Indeed, when the seabed reflectance is too high and waters too shallow, no estimate of the water column property is possible. Moreover, uncertainty exists in extremely turbid waters such as those of Laucala Bay (with shallow bottoms and turbidity of an order of magnitude higher than in New Caledonia) [38]. For these extremely turbid waters in enclosed bays, adapted algorithms using the NIR bands of sensors are still in test [66]. As soon as they propagate in oceans, particles and phytoplankton show dynamic coupling and decoupling patterns according to time and space, which are being explored in order to characterize the real impacts on coral reefs (Figure 14) [66].

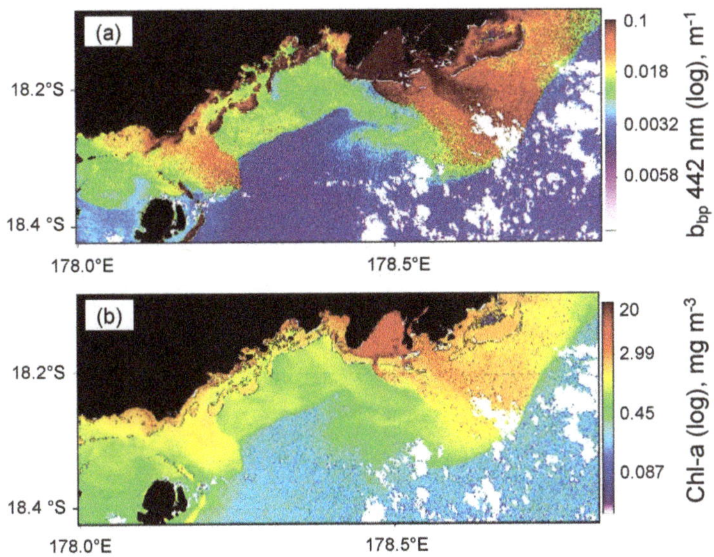

Figure 14. Maps of Sentinel 2 of Laucala Bay's (**a**) backscattering coefficient b_{bp} and (**b**) [Chl-a] after a sediment stirring event linked to the tropical cyclone Yasa, showing the plume extending to the ocean south of Fiji and the extent of the influence of land to the sea. Outputs were generated from Sentinel 2 immediately after tropical cyclone Yasa, which hit Fiji on 17 December 2020, and are a result of OC-SMART processing. Coral reefs and coral barriers south of Fiji and around the Beqa island are indicated. (Adapted from [66]).

4.2. *Scientific Accomplishments and Perspectives from Ocean Color Remote Sensing in the SWTP*

Ocean color has greatly helped to address questions about mesotrophy in the archipelagos of the SWTP through the identification of *Trichodesmium* blooms and tools adapted to the survey of tremendous variations in coastal areas around islands.

4.2.1. Novel Collection and Utilization of In Situ Optical Signatures

Long-term observation stations allowing bio-optical measurements are scarce in the SWTP (except on the Great Barrier Reef). Even Bio-Argo floats equipped with backscattering and fluorescence sensors are not numerous. Cruises have allowed gliders equipped with CDOM fluorometers, and many research cruises including bio-optical measurements together with carbon and nitrogen fluxes have provided an unprecedented view of mesotrophy in the SWTP. Linking in situ reflectance and IOP data is still a challenge due to the small amount of data. It is crucial to complete and archive data sets of R_{rs} and IOPs. An observation station was launched in 2012 in the southwestern lagoon of New Caledonia (MOISE, MOoring bIogeochemical Survey [52]), which could offer a means of observing trends in bio-optical parameters at 21°S together with coral reef surveys. The collection of *Trichodesmium* spectra should be multiplied to try to understand the link between abundances and spectral signatures and be able to better characterize their spatial patterns with new ocean color sensors.

4.2.2. Algal Blooms and the Detection of *Trichodesmium*

Algal blooms are indicators of the health of marine ecosystems, making their monitoring crucial for the effective management of coastal and oceanic resources. Although the maximal biomass of algal blooms in the SWTP is significantly lower than that in the Northern Hemisphere, their impact on carbon fixation in the ocean can still be substantial. Despite previous efforts, there remains a lack of comprehensive sampling of diazotrophs and assessment of nitrogen fixation levels in the ocean. Recent estimates indicate a significant increase in global oceanic nitrogen fixation, primarily driven by the Indian Ocean and the South Pacific Ocean, which contribute approximately 42% of the global nitrogen fixation. The South Pacific Ocean has been identified as a "hot spot" for diazotrophy due to iron and phosphate fertilization [67–71]. *Trichodesmium* blooms, detected through ocean color remote sensing, can introduce a substantial amount of nitrogen into the water column, and their impact should be further explored in comparison to monthly, seasonal, and yearly rates. Previous studies have successfully attributed large dendritic mesotrophic areas in the SWTP to *Trichodesmium* [9–13] and algorithms have been developed to retrieve their surface accumulations [20,26,63]. Additionally, satellite ocean color data and analyses of atmospheric dust could potentially explain large dendritic cross-tropical area blooms, especially if they are associated with the widespread transport of volcanic ash [72]. Volcanic eruptions in the Tonga trench may also contribute to the mesotrophy of the whole SWTP as, indeed, although the impact was minimal on [Chl-a] at the time of eruption [73,74], subsequent chlorophyll-a enrichment occurred with a significant temporal lag, and as eruptions are regular, this effect could provide nutrients usable by phytoplankton to the west of the Pacific region. The drift of material, including phytoplankton or *Trichodesmium* colonies, entrained by the rafts to the west, could act as a future source of productivity in more favorable zones around Fiji and other archipelagos [73]. Other physical environment changes favoring the supply of essential nutrients [75–77] or leading to deep trench degassing [78] and cyclones [79] could also be invoked.

4.2.3. Following Biogeochemistry of Pacific Island Lagoons

Ocean color remote sensing is the only means to estimate seasonal variability and responses to episodic events, as well as to assess trends in the health of tropical lagoon waters amidst global climate change and increasing anthropization [31,32]. These aspects can be better captured in the future with ocean color sensors providing more frequent coverage. Among the factors that significantly contribute to increased mesotrophy in archipelagos are intense runoffs resulting from tropical storms and cyclonic activity that modify the upper layer structure [80–82]. This mesotrophy is likely more pronounced in high islands, such as Vanuatu and New Caledonia (Figure 8). These effects have been observed extending from the rivers off Hienghène (Figure 9) or reef passages near the Rewa river south of Fiji, reaching hundreds of kilometers into the ocean [30]. As soon as they propagate in oceans, particles and phytoplankton show dynamic coupling and decoupling patterns according to time and space, which are being explored in order to characterize the real impacts on coral reefs (Figure 14) [66]. Runoffs impact not only [Chl-a] but also CDOM, which has been shown to contribute significantly to light absorption in lagoons after rainfall [50–52,83]. Large amounts of high [Chl-a] and other materials, such as mineral particles, can be transported effectively, transferring contaminants and nutrients far from the land. These phenomena should be monitored using ocean color sensors with better temporal resolutions. The understanding of river impacts on the pristine lagoons of the northeastern coast of New Caledonia, which are classified by UNESCO as World Heritage Sites, is of great value. Moreover, these findings reinforce the potential utility of OC-CCI products and satellite-based ocean color remote sensing in general for the monitoring of phytoplankton dynamics in all Pacific Island coasts. In a major way, [Chl-a] will be useful in validating 3D coupled physical–biogeochemical models, which are now able to reproduce the complex ocean–lagoon interface [84–86]. New empirical evidence has indicated that spatial gradients in nearshore primary production (PP) around reef islands can directly influence the nutritional status of corals on shallow reefs [87]. This hypothesis suggests that corals living in mesotrophic waters, which are relatively enriched with nutrients and plankton compared to oligotrophic waters, are more resilient to bleaching, as they can survive for longer periods without their symbionts [88]. As PP is strongly correlated with satellite-derived estimates of [Chl-a], monitoring [Chl-a] will help to test whether corals can compensate by relying on heterotrophic feeding until the environmental conditions improve and symbiont stability is reestablished [88]. *Trichodesmium* may be, in some tropical lagoons, a major component of phytoplankton composition (as in the ones of New Caledonia and Fiji), with slicks observed in summer [89–92], which should be followed with high-resolution sensors.

4.2.4. Long-Term Changes in the Ocean with Climate Change

One question arising is whether chlorophyll-a, diazotrophs, or runoffs will increase or decrease with climate change (sea surface temperature increase, salinity decrease, acidification) in the SWTP. The analysis of CMEMS data from the past two decades indicates an unexplained increasing trend in [Chl-a] in the Fiji–Tonga region, contrasting with a decreasing trend observed in the New Caledonia–Vanuatu region. It remains unclear whether these trends are associated with changes in phytoplankton composition [93], as satellite ocean color data have not yet provided documentation on this aspect. [Chl-a] in the Western Tropical Pacific Ocean appears to be sensitive to the Multivariate ENSO index and the Modoki El Nino index, although the effects in tropical regions and archipelagos are hardly detectable [94]. Reflectance anomaly approaches [95–97] applied on hyperspectral sensor reflectance will help in predicting the whole food chain changes in the ocean.

5. Conclusions

To address the questions of what governs chlorophyll variability in the southwest tropical archipelagos and lagoons and what are the predicted trends with climate change, it is necessary to look at [Chl-a] time series issued from satellite ocean color remote sensing as a function of environmental parameters and climatic indices, in addition to genetic analysis. These analyses require the discrimination of *Trichodesmium*, the major nitrogen-fixing species, from other phytoplankton, and applying regionally adapted algorithms to coastal ocean and coral lagoons including reefs.

Algorithms for the detection of *Trichodesmium*, as well as for the estimation of chlorophyll concentrations in tropical lagoons, have been developed or adapted for the oligo- to meso- South Pacific tropical waters, principally over New Caledonia and South Fiji. In order to validate them, we used a comprehensive data set of in situ bio-optical data representative of optical variability in Case 1 and Case 2 waters collected in various regions and over more than 20 years. This database has also the advantage of being the result of validated methodologies and obtained by the same operators and instruments during that period. The algorithms developed for *Trichodesmium* have been tested on different sensors with different spatial resolutions and the main improvement concerns the discrimination of *Trichodesmium* slicks from other phytoplankton patches in pixels obtained after atmospheric corrections, which was the main limitation for automatic detection. These promising results provide hope that future next-generation sensors, offering higher spectral resolutions and increased repetitiveness, both in low-Earth and geostationary orbits, will enhance efforts to remotely sense global *Trichodesmium* and other phytoplankton's abundance in the SWTP.

The algorithms developed for the better estimation of [Chl-a] in tropical lagoons result from a wide range of approaches, including quasi-empirical regression and classical optical inversion. The use of SVM was particularly well adapted to tropical lagoons, where in situ data were available, even if these are still restricted in number, compared to European waters. The non-parametric algorithms, although a priori adapted only to the studied areas where in situ measurements are available, may be tested for the same types of limitations encountered in other Pacific Island coastal areas. They proved to greatly improve the estimations of [Chla] and turbidity in shallower areas. Unsupervised clustering provided the seabed color at different scales and for some test areas of the New Caledonia lagoon. These methods are promising and will allow one to account effectively for the bottom effect (bathymetry + seabed color) in the next SVMs that will be used on other tropical lagoons and high-resolution sensors.

Both oceanic and coastal oceans are vital for the economy of Small Island Developing states, and multivariate correlative analysis will help to quantify ecosystem changes over spatial and temporal scales relevant to human activity and how they respond to natural and anthropogenic disturbances.

Author Contributions: Conceptualization, C.D. and R.F.; methodology, H.M., R.R., J.L., G.W., A.W. and H.M.; software, G.W. and H.M.; validation, C.D. and R.F.; formal analysis, G.W. and H.M.; investigation, C.D. and R.F.; resources, C.D., R.F. and G.W.; data curation, G.W., H.M. and J.L.; writing—original draft preparation, C.D. and J.T.; writing—review and editing, C.D., J.T. and R.F.; visualization, G.W. and R.A.; supervision, C.D. and R.F.; project administration, C.D. and A.S.; funding acquisition, C.D. All authors have read and agreed to the published version of the manuscript.

Funding: Projects were funded by PROOF FLUPAC, GOPS, ANR OUTPACE, INSU PNTS, INSU EC2CO, and LEFE CYBER GLI-FI GLIder in FIji, ReefTEMPS, OSU PYTHEAS, and IRD. Part of this research received external funding by the Centre de Recherche pour le Nickel et son Environment (CNRT) as a part of CNRT.IRD/DYNAMINE). The project benefited from the Fonds Pacifique Colour of the Ocean by Satellite Methods Exchanges in the Tropical Environment project (COMETE). R.F. and J.T. were supported by the National Aeronautics and Space Administration under various grants.

Data Availability Statement: The in-situ optical dataset presented in the study is available from the corresponding author upon reasonable request.

Acknowledgments: We thank all contributors for the in situ bio-optical cruise data in the South-western Tropical Pacific South Ocean as well in lagoons in New Caledonia and Fiji (during cruises Flupac, Zonal Flux, Ebene, Trichonesia—1998–1999, Valhybio—2008; Diapalis—2000–2001–2002–2003, Caliope—2011–2014–2016, Outpace—2015, Sokowasa—2022 South of Fiji, and the MOISE observation station). We thank the Flotte Océanographique Française (FOF) and particularly the crews of the R/V *Alis* and IRD UAR IMAGO. We gratefully acknowledge the NASA Ocean Biology Processing Group (OBPG) for making the MODIS ocean color imagery and products available, as well as the European Spatial Agency (ESA), the Centre National d'Etudes Spatiales (CNES), and Theia, which produce and provide Sentinel 2 images and satellite products. We thank the Japan Aerospace Exploration Agency (JAXA) for making available the TriOS radiometers.

Conflicts of Interest: The authors declare no conflict of interest.

References

1. Holland, E.; von Schuckmann, K.; Monier, M.; Legeais, J.-F.; Prado, S.; Sathyendranath, S.; Dupouy, C. Copernicus Marine Service Ocean State Report, Issue 3. *J. Oper. Oceanogr.* **2019**, *12* (Suppl. S1), s43–s48. [CrossRef]
2. Ganachaud, A.; von Schuckmann, K.; Whiteside, A.; Dupouy, C.; Le Meur, P.-Y.; Monier, M.; van Wynsberge, S.; De Ramon N'Yeurt, A.; Costa, M.; Aucan, J.; et al. CMEMS SST and Chl-a Indicators for Two Pacific Islands: A Co-Construction Monitoring Framework for an Integrated, Transdisciplinary, Multi-Scale Approach. *J. Oper. Oceanogr.* **2022**, *16*, 1–15.
3. Dandonneau, Y.; Gohin, F. Meridional and Seasonal Variations of the Sea Surface Chlorophyll Concentration in the Southwestern Tropical Pacific (14 to 32° S, 160 to 175° E). *Deep Sea Res.* **1984**, *31*, 1377–1393. [CrossRef]

4. Dandonneau, Y.; Lemasson, L. Water-Column Chlorophyll in an Oligotrophic Environment: Correction for the Sampling Depths and Variations of the Vertical Structure of Density, and Observation of a Growth Period. *J. Plankton Res.* **1987**, *9*, 215–234. [CrossRef]
5. Blanchot, J.; Rodier, M. Picophytoplankton Abundance and Biomass in the Western Tropical Pacific Ocean during the 1992 El Niño Year: Results from Flow Cytometry. *Deep Sea Res.* **1996**, *43*, 877–895. [CrossRef]
6. Le Bouteiller, A.; Blanchot, J.; Rodier, M. Size Distribution Patterns of Phytoplankton in the Western Pacific: Towards a Generalization for the Tropical Open Ocean. *Deep Sea Res.* **1992**, *39*, 805–823. [CrossRef]
7. Ceccarelli, D.M.; McKinnon, A.D.; Andréfouët, S.; Allain, V.; Young, J.; Gledhill, D.C.; Flynn, A.; Bax, N.J.; Beaman, R.; Borsa, P.; et al. Chapter Four—The Coral Sea: Physical Environment, Ecosystem Status and Biodiversity Assets. In *Advances in Marine Biology*; Lesser, M., Ed.; Academic Press: Cambridge, MA, USA, 2013; Volume 66, pp. 213–290.
8. Wilson, C.; Qiu, X. Global Distribution of Summer Chlorophyll Blooms in the Oligotrophic Gyres. *Prog. Oceanogr.* **2008**, *78*, 107–134. [CrossRef]
9. Dupouy, C.; Petit, M.; Dandonneau, Y. Satellite Detected Cyanobacteria Bloom in the Southwestern Tropical Pacific Implication for Oceanic Nitrogen Fixation. *Int. J. Remote Sens.* **1988**, *9*, 389–396. [CrossRef]
10. Dupouy, C. La chlorophylle de surface observée par le satellite NIMBUS-7 CZCS autour de la Nouvelle Calédonie et de ses dépendances. Une première analyse. *Bull. L'institut Océanographique Monaco* **1990**, *6*, 125–148.
11. Dupouy, C. Discoloured Waters in the Melanesian Archipelago (New Caledonia and Vanuatu). The Value of the Nimbus-7 Coastal Zone Colour Scanner Observations. In *Marine Pelagic Cyanobacteria: Trichodesmium and other Diazotrophs*; Carpenter, E.J., Capone, D.G., Rueter, J.G., Eds.; NATO ASI Series; Springer: Dordrecht, The Netherlands, 1992; pp. 177–191. ISBN 978-94-015-7977-3.
12. Dupouy, C.; Neveux, J.; Subramaniam, A.; Mulholland, M.R.; Montoya, J.P.; Campbell, L.; Carpenter, E.J.; Capone, D.G. Satellite Captures *Trichodesmium* Blooms in the Southwestern Tropical Pacific. *Eos Trans. Am. Geophys. Union* **2000**, *81*, 13–16. [CrossRef]
13. Dupouy, C.; Neveux, J.; Le Bouteiller, A. Spatial and temporal analysis of SeaWIFS sea surface chlorophyll, temperature, winds and sea level anomalies in the South Tropical Pacific Ocean (10° S–25° S, 150° E–180° E). In Proceedings of the 6ème Conférence PORSEC, Pan Ocean Remote Sensing Conference, Gayana, Conception, Chili, 29 November–3 December 2004; Gayana, Conception (Chili). Gayana 68(2) Suppl. I. Proc.. pp. 161–166. [CrossRef]
14. Dupouy, C.; Dirgerg, G.; Tenório, M.M.B.; Neveux, J.; Le Bouteiller, A. Surveillance des Trichodesmium autour de la Nouvelle-Calédonie, du Vanuatu, de Fidji et de Tonga 1998–2004. *Arch. Sci. Mer.* **2004**, *7*, 51. Available online: https://horizon.documentation.ird.fr/exl-doc/pleins_textes/divers14-03/010049853.pdf (accessed on 19 October 2023).
15. Campbell, L.; Carpenter, E.; Montoya, J.P.; Kutska, A.B.; Capone, D. Picoplankton community structure within and outside a *Trichodesmium* bloom in the Southwestern Pacific Ocean. *Vie Milieu* **2005**, *55*, 185–195.
16. Neveux, J.; Tenório, M.M.B.; Dupouy, C.; Villareal, T.A. Spectral diversity of phycoerythrins and diazotrophs abundance in tropical South Pacific. *Limnol. Oceanogr.* **2006**, *51*, 1689–1698. [CrossRef]
17. Moisander, P.H.; Beinart, R.A.; Hewson, I.; White, A.E.; Johnson, K.S.; Carlson, C.A.; Montoya, J.P.; Zehr, J.P. Unicellular Cyanobacterial Distributions Broaden the Oceanic N2 Fixation Domain. *Science* **2010**, *327*, 1512–1514. [CrossRef] [PubMed]
18. Shiozaki, T.; Kodama, T.; Furuya, K. Large-Scale Impact of the Island Mass Effect through Nitrogen Fixation in the Western South Pacific Ocean. *Geophy. Res. Lett.* **2014**, *41*, 2907–2913. [CrossRef]
19. Tenório, M.; Dupouy, C.; Rodier, M.; Neveux, J. *Trichodesmium* and other Filamentous Cyanobacteria in New Caledonian waters (South West Tropical Pacific) during an El Niño Episode. *Aquat. Microb. Ecol.* **2018**, *81*, 219–241. [CrossRef]
20. Subramaniam, A.; Carpenter, E.J.; Karentz, D.; Falkowski, P.G. Bio-Optical Properties of the Marine Diazotrophic Cyanobacteria *Trichodesmium* spp. I. Absorption and Photosynthetic Action Spectra. *Limnol. Oceanogr.* **1999**, *44*, 608–617. [CrossRef]
21. Subramaniam, A.; Brown, C.W.; Hood, R.R.; Carpenter, E.J.; Capone, D.G. Detecting *Trichodesmium* Blooms in SeaWiFS Imagery. *Deep Sea Res.* **2001**, *49*, 107–121. [CrossRef]
22. Hu, C.; Cannizzaro, J.; Carder, K.L.; Muller-Karger, F.E.; Hardy, R. Remote Detection of *Trichodesmium* Blooms in Optically Complex Coastal Waters: Examples with MODIS Full-Spectral Data. *Remote Sens. Environ.* **2010**, *114*, 2048–2058. [CrossRef]
23. Westberry, T.K.; Siegel, D.A.; Subramaniam, A. An Improved Bio-Optical Model for the Remote Sensing of *Trichodesmium* spp. Blooms. *J. Geophys. Res. Oceans* **2005**, *110*, C06012. [CrossRef]
24. Gower, J.; King, S.; Young, E. Global remote sensing of *Trichodesmium*. *Int. J. Remote Sens.* **2014**, *35*, 5459–5466. [CrossRef]
25. Dupouy, C.; Benielli-Gary, D.; Neveux, J.; Dandonneau, Y.; Westberry, T.K. An Algorithm for Detecting *Trichodesmium* Surface Blooms in the South Western Tropical Pacific. *Biogeosciences* **2011**, *8*, 3631–3647. [CrossRef]
26. McKinna, L.I.W. Three Decades of Ocean-Color Remote-Sensing *Trichodesmium* spp. in the World's Oceans: A Review. *Prog. Oceanogr.* **2015**, *131*, 177–199. [CrossRef]
27. Blondeau-Patissier, D.; Brando, V.E.; Lønborg, C.; Leahy, S.M.; Dekker, A.G. Phenology of *Trichodesmium* spp. Blooms in the Great Barrier Reef Lagoon, Australia, from the ESA-MERIS 10-Year Mission. *PLoS ONE* **2018**, *13*, e0208010. [CrossRef]
28. IOCCG. *Remote Sensing of Ocean Color in Coastal, and Other Optically-Complex Waters*; International Ocean-Colour Coordinating Group (IOCCG): Dartmouth, NS, Canada, 2000.
29. IOCCG. *Remote Sensing of Inherent Optical Properties: Fundamentals, Tests of Algorithms, and Applications*; International Ocean-Colour Coordinating Group (IOCCG): Dartmouth, NS, Canada, 2006.

30. Dekker, A.G.; Phinn, S.R.; Anstee, J.; Bissett, P.; Brando, V.E.; Casey, B.; Fearns, P.; Hedley, J.; Klonowski, W.; Lee, Z.P.; et al. Intercomparison of Shallow Water Bathymetry, Hydro-Optics, and Benthos Mapping Techniques in Australian and Caribbean Coastal Environments. *Limnol. Oceanogr. Methods* **2011**, *9*, 396–425. [CrossRef]
31. Blondeau-Patissier, D.; Gower, J.F.R.; Dekker, A.G.; Phinn, S.R.; Brando, V.E. A Review of Ocean Color Remote Sensing Methods and Statistical Techniques for the Detection, Mapping and Analysis of Phytoplankton Blooms in Coastal and Open Oceans. *Prog. Oceanogr.* **2014**, *123*, 123–144. [CrossRef]
32. Blondeau-Patissier, D.; Schroeder, T.; Brando, V.E.; Maier, S.W.; Dekker, A.G.; Phinn, S. ESA-MERIS 10-Year Mission Reveals Contrasting Phytoplankton Bloom Dynamics in Two Tropical Regions of Northern Australia. *Remote Sens.* **2014**, *6*, 2963–2988. [CrossRef]
33. Hedley, J.D.; Roelfsema, C.; Brando, V.; Giardino, C.; Kutser, T.; Phinn, S.; Mumby, P.J.; Barrilero, O.; Laporte, J.; Koetz, B. Coral Reef Applications of Sentinel-2: Coverage, Characteristics, Bathymetry and Benthic Mapping with Comparison to Landsat 8. *Remote Sens. Environ.* **2018**, *216*, 598–614. [CrossRef]
34. Zoffoli, M.L.; Frouin, R.; Kampel, M. Water Column Correction for Coral Reef Studies by Remote Sensing. *Sensors* **2014**, *14*, 16881–16931. [CrossRef]
35. Dupouy, C.; Neveux, J.; Dirberg, G.; Röttgers, R.; Tenório, M.M.B.; Ouillon, S. Bio-optical properties of the marine cyanobacteria *Trichodesmium* spp. *J. Appl. Remote Sens.* **2008**, *2*, 023503. [CrossRef]
36. Dupouy, C.; Neveux, J.; Ouillon, S.; Frouin, R.; Murakami, H.; Hochard, S.; Dirberg, G. Inherent Optical Properties and Satellite Retrieval of Chlorophyll Concentration in the Lagoon and Open Ocean Waters of New Caledonia. *Mar. Pollut. Bull.* **2010**, *61*, 503–518. [CrossRef]
37. Froidefond, J.-M.; Ouillon, S. Introducing a Mini-Catamaran to Perform Reflectance Measurements above and below the Water Surface. *Opt. Express* **2005**, *13*, 926–936. [CrossRef] [PubMed]
38. Ouillon, S.; Douillet, P.; Petrenko, A.; Neveux, J.; Dupouy, C.; Froidefond, J.-M.; Andréfouët, S.; Muñoz-Caravaca, A. Optical Algorithms at Satellite Wavelengths for Total Suspended Matter in Tropical Coastal Waters. *Sensors* **2008**, *8*, 4165–4185. [CrossRef] [PubMed]
39. Dupouy, C.; Frouin, R.; Tedetti, M.; Maillard, M.; Rodier, M.; Lombard, F.; Guidi, L.; Picheral, M.; Neveux, J.; Duhamel, S.; et al. Diazotrophic *Trichodesmium* Impact on UV–Vis Radiance and Pigment Composition in the Western Tropical South Pacific. *Biogeosciences* **2018**, *15*, 5249–5269. [CrossRef]
40. Ouillon, S.; Douillet, P.; Lefebvre, J.P.; Le Gendre, R.; Jouon, A.; Bonneton, P.; Fernandez, J.M.; Chevillon, C.; Magand, O.; Lefèvre, J.; et al. Circulation and Suspended Sediment Transport in a Coral Reef Lagoon: The South-West Lagoon of New Caledonia. *Mar. Pollut. Bull.* **2010**, *61*, 269–296. [CrossRef]
41. Röttgers, R.; Dupouy, C.; Taylor, B.B.; Bracher, A.; Woźniak, S.B. Mass-Specific Light Absorption Coefficients of Natural Aquatic Particles in the near-Infrared Spectral Region. *Limnol. Oceanogr.* **2014**, *59*, 1449–1460. [CrossRef]
42. Röttgers, R.; Doxaran, D.; Dupouy, C. Quantitative Filter Technique Measurements of Spectral Light Absorption by Aquatic Particles Using a Portable Integrating Cavity Absorption Meter (QFT-ICAM). *Opt. Express* **2016**, *24*, A1–A20. [CrossRef]
43. Dupouy, C.; Neveux, J.; Andre', J.M. Spectral Absorption Coefficient of Photosynthetically Active Pigments in the Equatorial Pacific Ocean (165°11–150° W). *Deep. Sea Res. Part II Top. Stud. Oceanogr.* **1997**, *44*, 1881–1906. [CrossRef]
44. Simeon, J.; Roesler, C.; Pegau, W.S.; Dupouy, C. Sources of Spatial Variability in Light Absorbing Components along an Equatorial Transect from 165° E to 150° W. *J. Geophys. Res. Oceans* **2003**, *108*, 3333. [CrossRef]
45. Gross, L.; Frouin, R.; Dupouy, C.; André, J.M.; Thiria, S. Reducing Variability That Is Due to Secondary Pigments in the Retrieval of Chlorophyll a Concentration from Marine Reflectance: A Case Study in the Western Equatorial Pacific Ocean. *Appl. Opt.* **2004**, *43*, 4041–4054. [CrossRef]
46. Dupouy, C.; Loisel, H.; Neveux, J.; Brown, S.L.; Moulin, C.; Blanchot, J.; Le Bouteiller, A.; Landry, M.R. Microbial Absorption and Backscattering Coefficients from in Situ and POLDER Satellite Data during an El Niño–Southern Oscillation Cold Phase in the Equatorial Pacific (180°). *J. Geophys. Res. Oceans* **2003**, *108*, 8138. [CrossRef]
47. Neveux, J.; Lefebvre, J.-P.; Le Gendre, R.; Dupouy, C.; Gallois, F.; Courties, C.; Gérard, P.; Fernandez, J.-M.; Ouillon, S. Phytoplankton Dynamics in the Southern New Caledonian Lagoon during a Southeast Trade Winds Event. *J. Mar. Syst.* **2010**, *82*, 230–244. [CrossRef]
48. Neveux, J.; Tenório, M.M.B.; Jacquet, S.; Torréton, J.-P.; Douillet, P.; Ouillon, S.; Dupouy, C. Chlorophylls and Phycoerythrins as Markers of Environmental Forcings Including Cyclone Erica Effect (March 2003) on Phytoplankton in the Southwest Lagoon of New Caledonia and Oceanic Adjacent Area. *Int. J. Oceanogr.* **2009**, *2009*, e232513. [CrossRef]
49. Favareto, L.R.; Rudorff, N.; Kampel, M.; Frouin, R.; Röttgers, R.; Doxaran, D.; Murakami, H.; Dupouy, C. Bio-Optical Characterization and Ocean Colour Inversion in the Eastern Lagoon of New Caledonia, South Tropical Pacific. *Remote Sens.* **2018**, *10*, 1043. [CrossRef]
50. Martias, C.; Tedetti, M.; Lantoine, F.; Jamet, L.; Dupouy, C. Characterization and Sources of Colored Dissolved Organic Matter in a Coral Reef Ecosystem Subject to Ultramafic Erosion Pressure (New Caledonia, Southwest Pacific). *Sci. Total Environ.* **2018**, *616–617*, 438–452. [CrossRef] [PubMed]
51. Dupouy, C.; Röttgers, R.; Tedetti, M.; Frouin, R.; Lantoine, F.; Rodier, M.; Martias, C.; Goutx, M. Impact of Contrasted Weather Conditions on CDOM Absorption/Fluorescence and Biogeochemistry in the Eastern Lagoon of New Caledonia. *Front. Earth Sci.* **2020**, *8*, 54. [CrossRef]

52. Dupouy, C.; Rodier, M. *MOISE Observing Station*; OSU PYTHEAS Report; OSU PYTHEAS: Marseille, France, 2014.
53. Fichez, R.D.P.; Chevillon, C.; Torreton, J.P.; Aung, T.H.; Chifflet, S.; Fernandez, J.M.; Gangaiya, P.; Garimella, S.; Gerard, P.; Ouillon, S.; et al. The Suva Lagoon Enviroment: An overview of a joint IRD Camelia Research Unit and USP Study. In *At the Cross Roads: Science and Management of the Suva Lagoon*; Morrison, R.J., Aalbersberg, W., Eds.; Institute of the Applied Sciences Press, University of the South Pacific: Suva, Fiji, 2006; Volume 1, pp. 93–105.
54. Koliyavu, T.; Martias, C.; Singh, A.; Mounier, S.; Gérard, P.; Dupouy, C. In-Situ Variability of DOM in Relation with Biogeochemical and Physical Parameters in December 2017 in Laucala Bay (Fiji Islands) after a Strong Rain Event. *J. Mar. Sci. Eng.* **2021**, *9*, 241. [CrossRef]
55. Lefèvre, J. *The MODIS Data Processing Chain*; Technical Document of the VALidation of Hyperspectral SATellite Data the Valhysat Project; IRD Nouméa: Nouméa, Nouvelle-Calédonie, 2010.
56. Murakami, H.; Dupouy, C. Coastal ocean atmospheric correction for AVNIR-2 high resolution images. In Proceedings of the SPIE, Remote Sensing of the Coastal Ocean, Land, and Atmosphere Environment, Incheon, Republic of Korea, 11–14 October 2010; Volume 7858, p. 785802. [CrossRef]
57. Murakami, H.; Dupouy, C.; Röttgers, R.; Frouin, R.J. Estimation of Inherent Optical Properties Using In Situ Hyperspectral Radiometer and MODIS Data along the East Coast of New Caledonia. In Proceedings of the SPIE, Remote Sensing of the Marine Environment II, Kyoto, Japan, 29 October–1 November 2012; Volume 8525, pp. 47–55.
58. Wattelez, G.; Dupouy, C.; Mangeas, M.; Lefèvre, J.; Touraivane; Frouin, R. A Statistical Algorithm for Estimating Chlorophyll Concentration in the New Caledonian Lagoon. *Remote Sens.* **2016**, *8*, 45. [CrossRef]
59. Wattelez, G.; Dupouy, C.; Lefèvre, J.; Ouillon, S.; Fernandez, J.-M.; Juillot, F. Application of the Support Vector Regression Method for Turbidity Assessment with MODIS on a Shallow Coral Reef Lagoon (Voh-Koné-Pouembout, New Caledonia). *Water* **2017**, *9*, 737. [CrossRef]
60. Minghelli-Roman, A.; Dupouy, C. Influence of Water Column Chlorophyll Concentration on Bathymetric Estimations in the Lagoon of New Caledonia, Using Several MERIS Images. *IEEE J. Sel. Top. Appl. Earth Obs. Remote Sens.* **2013**, *6*, 739–745. [CrossRef]
61. Minghelli-Roman, A.; Dupouy, C. Correction of the Water Column Attenuation: Application to the Seabed Mapping of the Lagoon of New Caledonia Using MERIS Images. *IEEE J. Sel. Top. Appl. Earth Obs. Remote Sens.* **2014**, *7*, 2619–2629. [CrossRef]
62. Wattelez, G.; Dupouy, C.; Juillot, F. Unsupervised Optical Classification of the Seabed Color in Shallow Oligotrophic Waters from Sentinel-2 Images: A Case Study in the Voh-Koné-Pouembout Lagoon (New Caledonia). *Remote Sens.* **2022**, *14*, 836. [CrossRef]
63. Rousset, G.; De Boissieu, F.; Menkes, C.E.; Lefèvre, J.; Frouin, R.; Rodier, M.; Ridoux, V.; Laran, S.; Bonnet, S.; Dupouy, C. Remote Sensing of *Trichodesmium* spp. Mats in the Western Tropical South Pacific. *Biogeosciences* **2018**, *15*, 5203–5219. [CrossRef]
64. Bazader, J.-F. Personal communication. 2023.
65. Benavides, M.; Conradt, L.; Bonnet, S.; Berman-Frank, I.; Barrillon, S.; Petrenko, A.; Doglioli, A. Fine-Scale Sampling Unveils Diazotroph Patchiness in the South Pacific Ocean. *ISME Commun.* **2021**, *1*, 3. [CrossRef]
66. Whiteside, A. *Ocean Color Plume Monitoring around the Fiji Islands from Remote Sensing through Anthropogenic and Climatic Influences. Estimation of the Impact of the Cyanobacteria Trichodesmium, Thèse de 3eme Cycle*; Aix-Marseille Université: Marseille, France, 7 November 2023.
67. Bonnet, S.; Caffin, M.; Berthelot, H.; Moutin, T. Hot Spot of N2 Fixation in the Western Tropical South Pacific Pleads for a Spatial Decoupling between N2 Fixation and Denitrification. *Proc. Natl. Acad. Sci. USA* **2017**, *114*, E2800. [CrossRef]
68. Messer, L.F.; Mahaffey, C.; M Robinson, C.; Jeffries, T.C.; Baker, K.G.; Bibiloni Isaksson, J.; Ostrowski, M.; Doblin, M.A.; Brown, M.V.; Seymour, J.R. High Levels of Heterogeneity in Diazotroph Diversity and Activity within a Putative Hotspot for Marine Nitrogen Fixation. *ISME J.* **2016**, *10*, 1499–1513. [CrossRef]
69. Bonnet, S.; Caffin, M.; Berthelot, H.; Grosso, O.; Benavides, M.; Helias-Nunige, S.; Guieu, C.; Stenegren, M.; Foster, R.A. In-Depth Characterization of Diazotroph Activity across the Western Tropical South Pacific Hotspot of N_2 Fixation (OUTPACE Cruise). *Biogeosciences* **2018**, *15*, 4215–4232. [CrossRef]
70. Moutin, T.; Karl, D.M.; Duhamel, S.; Rimmelin, P.; Raimbault, P.; Van Mooy, B.A.S.; Claustre, H. Phosphate Availability and the Ultimate Control of New Nitrogen Input by Nitrogen Fixation in the Tropical Pacific Ocean. *Biogeosciences* **2008**, *5*, 95–109. [CrossRef]
71. Bonnet, S.; Benavides, M.; Le Moigne, F.A.C.; Camps, M.; Torremocha, A.; Grosso, O.; Dimier, C.; Spungin, D.; Berman-Frank, I.; Garczarek, L.; et al. Diazotrophs Are Overlooked Contributors to Carbon and Nitrogen Export to the Deep Ocean. *ISME J.* **2023**, *17*, 47–58. [CrossRef]
72. Yoon, J.-E.; King, D.; Longman, J.; Cronin, S.J. Differential Response of Chlorophyll-a Concentrations to Explosive Volcanism in the Western South Pacific. *Front. Mar. Sci.* **2023**, *10*, 1072610. [CrossRef]
73. Whiteside, A.; Dupouy, C.; Singh, A.; Frouin, R.; Menkes, C.; Lefèvre, J. Automatic Detection of Optical Signatures within and around Floating Tonga-Fiji Pumice Rafts Using MODIS, VIIRS, and OLCI Satellite Sensors. *Remote Sens.* **2021**, *13*, 501. [CrossRef]
74. Whiteside, A.; Dupouy, C.; Singh, A.; Bani, P.; Tan, J.; Frouin, R. Impact of Ashes from the 2022 Tonga Volcanic Eruption on Satellite Ocean Color Signatures. *Front. Mar. Sci.* **2023**, *9*, 1028022. [CrossRef]
75. Martinez, E.; Antoine, D.; D'Ortenzio, F.; Gentili, B. Climate-Driven Basin-Scale Decadal Oscillations of Oceanic Phytoplankton. *Science* **2009**, *326*, 1253–1256. [CrossRef]

76. Messié, M.; Petrenko, A.; Doglioli, A.M.; Aldebert, C.; Martinez, E.; Koenig, G.; Bonnet, S.; Moutin, T. The Delayed Island Mass Effect: How Islands Can Remotely Trigger Blooms in the Oligotrophic Ocean. *Geophys. Res. Lett.* **2020**, *47*, e2019GL085282. [CrossRef]
77. Messié, M.; Petrenko, A.; Doglioli, A.M.; Martinez, E.; Alvain, S. Basin-Scale Biogeochemical and Ecological Impacts of Islands in the Tropical Pacific Ocean. *Nat. Geosci.* **2022**, *15*, 469–474. [CrossRef]
78. Bonnet, S.; Guieu, C.; Taillandier, V.; Boulart, C.; Bouruet-Aubertot, P.; Gazeau, F.; Scalabrin, C.; Bressac, M.; Knapp, A.N.; Cuypers, Y.; et al. Natural Iron Fertilization by Shallow Hydrothermal Sources Fuels Diazotroph Blooms in the Ocean. *Science* **2023**, *380*, 812–817. [CrossRef] [PubMed]
79. Russell, P.; Horvat, C. Extreme South Pacific Phytoplankton Blooms Induced by Tropical Cyclones. *Geophys. Res. Lett.* **2023**, *50*, e2022GL100821. [CrossRef]
80. Andrews, J.; Gentien, P. Upwelling as a Source of Nutrients for the Great Barrier Reef Ecosystems: A Solution to Darwin's Question? *Mar. Ecol. Prog. Ser.* **1982**, *8*, 257–269. [CrossRef]
81. Brodie, J.E.; Kroon, F.J.; Schaffelke, B.; Wolanski, E.C.; Lewis, S.E.; Devlin, M.J.; Bohnet, I.C.; Bainbridge, Z.T.; Waterhouse, J.; Davis, A.M. Terrestrial Pollutant Runoff to the Great Barrier Reef: An Update of Issues, Priorities and Management Responses. *Mar. Pollut. Bull.* **2012**, *65*, 81–100. [CrossRef]
82. Oubelkheir, K.; Ford, P.W.; Cherukuru, N.; Clementson, L.A.; Petus, C.; Devlin, M.; Schroeder, T.; Steven, A.D.L. Impact of a Tropical Cyclone on Terrestrial Inputs and Bio-Optical Properties in Princess Charlotte Bay (Great Barrier Reef Lagoon). *Remote Sens.* **2023**, *15*, 652. [CrossRef]
83. Fabricius, K.E.; Logan, M.; Weeks, S.; Brodie, J. The Effects of River Run-off on Water Clarity across the Central Great Barrier Reef. *Mar. Pollut. Bull.* **2014**, *84*, 191–200. [CrossRef]
84. Fuchs, R.; Dupouy, C.; Douillet, P.; Caillaud, M.; Mangin, A.; Pinazo, C. Modelling the Impact of a La Niña Event on a South West Pacific Lagoon. *Mar. Pollut. Bull.* **2012**, *64*, 1596–1613. [CrossRef]
85. Fuchs, R.; Pinazo, C.; Douillet, P.; Fraysse, M.; Grenz, C.; Mangin, A.; Dupouy, C. Modelling Ocean–Lagoon Interaction during Upwelling Processes in the Southwest of New Caledonia. *Estuar. Coast. Shelf Sci.* **2013**, *135*, 5–17. [CrossRef]
86. Chevalier, C.; Sous, D.; Devenon, J.-L.; Pagano, M.; Rougier, G.; Blanchot, J. Impact of Cross-Reef Water Fluxes on Lagoon Dynamics: A Simple Parameterization for Coral Lagoon Circulation Model, with Application to the Ouano Lagoon, New Caledonia. *Ocean Dyn.* **2015**, *65*, 1509–1534. [CrossRef]
87. Roche, R.C.; Heenan, A.; Taylor, B.M.; Schwarz, J.N.; Fox, M.D.; Southworth, L.K.; Williams, G.J.; Turner, J.R. Linking Variation in Planktonic Primary Production to Coral Reef Fish Growth and Condition. *R. Soc. Open Sci.* **2022**, *9*, 201012. [CrossRef]
88. Houlbrèque, F.; Ferrier-Pagès, C. Heterotrophy in Tropical Scleractinian Corals. *Biol. Rev. Camb. Philos. Soc.* **2009**, *84*, 1–17. [CrossRef] [PubMed]
89. Rodier, M.; Le Borgne, R. Population dynamics and environmental conditions affecting *Trichodesmium* spp. (filamentous cyanobacteria) blooms in the south-west lagoon of New Caledonia. *J. Exp. Mar. Biol. Ecol.* **2008**, *358*, 20–32. [CrossRef]
90. Rodier, M.; Le Borgne, R. Population and Trophic Dynamics of *Trichodesmium Thiebautii* in the SE Lagoon of New Caledonia. Comparison with *T. Erythraeum* in the SW Lagoon. *Mar. Pollut. Bull.* **2010**, *61*, 349–359. [CrossRef] [PubMed]
91. Saulia, E.; Benavides, M.; Henke, B.; Turk-Kubo, K.; Cooperguard, H.; Grosso, O.; Desnues, A.; Rodier, M.; Dupouy, C.; Riemann, L.; et al. Seasonal Shifts in Diazotrophs Players: Patterns Observed Over a Two-Year Time Series in the New Caledonian Lagoon (Western Tropical South Pacific Ocean). *Front. Mar. Sci.* **2020**, *7*, 581755. [CrossRef]
92. Boatman, T.G.; Upton, G.J.G.; Lawson, T.; Geider, R.J. Projected Expansion of *Trichodesmium*'s Geographical Distribution and Increase in Growth Potential in Response to Climate Change. *Glob. Chang. Biol.* **2020**, *26*, 6445–6456. [CrossRef]
93. De Monte, S.; Soccodato, A.; Alvain, S.; d'Ovidio, F. Can We Detect Oceanic Biodiversity Hotspots from Space? *ISME J.* **2013**, *7*, 2054–2056. [CrossRef] [PubMed]
94. Sharma, P.; Marinov, I.; Cabre, A.; Kostadinov, T.; Singh, A. Increasing Biomass in the Warm Oceans: Unexpected New Insights from SeaWiFS. *Geophys. Res. Lett.* **2019**, *46*, 3900–3910. [CrossRef]
95. Alvain, S.; Moulin, C.; Dandonneau, Y.; Bréon, F.M. Remote Sensing of Phytoplankton Groups in Case 1 Waters from Global SeaWiFS Imagery. *Deep Sea Res.* **2005**, *52*, 1989–2004. [CrossRef]
96. Alvain, S.; Moulin, C.; Dandonneau, Y.; Loisel, H. Seasonal Distribution and Succession of Dominant Phytoplankton Groups in the Global Ocean: A Satellite View. *Glob. Biogeochem. Cycles* **2008**, *22*, GB3001. [CrossRef]
97. De Boissieu, F.; Menkes, C.; Dupouy, C.; Rodier, M.; Bonnet, S.; Mangeas, M.; Frouin, R.J. Phytoplankton Global Mapping from Space with a Support Vector Machine Algorithm. In Proceedings of the SPIE, Ocean Remote Sensing and Monitoring from Space, Beijing, China, 13–16 October 2014; Volume 9261, pp. 354–367.

remote sensing

MDPI

Article

Satellite and High-Spatio-Temporal Resolution Data Collected by Southern Elephant Seals Allow an Unprecedented 3D View of the Argentine Continental Shelf

Melina M. Martinez [1,2,3,*], Laura A. Ruiz-Etcheverry [1,2,3], Martin Saraceno [1,2,3], Anatole Gros-Martial [4,5,6], Julieta Campagna [7], Baptiste Picard [6] and Christophe Guinet [6]

1 Departamento de Ciencias de la Atmósfera y los Océanos (DCAO), Facultad de Ciencias Exactas y Naturales, Universidad de Buenos Aires, Buenos Aires C1428EGA, Argentina; lruiz@cima.fcen.uba.ar (L.A.R.-E.); saraceno@cima.fcen.uba.ar (M.S.)
2 Centro de Investigaciones del Mar y la Atmósfera (CIMA), CONICET—Universidad de Buenos Aires, Buenos Aires C1428EGA, Argentina
3 Instituto Franco-Argentino de Estudios sobre el Clima y sus Impactos (IFAECI), IRL 3351, CNRS-CONICET-IRD-UBA, Buenos Aires C1428EGA, Argentina
4 ENSTA-Bretagne, UMR6285 Lab-STICC, F-29200 Brest, France; grosmartial.anatole@gmail.com
5 CNRS, Ifremer, IUEM, University Brest, UMR6538 Geo-Ocean, F-29280 Plouzané, France
6 Centre d'Etudes Biologiques de Chizé, UMR 7372, CNRS-La Rochelle Université, F-79360 Villiers en Bois, France; picard.baptiste@gmail.com (B.P.); christophe.guinet@cebc.cnrs.fr (C.G.)
7 Wildlife Conservation Society, Argentina, and Marine Programs, Amenábar 1595, Buenos Aires 1426, Argentina; jcampagna@wcs.org
* Correspondence: melina.martinez@cima.fcen.uba.ar

Abstract: High spatial and temporal resolution hydrographic data collected by Southern Elephant Seals (*Mirounga leonina,* SESs) and satellite remote sensing data allow a detailed oceanographic description of the Argentine Continental Shelf (ACS). In-situ data were obtained from the CTD (Conductivity, Temperature, and Depth), accelerometer, and hydrophone sensors attached to five SESs that crossed the ACS between the 17th and 31st of October 2019. The analysis of the temperature (T) and salinity (S) along the trajectories allowed us to identify two different regions: north and south of 42°S. Satellite Sea Surface Temperature (SST) data suggests that north of 42°S, warm waters are coming from the San Matias Gulf (SMG). The high spatio-temporal resolution of the in-situ data shows regions with intense gradients along the T and S sections that were associated with a seasonal front that develops north of Península Valdés in winter due to the entrance of cold and fresh water to the SMG. The speed of the SESs is correlated with tidal currents in the coastal portion of the northern region, which is in good agreement with the macrotidal regime observed. A large number of Prey Catch Attempts (PCA), a measure obtained from the accelerometer sensor, indicates that SESs also feed in this region, contradicting suggestions from previous works. The analysis of wind intensity estimated from acoustic sensors allowed us to rule out the local wind as the cause of fast thermocline breakups observed along the SESs trajectories. Finally, we show that the maximum depth reached by the elephant seals can be used to detect errors in the bathymetry charts.

Keywords: elephant seal; *Mirounga leonina*; Península Valdés (PV); Argentina; seasonal front in the north of PV; water masses

1. Introduction

In the southern hemisphere (SH), one of the main limitations to understanding the marine environment is the scarce availability of in-situ data. Traditionally, in-situ observations of the marine environment are collected from research vessels, drifting buoys, and moorings. In recent decades, several animals have been used as autonomous platforms to collect oceanographic data, contributing to better sampling of the SH [1]. The use of diving animals allows for the best spatial and temporal resolution in the SH [2,3].

Thanks to the fact that Southern Elephant Seals (SESs) dive almost continuously [4] to deep waters (up to 2000 m) [5,6], on the order of 60 times per day [3,7], and travel thousands of kilometers [8,9], they represent excellent platforms to measure data in the water column of the ocean. SESs show strong site fidelity, returning reliably to natal breeding grounds twice a year, allowing for both attachment and recovery of instruments. The in-situ data collected allows us to study in detail the behavior of those animals and, simultaneously, allows us to evaluate the seasonal and interannual variability of key oceanographic parameters [9].

SESs colonies are located within the Antarctic Circle, except for the one located at Península Valdés (PV), in South America (Figure 1). While at sea, the spatial distribution of females and males is quite different [10]. Males forage mostly on the Patagonian shelf [11] and females mainly along the Patagonian shelf slope, shelf break, or Argentine Basin [12–15]. SESs females breed between late August and November and molt between December and January along ca. 200 km of open ocean coastline at PV [16,17]. SESs have been instrumented in PV since 1992 [14]. The oceanographic data collected by the PV colony have been used to study mesoscale structures in the open ocean [18], identify foraging places [12], and relate the species that the SESs eat to different water masses along the Patagonian shelf slope [13]. Based on the SESs path, speed, and swimming pattern, Campagna et al. [14] suggested that the shelf area is mainly a transit zone and is not the preferred feeding ground of the SESs females of PV. This was recently confirmed by McGovern et al. [15]. Here we propose to use T and S data to characterize the oceanographic conditions along the trajectories of the SESs within the continental shelf, taking advantage of the fact that they crossed it very fast. Indeed, the high-resolution profile data collected by SESs will provide a synoptic description of the oceanographic conditions of the Argentine Continental Shelf (ACS) adjacent to PV.

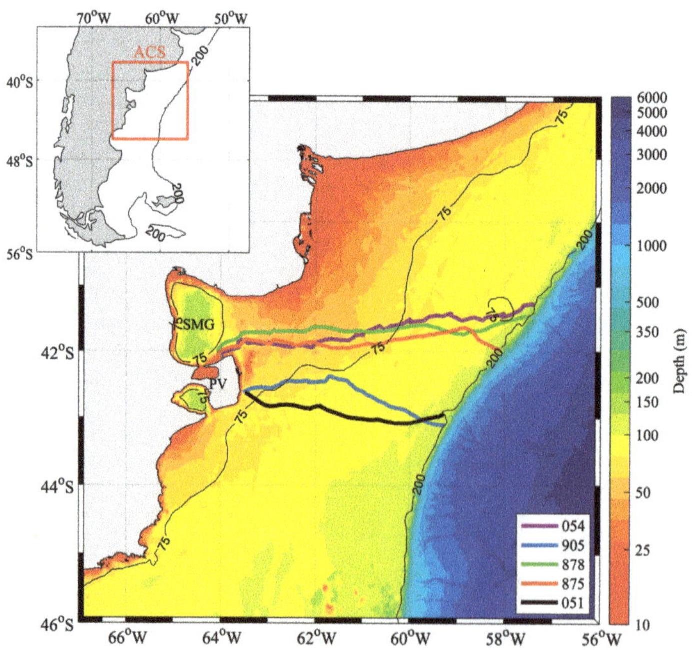

Figure 1. Thick color lines represent the trajectories of the five SESs that crossed the Argentine Continental Shelf (ACS) in October 2019. The colors in the background represent the bathymetry of GEBCO 2021. The thin black lines represent the 75 and 200 m isobaths. SMG: San Matias Gulf. PV: Península Valdés.

We focus our study on the region from the PV coast to the edge of the Patagonian shelf slope (defined by the 200 m isobath) from 55°W to 65°W and from 45°S to 41°S (Figure 1). The portion of ACS adjacent to PV is subject to different dynamics: In the eastern boundary, the Malvinas Current (MC) flows northward over the continental slope, carrying cold, nutrient-rich water. At 41°S, the slope water signal can be observed up to the mid-shelf due to MC intrusions [19]. In the western boundary, the PV tidal front is the main characteristic. The PV front is associated with transitioning between homogeneous coastal waters and vertically stratified mid-shelf waters during spring and summer [20,21]. The homogeneous coastal waters are the product of vertical mixing generated by energetic tidal currents, while the seasonal stratification on the mid-shelf is due to surface warming during spring and summer. The position of the PV front is strongly linked to the bottom topography [22,23], with a NE-SW alignment, following approximately the 75 m isobath between 41°S and 45°S [24,25]. This study area is affected by the MC and by the strong tides in the coastal region, as well as by the circulation of the San Matias Gulf (SMG). This gulf, a semi-enclosed coastal basin, is a source of salty and warm water due to the bathymetry and the current dynamics. Cold and relatively fresh shelf waters enter through the southeastern portion of the mouth mainly in winter and spring [26] and re-circulate within the SMG, gaining heat and salinity through solar heating and evaporation, respectively [27]. The warm and salty waters exported through the northern portion of the gulf mouth to the inner shelf are known as High-Salinity Coastal Waters (>34) [20,28,29]. The other water masses present in the ACS are Mid-Continental Shelf Water, which has a salinity between 33.4 and 33.8 [29,30], Coastal Water (Low Salinity, <33.4), and Malvinas Water (>33.8). Salinity is usually used in ACS to classify water masses [29,31].

The brief oceanographic description of the region presented here is based on the use of remote sensing data, numerical model outputs, and a few in-situ data obtained during oceanographic cruises. As we show below, the high-resolution in-situ data obtained with the instruments mounted on the SESs offer new, three-dimensional observations at a very fine scale resolution in the region, allowing to distinguish small fronts almost not described previously in the literature, which might be crucial in the CO_2 flux dynamics. Thus, the aim of this study is to provide an updated and detailed oceanographic description of the PV region using this unique dataset in combination with satellite data. This work also aims to relate the environmental changes to SESs foraging behavior. The article is organized as follows: the data and methods are described in Section 2. Section 3 presents the results. Finally, Section 4 present the discussion and the main conclusions.

2. Materials and Methods

2.1. In-Situ Data

In-situ data used in this work have been collected as part of an international collaboration between the Centre National de la Recherche Scientifique (CNRS), the Wildlife Conservation Society (WCS), the University of Tasmania (UTAS), and the University of Buenos Aires (UBA). We analyzed data collected by 5 Southern elephant seals (SESs hereafter, *Mirounga leonina*) females that were equipped with Conductivity-Temperature-Depth Satellite Relay Data Loggers (CTD-SRDLs) between October 17 and 31, 2019 in PV, Argentina. Three biologging devices were glued to the animals before their departure to the sea for their post-breeding foraging trip (Tables 1 and 2, Figure 2): a head-mounted DTAG-4 tag (configured as either a sound and movement tag, 97 × 55 × 33 mm, 200 g in air, n = 10, or a sonar and movement tag, 95 × 55 × 37 mm, 200 g in the air), a back-mounted Satellite Relayed Data Logger (CTD-SRDL, Sea Mammal Research Unit (SMRU), St Andrews, UK, 115 × 100 × 40 mm, 680 g in the air), and a neck-mounted Argos location-only satellite tag (SPOT, Wildlife Computers, Redmond, WA, USA, 72 × 54 × 24 mm, 119 g in the air).

Table 1. The location where the instruments have been placed within the body of each SESs.

Id SESs	Head	Back
054		CTD SMRU
875	Dtag-4 sonar	CTD SMRU
878	Dtag-4 sonar	CTD SMRU
905	Dtag-4 sonar	CTD SMRU
051	Dtag-4- Acoustic	CTD SMRU

Table 2. Main characteristics of the instruments deployed.

Tag Name	Variables Measured	Range	Accuracy	Precision	Frequency	Range
CTD	Temperature	−5 to 35 °C	0.0005 °C	0.0001 °C	0.5 Hz	−5 to 35 °C
	Geolocation (ARGOS)		5km			
	Salinity	0 to 50	0.01	0.002	0.5Hz	0 to 50
	Pressure	0 to 2000 dBar	2 dbar	0.05 dBar	0.5Hz	0 to 2000 dBar
SPOT	Geolocation (ARGOS)		5 km			
Sonar	Pressure	0 to 2000 dBar	2 dbar	0.05 dBar	50 Hz	0 to 2000 dBar
	Triaxial acceleration			0.03 ms -2	200 Hz	
	Magnetic field			0.5 μT	50 Hz	
	Geolocation (GPS)		50 m			
Dtag-4	Pressure	0 to 2000 dBar	2 dbar	0.05 dBar	50 Hz	0 to 2000 dBar
	Triaxial acceleration			0.03 ms -2	200 Hz	
	Magnetic field			0.5 μT	50 Hz	
	Geolocation (GPS)		50 m			
	Sound (Acoustic sensor)				20kHz	

The CTD-SRDL records salinity derived from conductivity, temperature, and pressure in hydrographic profiles. When the animals reach the surface, the data collected along the profiles is compressed and sent via messages, along with the positional information, through the Advanced Research and Global Observation Satellite (ARGOS) (http://www.argos-system.org/, accessed on 3 May 2021) [32]. The transmission occurs several times per day. The CTD-SRDL archives pressure, temperature, and salinity data recorded continuously throughout the foraging trip at 0.5 Hz for internal memory. This is far more frequent than the transmitted data (i.e., 3 to 4 24-T/S point profiles per day), but this requires the tags to be recovered when SESs come back to shore to molt in the summer (December–January) [1]. The typical accuracies are ±0.02 °C for temperature and ±0.05 or better for salinity [33]. Additionally, the SPOT tag was included to facilitate the recovery of the instruments when the animal returned to land in case the CTD-SDRL had stopped transmitting.

Acceleration measurements associated with the head-mounted DTAG-4 movement tag provide an index of prey capture attempts (PCA) [34]. This variable is considered the most reliable proxy for foraging success [35,36]. It offers the opportunity to link variations in elephant seal movements, diving, and foraging behavior with environmental conditions. The wind estimation can be obtained from the sound tag, which is derived from the absolute sound pressure following [37]. The DTAG-4 is also equipped with accurate location sensors that provide depth, longitude, and latitude data. The depth data were sampled at a frequency of 0.2 s, while the latitude and longitude data were sampled intermittently, with a variable sampling rate ranging from 1e-5 to 0.1 s. GPS (Global Positioning System) positions were obtained at each surfacing. At-sea movements were monitored using the ARGOS satellite-tracking system, and the loggers were recovered as soon as the seals hauled out for molting. The geolocation accuracy provided by satellite using ARGOS is

±5 km [38,39] and using GPS is about 50 m [40,41]. The CTD and sonar tags were synchronized by matching the depth data sampled by their respective pressure sensors, according to Le Bras et al. [42]. The environmental data are collected up to 50 Hz (Table 2). To compute the vertical profile data, we followed the procedure described in Siegelman et al. [43]: First, temperature and conductivity are corrected for a thermal mass effect; then salinity spiking and density inversion are removed by adjusting salinity while leaving temperature unchanged.

Figure 2. A female SES with three biologging devices glued: a head-mounted DTAG-4, a back-mounted CTD-SRDL, and a neck-mounted SPOT. The main characteristics of the devices are listed in Table 2.

The methodology used for the colocation of instruments on SESs is described by Campagna et al. [44]. The loggers are noninvasive [45]; they are glued to the animal's hair and fall off when it molts. Furthermore, elephant seals' performances do not appear to be impacted by loggers [46]. Thanks to the satellite tag, the tools are recovered when the animals return to shore. After recovering, the data collected were calibrated following the methodology described by Roquet et al. [45] and Sigelman et al. [43].

2.2. Satellite Data

Sea Surface Temperature (SST) from the Multi-scale Ultra-High Resolution L4 product (MUR—https://podaac.jpl.nasa.gov/MEaSUREs-MUR, accessed on 31 August 2021) with daily resolution and spatial resolution of 0.011° × 0.011° is used to compare the surface temperature obtained by the SESs. This high spatial resolution product is generated by combining MODIS, AVHRR, GOES, AMSR-E, and TMI measurements through the technique known as multi-resolution variational analysis [47]. The choice of this product is due to its high spatial resolution, and it has been successfully compared with independent in-situ data in this study region [48,49].

2.3. Reanalysis Data

Surface wind speed values are obtained from ECMWFs reanalysis model, ERA5. Reanalysis data are the product of a physics-based model combined with in-situ and satellite data [50]. More precisely, the integrated forecasting system has a 4D-variational assimilation with 12 h long windows and 12 min time steps [51] and uses data such as French RADOME, Land SYNOP, quickSCAT, and ASCAT as observations [52]. It gives hourly data on a 0.25° grid. The model provides an estimation of the 10 m wind, which is the wind speed ten meters above the surface.

2.4. Bathymetry Data

We used three gridded bathymetry dataset versions (2019, 2020, and 2021) of the General Bathymetric Chart of the Oceans (GEBCO) that are available at https://www.gebco.net/, accessed on 1 July 2021. All versions have been developed through the Nippon Foundation-GEBCO Seabed 2030 Project, which brings together all available bathymetric data to produce a map of the world ocean floor. The GEBCO 2019 portion of the ACS that we used has been corrected, considering bathymetry measurements provided by the Servicio de Hidrografia Naval. This version is currently the most used for regional numerical models in the region [26,53,54].

2.5. Tidal Model

TPXO is a global tidal model that provides a spatial resolution of 1/30°. The model assimilates satellite data from Topex Poseidon, Topex Tandem, ERS, GFO, and other data sets (i.e., tide gauges, ship-born ADCP). The model has a toolbox for accessing the coefficients for tidal harmonic constituents in barotropic tide models and for making predictions of tide height and currents [55]. TPXO8 is used to compare the zonal component velocity obtained by the SESs. The ACS presents a complex tidal region that is largely dominated by the semi-diurnal lunar constituent M2 [20,31]. TPXO8 represents tides very well in the region [56]. It has also been used to implement regional models [26,54].

2.6. Methodology

Several considerations must be considered to compare satellite SST with in-situ data. First, in-situ data are reliable at 15 m depth [35], not at the sea surface, where satellite data indicate SST. Second, satellite images of SST are typically available only once a day, while SESs conduct a vertical profile every 16 min and every 1.1 km on average over the continental shelf and therefore offer about 90 values every day in a large area. Hence, we extracted the nearest value, in time and space, of the satellite images to the 15 m in-situ data. Pearson correlations between both data sets are computed at a 95% confidence level. To evaluate the dispersion between both datasets, biases were calculated for each trajectory as follows:

$$BIAS = \mathrm{T_{Mur}} - \mathrm{T_{in\text{-}situ}} \quad (1)$$

The potential temperature and potential density were calculated using the Gibbs Sea Water (GSW) oceanographic toolbox [57] that considers the Thermodynamic Equation of Seawater 2010 (TEOS-2010).

3. Results

3.1. SESs Trajectories

Figure 1 displays the trajectories of the five SESs analyzed in this work. SESs moved mainly towards the east from PV, reaching the open ocean in three to six days. The average number of dives per day on the continental shelf was 92.6 ± 10.2, and the average distance traveled per day was 110.1 ± 13.8 km (Table 3). The five SESs left PV and returned to molt to sites located very close to the place of departure in PV, between 21 December 2019, and 3 January 2020.

Table 3. Date of departure, date of arrival at isobath 200 m, mean number of dives per day within the ACS, number of days to cross the ACS, distance traveled (km), and mean distance traveled per day (km/day).

SESs ID	Date and Time of Departure	Date and Time of Arrival at the 200 m Isobath	Mean Number of Dives per Day within the ACS	Number of Days to Cross the ACS	Distance Traveled (km)	Mean Distance Traveled per Day (km)
054	25-October 11:29:54	01-November 03:08:00	92.8	6.7	667.2	100.3
878	17-October 23:32:56	23-October 02:41:46	85.4	5.1	586.8	114.4
875	18-October 01:09:10	23-October 19:31:48	83.1	5.8	540.4	93.7
905	20-October 15:44:22	23-October 22:19:10	109.0	3.4	381.1	116.4
051	21-October 06:48:48	24-October 01:52:00	92.7	2.8	359.4	128.6
Mean			92.6 ± 10.2	4.7 ± 1.6	507.0 ± 133.0	110.1 ± 13.8

Table 3 summarizes basic statistics for the trajectories of the five elephant seals. All the SESs traveled a similar distance per day. The SESs that left from the north of PV (SESs 054, 878, 875) took more days to arrive at the 200 m isobath compared to the SESs that went from the south of PV (SESs 905, 051) because they traveled a longer distance to reach the eastern limit of the ACS, the 200 m isobath (Figure 1). In the following section, we analyze the speed of SESs along their trajectory in more detail.

3.2. SESs Speeds

Figure 3 shows the amplitude of the M2 tidal component and the trajectory of the SESs. M2 is the largest tidal component in the region, leading other components by an order of magnitude almost everywhere [31]. The three SESs left from the northern coast of PV crossed a portion of the ACS where the M2 tidal amplitude values were much greater than those along the trajectories of the SESs that left PV from the south (Figure 3). The speed of the zonal component of the tidal currents (considering all the tidal components of the TPXO8 model) along the trajectory of each SES is compared with the zonal speed of the animals in Figure 4. Both time series present similar variability during the three days after departure for the three northern elephants (054, 878, and 875); this feature is not observed in the two southern elephants (905 and 051). We can therefore observe that SES speed is largely correlated to tidal currents when the latter are large (>150 cm/s). It is an open question whether the SESs are taking advantage of the tidal currents or are affected by them.

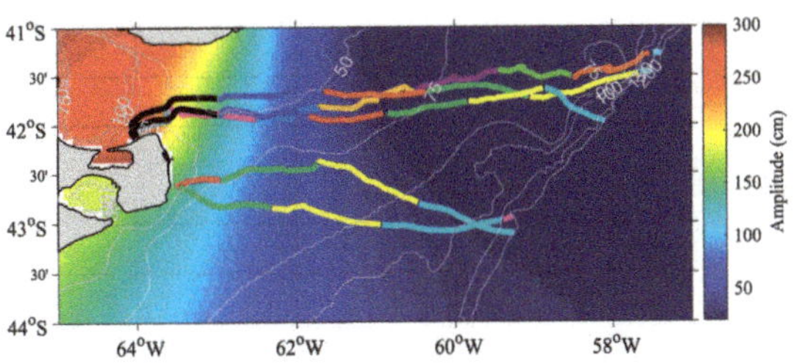

Figure 3. Amplitude (cm) for the M2 constituent as obtained with the TPX0 tide model. Thick lines represent the trajectories of the five SESs in October 2019. Colors along the trajectories correspond to the different days, as detailed in Figure 4. Thin gray lines represent the 50, 75, 90, 100, 150, and 200 m isobaths.

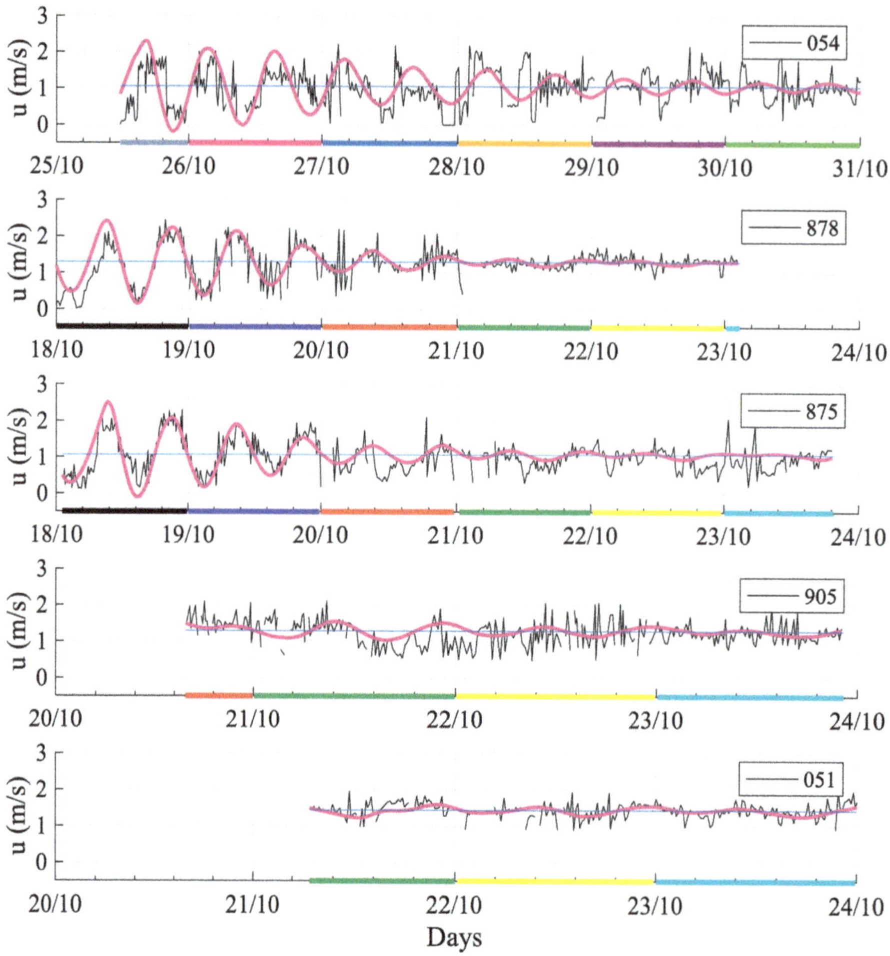

Figure 4. The zonal component of the SES swimming velocity (black) and the tidal current velocity derived from TPXO (magenta) along the five trajectories. The mean value along each trajectory of the in-situ velocity (blue line) was added to the tidal currents. A threshold of two standard deviations was used to eliminate the outliers from the in-situ velocity data. The colors on the x-axis match the colors along the trajectories in Figure 3.

3.3. Comparison between Near-Surface SES Temperatures and Satellite SST

Figure 5 shows the scatterplot and the spatial distribution of the difference between the satellite and in-situ temperature datasets. The correlation between the near-surface temperature (15 m depth) collected by the SESs and the time-space matching SST is 0.97, significant at the 95% confidence level (CL95%).

Despite the good correlation between in-situ and satellite SST, the two datasets show a moderate dispersion in temperature values lower than 8 °C along the linear regression (Figure 5a). The differences are reflected in the bias values (Equation (1)) that, in general, are less than 0.5 °C, except in some parts of the trajectories, mainly near the shelf-break. (Figure 5b). Note that the bias values in all trajectories are positive. This means that the in-situ temperatures are lower than the SST-MUR. At first glance, this is an expected result since in-situ data are collected at 15 dbar, while the satellite sensor represents the skin layer of the ocean. Yet the differences are very small in most of the places, suggesting that the top 15 m of the water column are very well mixed in this portion of the ocean.

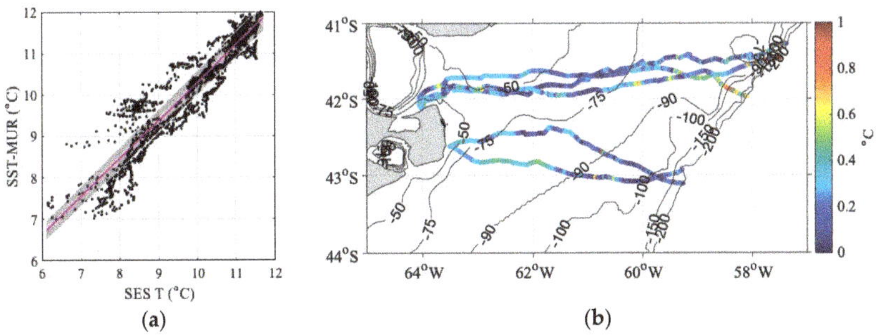

Figure 5. (**a**) A scatter plot of the in-situ temperature of SESs and SST-MUR was extracted along the trajectories (black dots). Linear regression and 95% confidence level intervals are indicated with a magenta line and a shaded gray region. (**b**) Temperature bias (SST MUR—T at 15 dbar) along trajectories. Black lines represent the 50, 75, 90, 100, 150, and 200 m isobaths.

3.4. Analysis of the Temperature and Salinity Spatial Distributions

The spatial distribution of the temperature along the trajectories displays a zonal gradient, observing warm (warmer than 11 °C) water in the inner shelf and cold (colder than 6 °C) water in the outer shelf (Figure 6, left column). This is in good agreement with similar in-situ observations presented by Campagna et al. [11]. The zonal temperature gradient is evident in the three pressure levels (Figure 6, left column). The cold water close to the shelfbreak is associated with MC waters. A meridional gradient in the inner and middle shelves is also evident, showing that, north of 42°S, the surface temperature is 1 °C warmer than in the south (Figure 6).

In the northern trajectories, a maximum salinity (34 < S < 34.2) is observed when the depth is less than 60 m (Figure 6, right column). This region also has the warmest water (~12 °C). Near the 75 m isobath, the salinity decreases by 1 unit (S = 33.3) and increases again until it reaches 33.9 on the outer shelf. The temperature and salinity observed north of 42°S and west of the 75 m isobath are likely due to the mixing with waters coming from the SMG. The 75 m isobath coincides with the limit between the colder and fresher water from the south that moves towards the outer portion of the continental shelf and the hotter and saltier water closer to the coast, north of 42°S. The influence of the SMG waters over the adjacent portion of the continental shelf is in good agreement with previous observations [27].

Along the southern trajectories (south of 42°S), the salinity increases from 33.2 near the coast to 33.9 at the shelf-break in the three depth layers displayed (Figure 6).

Figure 6 also shows different values at the different depths in the temperature record, particularly in the mid and outer shelf, suggesting that waters are more stratified in those regions than in the inner shelf. Considering the large tidal amplitude in the region (Figure 3), we computed the Simpson parameter [58] to quantify in which regions tides are responsible for mixing waters and preventing seasonal stratification. We estimated the Simpson parameter (ϕ) as in Kahl et al. [59] using the in-situ data collected by the SESs. The spatial distribution of ϕ displays low values (<10 J/m^3) in the inner shelf and larger values (10–50 J/m^3) in the middle and outer shelves (not shown). The most significant ϕ (critical value of 10 J/m^3) occurs along the 75 m isobath, suggesting that well-mixed waters are located between such isobath and the coast and stratified waters are in waters deeper than 75 m. This result is in very good agreement with our observations in the following sections. In this study region, Lucas et al. [27] and Kahl et al. [59] showed the spatial distribution of ϕ for spring/summer with a critical value of 40–50 J/m^3 that followed the 50–60 m isobath. The values obtained in this work are low because October is a transition period toward strong stratification due to the solar radiation annual cycle. A similar result was found between 47° and 55°S in October 2005 by Sabatini et al. [60].

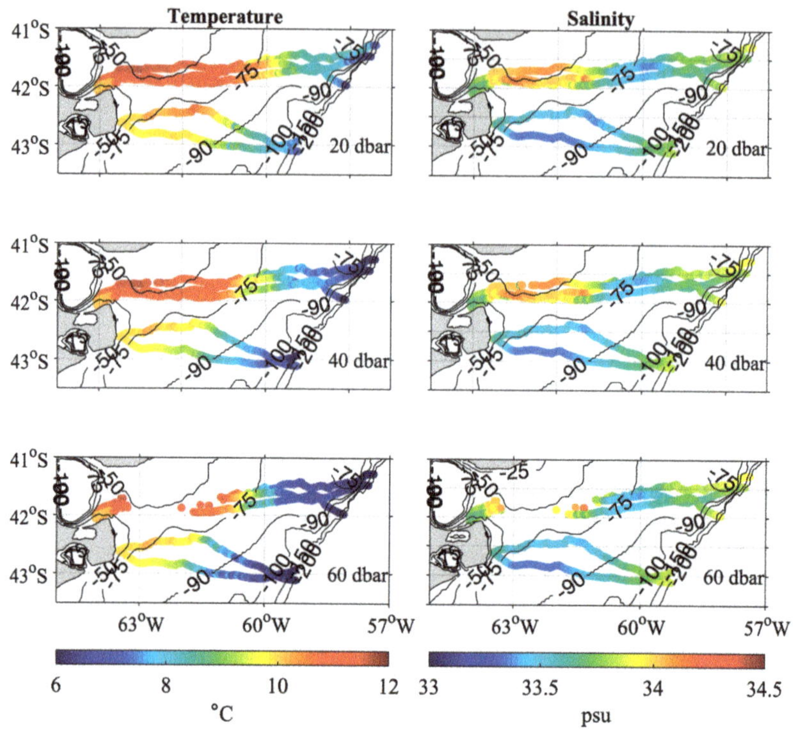

Figure 6. Temperature (°C) (**left** column) and Salinity (**right** column) were measured by sensors placed on 5 elephant seals during October 2019. The variables are presented at 3 pressure levels: 20 dbar (upper panel), 40 dbar (middle panel), and 60 dbar (bottom panel). Black lines represent the 50, 75, 90, 100, 150, and 200 m isobaths.

3.5. Water Mass Characterization

In Section 3.4, we showed that the data collected in the northern trajectories differs quite a bit from the data collected in the southern trajectories. In what follows, we analyze the water masses north and south of 42°S, selecting one representative trajectory for each region. We used trajectory 051 to represent the southern ones and 875 for the northern ones (Figure 1).

Both Northern and Southern SESs swam across three distinctive water masses that were identified in the TS diagrams: Low Salinity Coastal Waters (LSCW, S < 33.4), Mid-Shelf Waters (MSW, 33.4 < S < 33.8), and Malvinas Waters (MW, S > 33.8) (Figure 7b,c). The Northern elephant also crossed the High Salinity Coastal Waters (HSCW, S > 34) at the beginning of its trip (18, 19, and 20 of October), which contain waters from SMG (Figure 7b). SMG has higher temperatures and salinity values relative to the continental shelf because it is an evaporation basin, i.e., precipitation is lower than evaporation [28,61,62]. As described above, the waters adjacent to the SMG are affected by the waters from the gulf. The HSCW is not present south of 42°S.

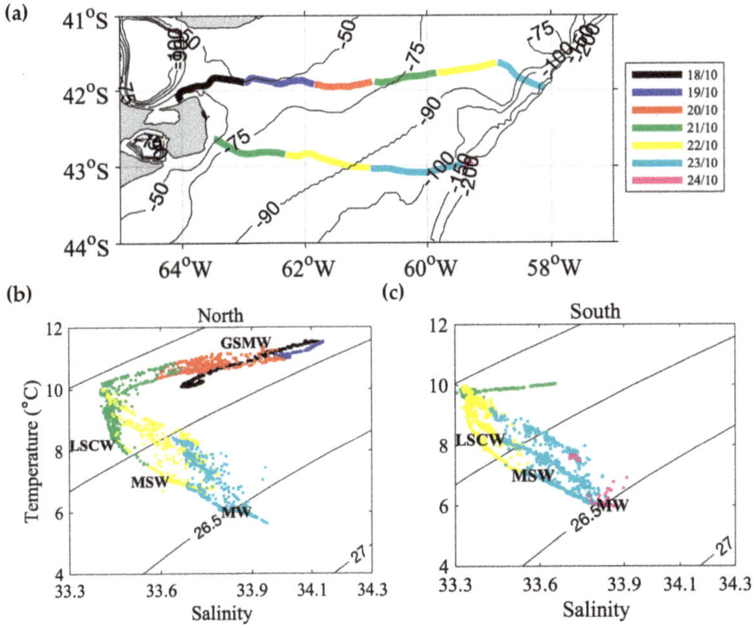

Figure 7. (**a**) Trajectories represent the northern (875) and southern (051) regions. The corresponding TS diagrams are presented in panels (**b**,**c**). The colors in panels (**a**–**c**) correspond to the segment of trajectory traveled per day by each elephant. The black lines represent isolines of σ in kg/m^3.

The LSCW and MSW were sampled by the southern and northern SESs on the 22 and 23 of October (Figure 7). These water masses originate in the coastal Pacific Ocean and are transported by the Patagonian current [63].

3.6. Vertical Profiles of Temperature, Salinity, and Density

3.6.1. Northern Trajectory

The northern elephant had 479 profiles, with an average distance between each profile of 1.1 km and an average profile depth of 67.3 m.

The vertical structure of the temperature is homogeneous from the surface to the bottom in the first 300 km from the coast (Figure 8, upper panel). During the first day after departure from the north of PV (19 October 2019), the potential temperature profile displays several increases and decreases in temperature (between 10 °C and 11 °C). From km 300 towards the shelf break, a vertical stratification begins, which is accentuated towards the slope with temperatures of 8 °C on the surface and 6 °C on the bottom (Figure 8 upper panel).

The vertical salinity structure along the trajectory also presents an oscillating pattern during the first day after departure, with salinity values between 33.6 and 34. Between 300 and 400 km from the coast, on the middle continental shelf, the salinity shows minimal values of 33.4 from surface to bottom. Then, the salinity increases towards the east, reaching values of 33.6–33.9 (Figure 8 middle panel).

The vertical density structure depends on temperature and salinity; homogeneous values are observed from surface to bottom and from the coast up to about 320 km (Figure 8 bottom panel). The spatial density distribution also shows the strip pattern identified in temperature and salinity during 19th of October. In the middle continental shelf (~350 km), a density minimum modulated mainly by salinity is observed. From 400 km to the edge of the shelf, the density is dominated by T and S. In this section, a weak vertical stratification begins, which increases towards the slope. An increase in salinity (33.6 < S < 33.9) and a decrease in temperature (<8 °C) in the bottom are observed in this portion of the section in

Figure 8 (upper and middle panels), clearly suggesting intrusions of the saltier and colder MC water. Such intrusions have been previously reported at 41°S by Piola et al. [19].

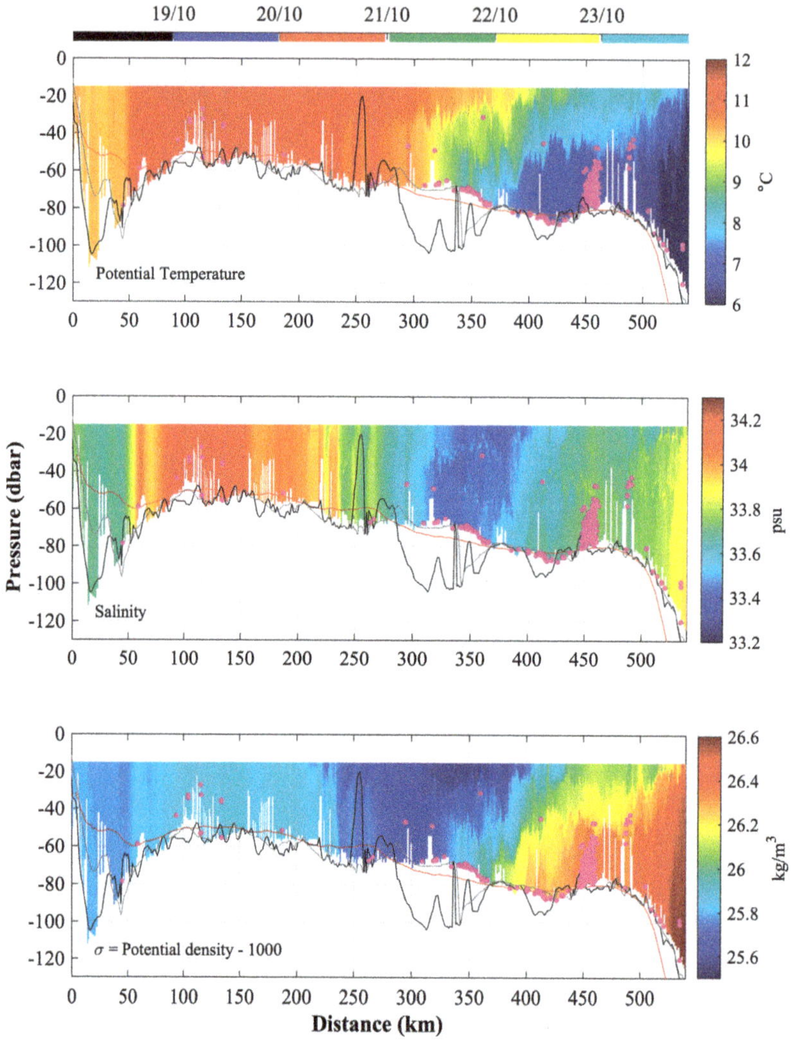

Figure 8. Temperature (**top**), salinity (**middle**), and density (**bottom**) profiles for SES 875 after departing north of PV. The horizontal axis is the distance in km from PV. The red, gray, and black lines show the seafloor using the GEBCO-SHN 2019, GEBCO 2020, and GEBCO 2021 bathymetries, respectively. The top colorbar indicates the day and month of 2019; the colors are the same as in Figure 7. Magenta dots are Prey Catch Attempts.

PCA values are indicated with magenta dots in the three panels of Figure 8. A few values are observed at the sea bottom along the section. A large concentration of PCAs is observed in a particular region on the outer shelf, between 450 and 480 km from the coast. A careful inspection of the T and S values shows an increase in salinity in this region, which coincides with a decrease in temperature of 0.5 °C and an increase in density (from 26.3 to 26.4 kg/m^3). As discussed above, this water mass can only originate from an intrusion of the MC. It is very interesting to note the correspondence between the large number of

PCAs and the T and S values that most closely resemble the MC waters. The fact that many PCAs are observed in a particular region where the MC rested isolated suggests that the MC waters bring food the SESs like.

The analysis of Figure 8 in conjunction with the analysis of the spatial distribution of T and S (Section 3.4) and the TS diagrams (Section 3.5) allows us to identify that the warm and salty region (between km 80 and km 250, Figure 8) corresponds to the HSCW from SMG. The strip pattern observed in T and S the first day after its departure indicates that the SES swam across different water masses at the beginning of its trip. To further understand these features, we used satellite SST images that showed excellent agreement with the in-situ data (Section 3.3). On the 18th of October 2019, SST agreed to an alternating pattern of increasing and decreasing values (Figure 9a) that agreed with the pattern observed in Figure 8. The spatial pattern of SST indicates that the relative cold water is advected from the south into the SMG and that the relative warm water corresponds to HSCW. Moreover, the monthly average of 17 years of SST gradient reveals that in October there is a strong and persistent front (0.06 °C/km) at 41.9°S and between 63.3° and 63.6°W (Figure 9b). The front separates warm and salty water from SMG and cold and fresh water from the southern portion of the adjacent shelf. The Northern SES crossed this frontal area, explaining the abrupt temperature changes observed in Figure 8. Figure 9 also indicates that the other two northern SESs crossed the frontal area. The frontal area north of PV has been reported in Pisoni [64]. Using SST satellite images, they show that it is a seasonal front that exists during winter and the beginning of spring. The TS values recorded by the SESs in conjunction with SST images show that the front separates fresh and cold waters that enter the SMG in the narrow region between the coast and the front from warm and salty water created in the SMG, north of the front. The water entering the SMG comes from the coastal region east of PV, where a strong upwelling took place [26].

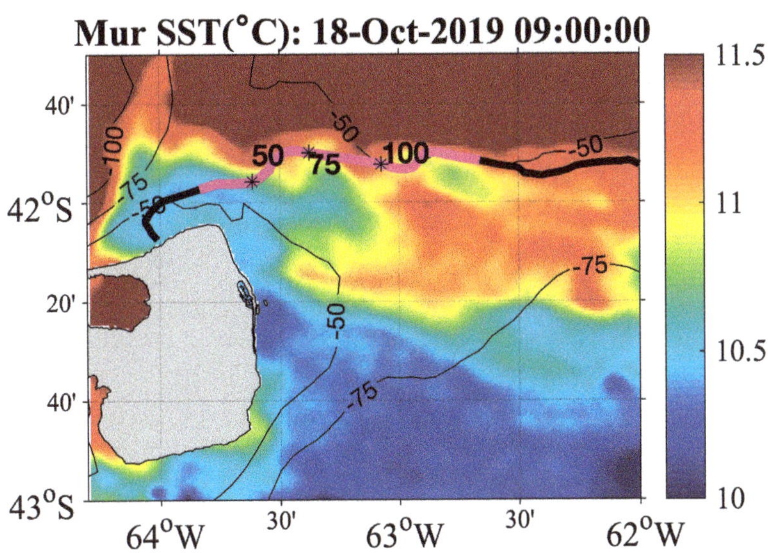

(a)

Figure 9. *Cont.*

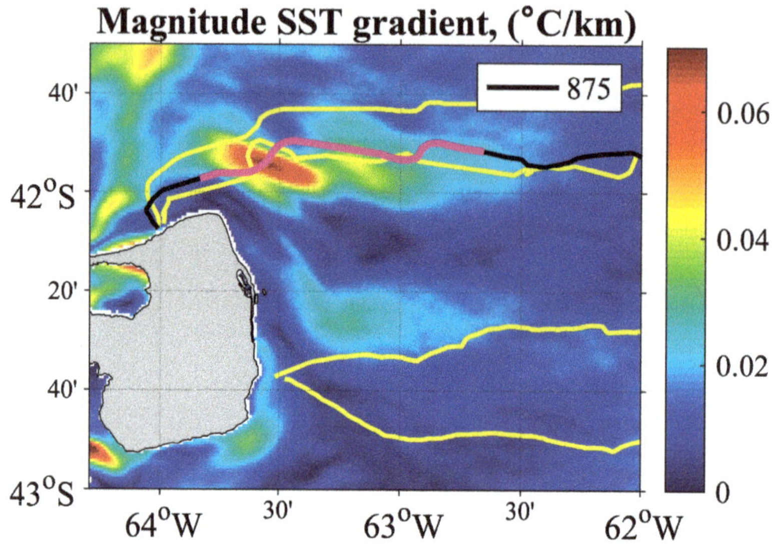

(b)

Figure 9. (**a**) Sea surface temperature (SST) is derived from the Multi-Scale Ultra-High Resolution L4 (MUR) satellite product for 18 October 2019. The stars are kilometers from the coast (Figure 8). Black lines represent the 50, 75, and 100 m isobaths. (**b**) Average magnitude of Sea surface temperature (SST) gradient (°C/km) derived from the Multi-scale Ultra-High Resolution L4 (MUR) satellite product for all months of October from 2002 to 2019. The yellow lines represent the trajectories of SESs. The trajectory of the SES 875 is marked with the black-magenta line on both panels. The magenta color corresponds to the segment of trajectory traveled on October 18 by the SES.

3.6.2. Southern Trajectory

Potential temperature, salinity, and potential density for the Southern elephant seal are presented in Figure 10. The southern elephant conducted 259 profiles, with an average distance between each profile of 1.4 km and an average profile depth of 82.8 m. The temperature shows homogeneous values in the vertical of about 10 °C in the first 50 km of the trajectory. Warmer values in the upper 30 m of the water column are observed from km 50 to the shelf-break, suggesting the presence of seasonal stratification (Figure 10 upper panel). Cold waters (less than 8 °C) are observed close to the sea floor from km 180 towards the edge of the shelf-break (Figure 10), as observed in the northern trajectory.

Salinity along the southern trajectory (Figure 10 middle panel) is vertically homogeneous in the first 180 km, with relative salty values close to the coast (33.6) that decrease up to 33.3 in the east. The lowest salinity values (33.3) occupy a large portion of the section, from 50 to 180 km from the shore. These values are characteristic of the diluted water coming from the Chilean coast through the Magellan Strait. From km 180 to the shelf-break, the salinity increases to 33.8.

The density section (Figure 10, bottom panel) shows homogeneous values in the vertical range of 25.7 kg/m^3 in the first 50 km of the trajectory. The density pattern is similar to the salinity pattern between 0 and 150 km from the shore, suggesting that density is dominated by S. Between 150 km and the shelf break, a comparison of the three panels of Figure 10 indicates that the density is dominated by the contributions of both T and S. The cold and salty waters (T < 7 °C and S > 33.8) occupy most of the mid and outer shelf in the deepest portion of the water column between 180 km and the shelf break. These values are due to the mixing between MW and MSW.

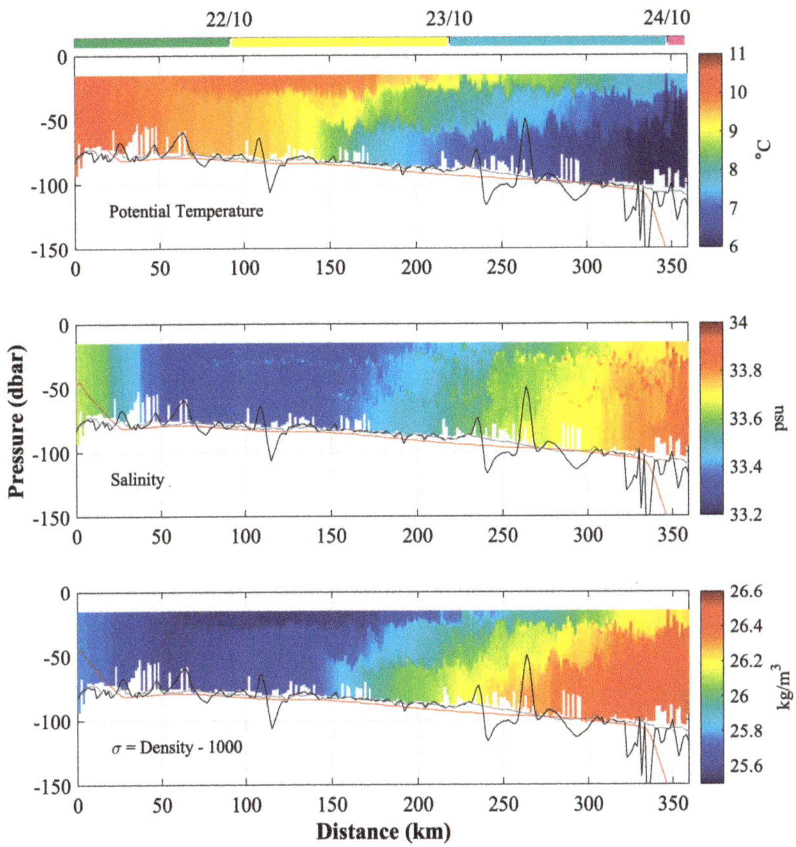

Figure 10. Temperature (**top**), salinity (**middle**), and density (**bottom**) profiles along the southern trajectory 051. The horizontal axis is the distance in km from PV. The red, gray, and black lines show the seafloor using the GEBCO-SHN 2019, GEBCO 2020, and GEBCO 2021 bathymetries, respectively. The top colorbar indicates the day and month of 2019; the colors are the same as in Figure 7.

A particular feature is observed in three consecutive temperature profiles of the Southern elephant, between 15 and 35 dbar at ~225 km from the coast (Figure 10 top panel). To better study this event, we plotted the potential temperature and salinity at 15 dbar for all the SESs. The cooling event detected by SES 051 was also detected in SES 905, with a time difference of 6 h (Figure 11). The potential temperature drop of 1 °C in both SESs was accompanied by an increase in salinity of ~0.2 psu (Figure 11).

Similar cooling events have been observed in the ACS and have been attributed to strong winds and/or rapid changes in the wind stress direction, generally due to the passage of a low-pressure system [49]. Such low-pressure systems intensify the sea surface wind and temporarily disrupt vertical stratification. The sea level pressure analysis provided by ERA5 (ECMWF Reanalysis v5) shows that a low-pressure system over the shelf coincided in time and space with the anomalies along the SESs 905 and 051 discussed here (Figure 11). Yet, wind estimation from ERA5 does not show an abrupt change in the wind stress intensity at the time when the anomalies were observed on the 22 and 23 of October 2019 (Figure 11). Figure 11b also shows an excellent correlation (r = 0.81, significant at the 95% confidence level (CL95%)) between the DTAG-4 wind estimates and ERA5, further supporting the capacity of the DTAG-4 logger to estimate wind and providing confidence in the ERA5 wind data [65].

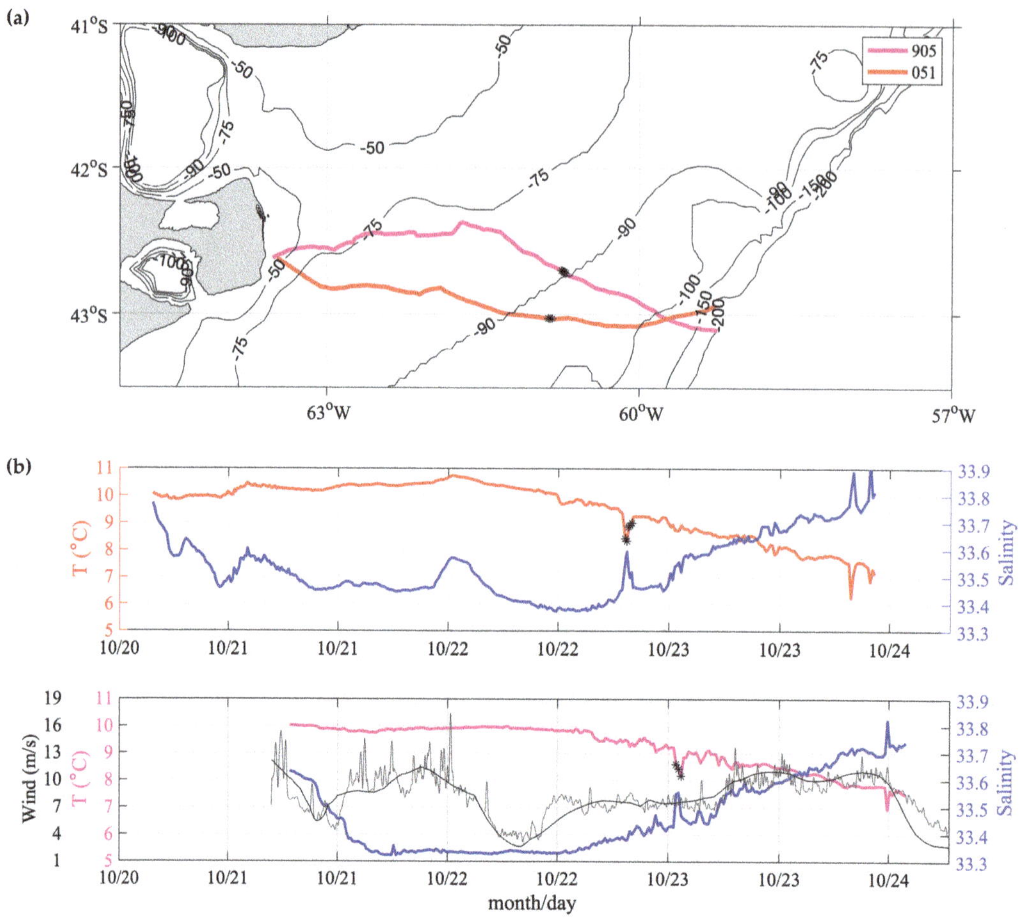

Figure 11. (**a**) Trajectories of SESs 905 (magenta) and 051 (red). (**b**) In-situ temperature (magenta—red), salinity (blue), and wind speed from ERA5 (grey) and DTAG-4 (black) along the two trajectories. DTAG-4 values were low-pass filtered with a 20 min cut-off period using a Loess filter. The black dots in panel (**a**) and the black stars in panel (**b**) indicate the cooling event in both trajectories.

We therefore conclude that local wind was not responsible for disrupting the stratification of the water column observed on the 22 and 23 of October 2019. The disruption of the stratification could also be due to the passage of internal waves, a possibility that we will investigate in future work.

3.7. Comparison between Depths Reached by SESs and GEBCO Bathymetric Charts

The maximum depth reached by the SESs is higher than reported in bathymetric charts from GEBCO (2019 + SHN, 2020, and 2021) along several portions of the trajectories (Figures 8 and 10). Thus, the depths reached by the SESs may be a useful variable to calibrate/validate the bathymetric charts. Here, a quantitative comparison is presented between the maximum depth reached by SESs and the corresponding value provided by bathymetric charts (Figure 10). Note that we neglected the cases where the depth reached by the SESs was lower than the GEBCO products since we cannot guarantee that the SESs reached the sea floor. Most of the discrepancies are observed near the coast for the three datasets, especially in the northern trajectories, reaching values larger than 10 m (Figure 12).

The lowest bias values (< 9 m) are observed on the middle shelf for GEBCO 2019 + SHN and GEBCO 2021. GEBCO 2019 + SHN underestimates the depth on the outer shelf with values larger than 12 m. Figure 12 also revealed that GEBCO 2020 undervalues the depth throughout the trajectories. The histogram of the bathymetry differences higher than 5 m (Figure 13) confirms that GEBCO 2021 is the bathymetry chart that presented the lowest number of cases where the depth reached by SESs is higher, suggesting that it is the best option available to represent the bathymetry in the region.

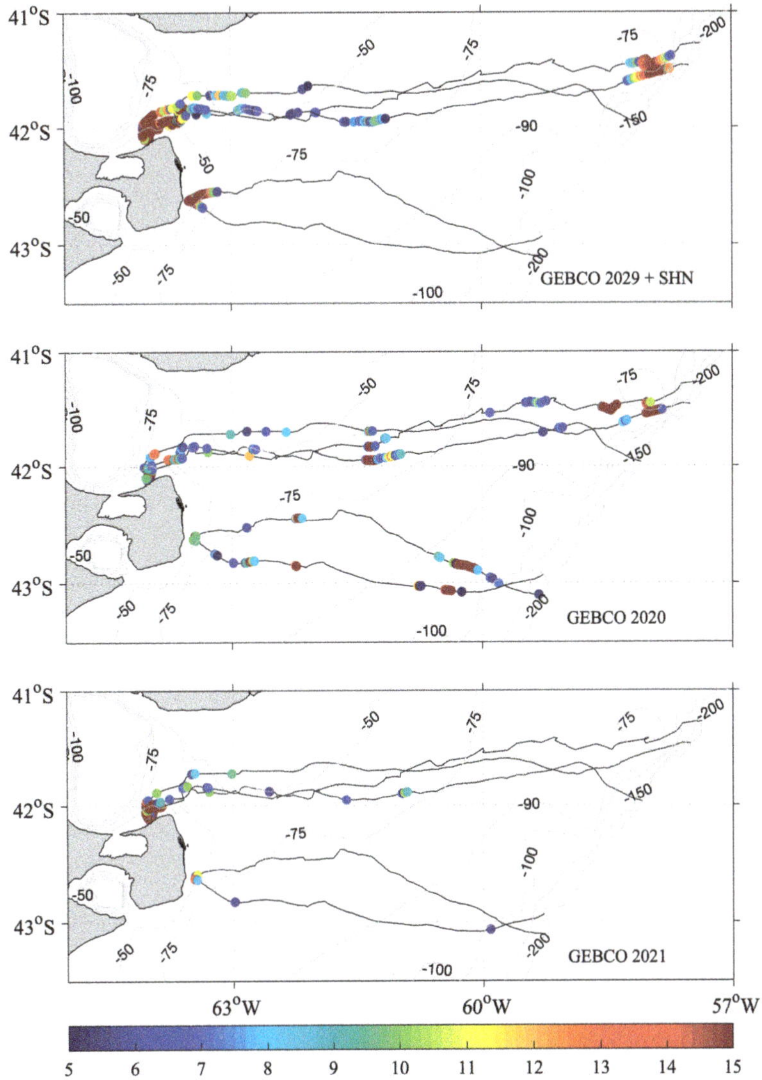

Figure 12. Bathymetry differences higher than 5 m between the depth reached by SESs and the depth reported by GEBCO 2019 + SHN (**upper** panel), GEBCO 2020 (**middle** panel), and GEBCO 2021 (**bottom** panel). The black lines represent the five SES trajectories, and the gray contours represent the 25, 50, 75, 100, 150, and 200 m isobaths.

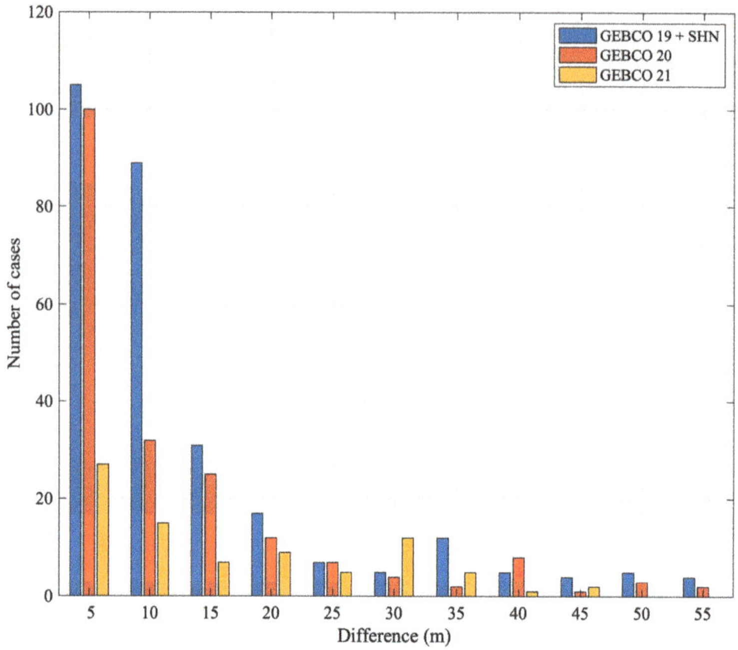

Figure 13. Histogram of the bathymetry differences higher than 5 m between depths reached by SESs and bathymetric charts: GEBCO 2019 + SHN (dark blue), GEBCO 2020 (red), and GEBCO 2021 (dark yellow).

4. Summary and Discussion

In this study, we analyzed high-spatial and temporal-resolution hydrographic data collected by five SESs that crossed the ACS between 44°S and 41°S in October 2019.

The spatial distribution of temperature and salinity revealed two regions, north and south of 42°S. The northern region is 1 °C warmer and 0.5 psu saltier than the southern region due to the fact that in the latter, the diluted water coming from the south mixes with the High-Salinity Coastal Water coming from SMG.

The spatial patterns of T and S at different pressure levels also indicated that the 75 m isobath delimited the vertically homogeneous region from the stratified one. Indeed, the 75 m isobath corresponds with the position of the largest increment of the Simpson parameter computed along the SESs trajectories. The Simpson parameter at the 75 m isobath is 10 J/m^3, a rather low value compared with values commonly found in the region to separate coastal mixed waters from open ocean stratified waters [27,59,66]. This is probably due to the fact that our estimation of the Simpson parameter has been conducted in austral spring (mid-October), a period of the year when the vertical stratification over the continental shelf is much lower than in austral summer, when the other estimations have been conducted [24,66].

SST from MUR correlates very well (r = 0.97) with the T measured at 15 m depth by SESs. Both datasets show the presence of an advective front in the northern region of PV, between 63.3°W and 63.6°W, at 41.9°S. Pisoni [64] described such a front using only satellite SST data and discussed that an intense coastal jet carries cold and fresh waters from the western coast of PV, causing the formation of the so-called advective front. T and S data provided by the three SESs that sampled the waters on both sides of the front allowed showing that there is a gradient of salinity also in the region.

The SESs crossed the ACS in a few days with a mean speed of 1–1.5 m/s, similar to the values documented by Campagna et al. [11] and one order of magnitude higher than the geostrophic velocities in the area [49]. We show for the first time that SESs speed might be modulated by tidal currents: the speed of the three northern SESs showed a semi-diurnal cycle that correlates with the macro-tidal regime observed close to the coast, north of 42°S. South of 42°S, this oscillation is not observed since the tidal velocities are not as large as in the northern region.

Temperature and salinity along the trajectories might present large variations (Figure 11). Local wind stress estimated from an acoustic sensor mounted on some of the SESs and as computed by ERA5 reanalysis data does not show large values related to T and S abrupt changes observed. We argue that the passage of internal waves originating in a remote place could be the force that explains the anomalies observed.

PCA values suggest that SESs feed mostly at the bottom of the sea when crossing the ACS. Only SES 875 fed intensively in a particular region located on the outer shelf, between 50 m and the sea bottom, located at 80 m. The T and S analyses showed that such a region was occupied by an intrusion of MC waters. We suggest that MC carried prey that SES 875 liked.

Lastly, the records of the maximum depth reached by the SESs provide valuable information to calibrate/validate bathymetric charts. Our results pointed out the regions along the trajectories where the bathymetry from three GEBCO products underestimated the depth of the sea. A reliable topography is crucial for ocean numerical models to reproduce a realistic circulation [67].

The SES colony of PV has been demonstrated to be an excellent platform for measuring oceanographic variables at high spatial and temporal resolution over the ACS, providing an average of 92.6 dives per day and one profile every 1.1km. This unique dataset opens the opportunity to explore physical, small-scale processes that are not yet well understood but can have a significant impact on the air-sea fluxes of CO_2 and heat advection. The continuity of high-frequency, high-spatial-resolution hydrographic data are crucial for studying these processes and their variability. Hence, maintaining the deployment and collection of data collected through SESs could enhance the comprehension of the climate change process in the region. Variations in temperatures, currents, and salinity can deeply impact marine ecosystems, biodiversity, and biological productivity [33,68]. Finally, this work highlights the advantages of the utilization of mammal-based sensors in comparison to gliders and Argo floats. Many of the regions sampled by the SESs in the ACS are shallow (less than 50 m) and are characterized by strong tidal currents (up to 2.5 m/s), largely limiting the operability of gliders and Argo floats. The mammal-based sensors, through their unique capabilities, allow more effective data collection under such conditions.

Author Contributions: Conceptualization, L.A.R.-E. and M.S.; Methodology, M.M.M., L.A.R.-E. and M.S.; Software, M.M.M.; Validation, M.M.M., A.G.-M., B.P. and C.G.; Formal analysis, M.M.M.; Investigation, M.M.M., L.A.R.-E. and M.S.; Resources, M.S. and C.G.; Data curation, B.P.; Writing—original draft, M.M.M.; Writing—review & editing, M.M.M., L.A.R.-E., M.S., A.G.-M., J.C., B.P. and C.G.; Visualization, M.M.M. and L.A.R.-E.; Supervision, M.S.; Project administration, M.S. All authors have read and agreed to the published version of this manuscript.

Funding: This research was funded by European Organization for the Exploitation of Meteorological Satellites (EUMETSAT)/ Centre National de la Recherche Spatial (CNES) through Project Studying physical processes in the Southwestern Atlantic to understand BIOlogical productivity & regional Ecosystems (SABIO) and Wildlife Conservation Society (WCS).

Data Availability Statement: The marine mammal data were collected and made freely available by the International MEOP Consortium and the national programs that contribute to it. (http://www.meop.net, accessed on 3 May 2021).

Acknowledgments: We are grateful to all the field personnel that assisted in the deployment and recoveries of data loggers: Claudio Campagna, Loic Le Ster, Mark Hindell and Ricardo Vera. We also thank the institutions that provided additional support: Mincyt Pampa Azul Project B6 and PICT 2020-SERIEA-01468. The manuscript was improved by comments from the four anonymous reviewers.

Conflicts of Interest: The authors declare no conflict of interest.

References

1. Harcourt, R.; Sequeira, A.M.M.; Zhang, X.; Roquet, F.; Komatsu, K.; Heupel, M.; McMahon, C.; Whoriskey, F.; Meekan, M.; Carroll, G.; et al. Animal-Borne Telemetry: An Integral Component of the Ocean Observing Toolkit. *Front. Mar. Sci.* **2019**, *6*, 326. [CrossRef]
2. Chapman, C.C.; Lea, M.-A.; Meyer, A.; Sallée, J.-B.; Hindell, M. Defining Southern Ocean fronts and their influence on biological and physical processes in a changing climate. *Nat. Clim. Chang.* **2020**, *10*, 209–219. [CrossRef]
3. Treasure, A.M.; Roquet, F.; Ansorge, I.J.; Bester, M.N.; Boehme, L.; Bornemann, H.; Charrassin, J.-B.; Chevallier, D.; Costa, D.P.; Fedak, M.A.; et al. Marine Mammals Exploring the Oceans Pole to Pole: A Review of the MEOP Consortium. *Oceanography* **2017**, *30*, 132–138. [CrossRef]
4. McIntyre, T.; de Bruyn, P.J.N.; Ansorge, I.J.; Bester, M.N.; Bornemann, H.; Plötz, J.; Tosh, C.A. A lifetime at depth: Vertical distribution of southern elephant seals in the water column. *Polar Biol.* **2010**, *33*, 1037–1048. [CrossRef]
5. Boehme, L.; Lovell, P.; Biuw, M.; Roquet, F.; Nicholson, J.; Thorpe, S.E.; Meredith, M.P.; Fedak, M. Technical Note: Animal-borne CTD-Satellite Relay Data Loggers for real-time oceanographic data collection. *Ocean Sci.* **2009**, *5*, 685–695. Available online: www.ocean-sci.net/5/685/2009/ (accessed on 30 June 2021). [CrossRef]
6. Roquet, F.; Park, Y.-H.; Guinet, C.; Bailleul, F.; Charrassin, J.-B. Observations of the Fawn Trough Current over the Kerguelen Plateau from instrumented elephant seals. *J. Mar. Syst.* **2009**, *78*, 377–393. [CrossRef]
7. Hindell, M.A.; McMahon, C.R.; Jonsen, I.; Harcourt, R.; Arce, F.; Guinet, C. Inter- and intrasex habitat partitioning in the highly dimorphic southern elephant seal. *Ecol. Evol.* **2021**, *11*, 1620–1633. [CrossRef]
8. Bailleul, F.; Charrassin, J.-B.; Monestiez, P.; Roquet, F.; Biuw, M.; Guinet, C. Successful foraging zones of southern elephant seals from the Kerguelen Islands in relation to oceanographic conditions. *Philos. Trans. R. Soc. B Biol. Sci.* **2007**, *362*, 2169–2181. [CrossRef]
9. Dragon, A.-C.; Monestiez, P.; Bar-Hen, A.; Guinet, C. Linking foraging behaviour to physical oceanographic structures: Southern elephant seals and mesoscale eddies east of Kerguelen Islands. *Prog. Oceanogr.* **2010**, *87*, 61–71. [CrossRef]
10. Hindell, M.A.; McMahon, C.R.; Bester, M.N.; Boehme, L.; Costa, D.; Fedak, M.A.; Guinet, C.; Herraiz-Borreguero, L.; Harcourt, R.G.; Huckstadt, L.; et al. Circumpolar habitat use in the southern elephant seal: Implications for foraging success and population trajectories. *Ecosphere* **2016**, *7*, e01213. [CrossRef]
11. Campagna, C.; Piola, A.R.; Marin, M.R.; Lewis, M.; Fernández, T. Southern elephant seal trajectories, fronts and eddies in the Brazil/Malvinas Confluence. *Deep Sea Res. Part I Oceanogr. Res. Pap.* **2006**, *53*, 1907–1924. [CrossRef]
12. Campagna, J.; Lewis, M.N.; González Carman, V.; Campagna, C.; Guinet, C.; Johnson, M.; Randall WDavis Rodríguez, D.H.; Hindell, M.A. Ontogenetic niche partitioning in southern elephant seals from Argentine Patagonia. *Mar. Mammal Sci.* **2020**, *37*, 631–651. [CrossRef]
13. Aubone, N.; Saraceno, M.; Alberto, M.T.; Campagna, J.; Le Ster, L.; Picard, B.; Hindell, M.; Campagna, C.; Guinet, C. Physical changes recorded by a deep diving seal on the Patagonian slope drive large ecological changes. *J. Mar. Syst.* **2021**, *223*, 103612. [CrossRef]
14. Campagna, C.; LE Boeuf, B.J.; Blackwell, S.B.; Crocker, D.E.; Quintana, F. Diving behaviour and foraging location of female southern elephant seals from Patagonia. *J. Zool.* **1995**, *236*, 55–71. [CrossRef]
15. McGovern, K.; Rodríguez, D.; Lewis, M.; Eder, E.; Piola, A.; Davis, R. Habitat associations of post-breeding female southern elephant seals (*Mirounga leonina*) from Península Valdés, Argentina. *Deep Sea Res. Part I Oceanogr. Res. Pap.* **2022**, *185*, 103789. [CrossRef]
16. Campagna, C.; Lewis, M. Growth and Distribution of A Southern Elephant Seal Colony. *Mar. Mammal Sci.* **1992**, *8*, 387–396. [CrossRef]
17. Laws, R.M. History and present status of southern elephant seal populations. In *Elephant Seals: Population ecology, Behavior and Physiology*; Le Boeuf, B.J., Laws, R.M., Eds.; University of California Press: Berkeley, CA, USA, 1994; pp. 49–65.
18. Campagna, C.; Piola, A.R.; Marin, M.R.; Lewis, M.; Zajaczkovski, U.; Fernández, T. Deep divers in shallow seas: Southern elephant seals on the Patagonian shelf. *Deep Sea Res. Part I Oceanogr. Res. Pap.* **2007**, *54*, 1792–1814. [CrossRef]
19. Piola, A.R.; Avellaneda, N.M.; Guerrero, R.A.; Jardón, F.P.; Palma, E.D.; Romero, S.I. Malvinas-slope water intrusions on the northern Patagonia continental shelf. *Ocean Sci.* **2010**, *6*, 345–359. [CrossRef]
20. Glorioso, P.D. Temperature distribution related to shelf-sea fronts on the Patagonian Shelf. *Cont. Shelf Res.* **1987**, *7*, 27–34. [CrossRef]
21. Sabatini, M.; Martos, P. SCIENTIA MARINA Mesozooplankton features in a frontal area off northern Patagonia (Argentina) during spring 1995 and 1998. *Sci. Mar.* **1995**, *66*, 215–232. [CrossRef]
22. Romero, S.I.; Piola, A.R.; Charo, M.; Garcia, C.A.E. Chlorophyll-a variability off Patagonia based on SeaWiFS data. *J. Geophys. Res. Ocean* **2006**, *111*. [CrossRef]
23. Rivas, A.L.; Pisoni, J.P. Identification, characteristics and seasonal evolution of surface thermal fronts in the Argentinean Continental Shelf. *J. Mar. Syst.* **2010**, *79*, 134–143. [CrossRef]

24. Bogazzi, E.; Baldoni, A.; Rivas, A.; Martos, P.; Reta, R.; Orensanz, J.M.; Lasta, M.; Dell'Arciprete, P.; Werner, F. Spatial correspondence between areas of concentration of Patagonian scallop (*Zygochlamys patagonica*) and frontal systems in the southwestern Atlantic. *Fish. Oceanogr.* **2005**, *14*, 359–376. [CrossRef]
25. Franco, B.C.; Palma, E.D.; Tonini, M.H. Benthic-pelagic uncoupling between the Northern Patagonian Frontal System and Patagonian scallop beds. *Estuar. Coast. Shelf Sci.* **2015**, *153*, 145–155. [CrossRef]
26. Tonini, M.H.; Palma, E.D.; Pisoni, J.P. Modeling the seasonal circulation and connectivity in the North Patagonian Gulfs, Argentina. *Estuar. Coast. Shelf Sci.* **2022**, *271*, 107868. [CrossRef]
27. Lucas, A.J.; Guerrero, R.A.; Mianzán, H.W.; Acha, E.M.; Lasta, C.A. Coastal oceanographic regimes of the Northern Argentine Continental Shelf (34–43°S). *Estuar. Coast. Shelf Sci.* **2005**, *65*, 405–420. [CrossRef]
28. Piola, A.R.; Scasso, L.M. Circulación en el golfo San Matías. *Geoacta* **1988**, *15*, 33–51.
29. Bianchi, A.A.; Bianucci, L.; Piola, A.R.; Pino, D.R.; Schloss, I.; Poisson, A.; Balestrini, C.F. Vertical stratification and air-sea CO_2 fluxes in the Patagonian shelf. *J. Geophys. Res. Ocean* **2005**, *110*, 1–10. [CrossRef]
30. Lusquiños, A.J.; Valdés, A.J. *Aportes al Conocimiento de las Masas de Agua del Atlántico Sudoccidental*; Rep. H659; Serv. de Hidrografía Nav.: Buenos Aires, Argentina, 1971; p. 48.
31. Saraceno, M.; Martín, J.; Moreira, D.; Pisoni, J.P.; Tonini, M.H. Physical Changes in the Patagonian Shelf. In *Global Change in Atlantic Coastal Patagonian Ecosystems: A Journey Through Time*; Helbling, E.W., Narvarte, M.A., González, R.A., Villafañe, V.E., Eds.; Springer International Publishing: Berlin/Heidelberg, Germany, 2022; pp. 43–71. [CrossRef]
32. Photopoulou, T.; Fedak, M.A.; Matthiopoulos, J.; McConnell, B.; Lovell, P. The generalized data management and collection protocol for Conductivity-Temperature-Depth Satellite Relay Data Loggers. *Anim. Biotelemetry* **2015**, *3*, 21. [CrossRef]
33. Roquet, F.; Boehme, L.; Block, B.; Charrasin, J.B.; Costa, D.; Guinet, C.; Harcourt, R.G.; Hindell, M.A.; Hückstädt, L.A.; McMahon, C.R.; et al. Ocean Observations Using Tagged Animals. *Oceanography* **2017**, *30*, 139. [CrossRef]
34. Goulet, P.; Guinet, C.; Swift, R.; Madsen, P.T.; Johnson, M. A miniature biomimetic sonar and movement tag to study the biotic environment and predator-prey interactions in aquatic animals. *Deep Sea Res. Part I Oceanogr. Res. Pap.* **2019**, *148*, 1–11. [CrossRef]
35. Guinet, C.; Vacquié-Garcia, J.; Picard, B.; Bessigneul, G.; Lebras, Y.; Dragon, A.; Viviant, M.; Arnould, J.; Bailleul, F. Southern elephant seal foraging success in relation to temperature and light conditions: Insight into prey distribution. *Mar. Ecol. Prog. Ser.* **2014**, *499*, 285–301. [CrossRef]
36. Vacquié-Garcia, J.; Guinet, C.; Laurent, C.; Bailleul, F. Delineation of the southern elephant seal's main foraging environments defined by temperature and light conditions. *Deep Sea Res. Part II Top. Stud. Oceanogr.* **2015**, *113*, 145–153. [CrossRef]
37. Merchant, N.D.; Fristrup, K.M.; Johnson, M.P.; Tyack, P.L.; Witt, M.J.; Blondel, P.; Parks, S.E. Measuring acoustic habitats. *Methods Ecol. Evol.* **2015**, *6*, 257–265. [CrossRef] [PubMed]
38. Lopez, R.; Malardé, J.-P.; Danès, P.; Gaspar, P. Improving Argos Doppler location using multiple-model smoothing. *Anim. Biotelemetry* **2015**, *3*, 32. [CrossRef]
39. Dragon, A.-C.; Bar-Hen, A.; Monestiez, P.; Guinet, C. Horizontal and vertical movements as predictors of foraging success in a marine predator. *Mar. Ecol. Prog. Ser.* **2012**, *447*, 243–257. [CrossRef]
40. Lowther, A.D.; Lydersen, C.; Fedak, M.A.; Lovell, P.; Kovacs, K.M. The Argos-CLS Kalman Filter: Error Structures and State-Space Modelling Relative to Fastloc GPS Data. *PLoS ONE* **2015**, *10*, e0124754. [CrossRef] [PubMed]
41. Ryan, P.G.; Petersen, S.L.; Peters, G. GPS tracking a marine predator: The effects of precision, resolution and sampling rate on foraging tracks of African Penguins. *Mar. Biol.* **2004**, *145*, 215–223. [CrossRef]
42. Le Bras, Y.; Jouma'a, J.; Picard, B.; Guinet, C. How Elephant Seals (*Mirounga leonina*) Adjust Their Fine Scale Horizontal Movement and Diving Behaviour in Relation to Prey Encounter Rate. *PLoS ONE* **2016**, *11*, e0167226. [CrossRef]
43. Siegelman, L.; Roquet, F.; Mensah, V.; Rivière, P.; Pauthenet, E.; Picard, B.; Guinet, C. Correction and Accuracy of High- and Low-Resolution CTD Data from Animal-Borne Instruments. *J. Atmos. Ocean. Technol.* **2019**, *36*, 745–760. [CrossRef]
44. Campagna, C.; Fedak, M.A.; McConnell, B.J. Post-Breeding Distribution and Diving Behavior of Adult Male Southern Elephant Seals from Patagonia. *J. Mammal.* **1999**, *80*, 1341–1352. Available online: https://academic.oup.com/jmammal/article/80/4/1341/851974 (accessed on 20 April 2022). [CrossRef]
45. Roquet, F.; Charrassin, J.-B.; Marchand, S.; Boehme, L.; Fedak, M.; Reverdin, G.; Guinet, C. Delayed-Mode Calibration of Hydrographic Data Obtained from Animal-Borne Satellite Relay Data Loggers. *J. Atmos. Ocean. Technol.* **2011**, *28*, 787–801. [CrossRef]
46. McMahon, C.R.; Field, I.C.; Bradshaw, C.J.; White, G.C.; Hindell, M.A. Tracking and data–logging devices attached to elephant seals do not affect individual mass gain or survival. *J. Exp. Mar. Biol. Ecol.* **2008**, *360*, 71–77. [CrossRef]
47. Chin, T.M. Multi-Resolution Variational Analysis (MRVA): High-resolution data fusion over global surface. *AGU Fall Meet. Abstr.* **2012**, *2012*, NG31A-1574.
48. Valla, D.; Piola, A.R. Evidence of upwelling events at the northern Patagonian shelf break. *J. Geophys. Res. Oceans* **2015**, *120*, 7635–7656. [CrossRef]
49. Lago, L.S.; Saraceno, M.; Martos, P.; Guerrero, R.A.; Piola, A.R.; Paniagua, G.F.; Ferrari, R.; Artana, C.I.; Provost, C. On the Wind Contribution to the Variability of Ocean Currents Over Wide Continental Shelves: A Case Study on the Northern Argentine Continental Shelf. *J. Geophys. Res. Oceans* **2019**, *124*, 7457–7472. [CrossRef]

50. Hersbach, H.; Bell, B.; Berrisford, P.; Horányi, A.; Muñoz-Sabater, J.; Nicolas, J.; Peubey, C.; Radu, R.; Schepers, D.; Simmons, A.; et al. *ERA5 Hourly Data on Single Levels from 1940 to Present. Copernicus Climate Change Service (c3s) Climate Data Store (CDS)*; European Commission: Luxembourg, 2018; p. 224. [CrossRef]
51. ECMWF. *IFS Documentation CY41r2—Part I: Observations*; ECMWF: Reading, UK, 2016; p. 223.
52. ECMWF. *IFS Documentation CY41r2—Part II: Data Assimilation*; ECMWF: Reading, UK, 2016.
53. Pisoni, J.P.; Rivas, A.L.; Tonini, M.H. Coastal upwelling in the San Jorge Gulf (Southwestern Atlantic) from remote sensing, modelling and hydrographic data. *Estuar. Coast. Shelf Sci.* **2020**, *245*, 106919. [CrossRef]
54. Dinápoli, M.G.; Simionato, C.G.; Moreira, D. Nonlinear Interaction Between the Tide and the Storm Surge with the Current due to the River Flow in the Río de la Plata. *Estuaries Coasts* **2021**, *44*, 939–959. [CrossRef]
55. Egbert, G.D.; Erofeeva, S.Y. Efficient Inverse Modeling of Barotropic Ocean Tides. *J. Atmos. Ocean. Technol.* **2002**, *19*, 183–204. [CrossRef]
56. Lago, L.S.; Saraceno, M.; Ruiz-Etcheverry, L.A.; Passaro, M.; Oreiro, F.A.; Donofrio, E.E.; Gonzalez, R.A. Improved Sea Surface Height From Satellite Altimetry in Coastal Zones: A Case Study in Southern Patagonia. *IEEE J. Sel. Top. Appl. Earth Obs. Remote Sens.* **2017**, *10*, 3493–3503. [CrossRef]
57. McDougall, T.J.; Barker, P.M. *Getting Started with TEOS-10 and the Gibbs Seawater (GSW) Oceanographic Toolbox*; SCOR/IAPSO WG: Arnhem, The Netherlands, 2011.
58. Simpson, J.H. Simpson_1981s.pdf. *Phil. Trans. R. Soc. Lond.* **1981**, *302*, 531–546.
59. Kahl, L.C.; Bianchi, A.A.; Osiroff, A.P.; Pino, D.R.; Piola, A.R. Distribution of sea-air CO_2 fluxes in the Patagonian Sea: Seasonal, biological and thermal effects. *Cont. Shelf Res.* **2017**, *143*, 18–28. [CrossRef]
60. Sabatini, M.; Akselman, R.; Reta, R.; Negri, R.; Lutz, V.; Silva, R.; Segura, V.; Gil, M.; Santinelli, N.; Sastre, A.; et al. Spring plankton communities in the southern Patagonian shelf: Hydrography, mesozooplankton patterns and trophic relationships. *J. Mar. Syst.* **2012**, *94*, 33–51. [CrossRef]
61. Palma, E.D.; Matano, R.P. A numerical study of the Magellan Plume. *J. Geophys. Res. Ocean.* **2012**, *117*. [CrossRef]
62. Tonini, M.H.; Palma, E.D.; Piola, A.R. A numerical study of gyres, thermal fronts and seasonal circulation in austral semi-enclosed gulfs. *Cont. Shelf Res.* **2013**, *65*, 97–110. [CrossRef]
63. Brandhorst, W.; Castello, J.P. *Evaluación de los Recursos de Anchoita (Engraulis anchoita) Frente a la Argentina y Uruguay*; Proyecto de Desarrollo Pesquero; Ser. Inf. Tee. Publicación: Mar del Plata, Argentina, 1971.
64. Pisoni, J.P. Los Sistemas Frontales y la Circulación en las Inmediaciones de los Golfos Norpatagónicos. Ph.D. Thesis, Universidad de Buenos Aires, Facultad de Ciencias Exactas y Naturales, Buenos Aires, Argentina, 2012. Available online: https://hdl.handle.net/20.500.12110/tesis_n5193_Pisoni (accessed on 12 July 2022).
65. Cazau, D.; Bonnel, J.; Jouma'a, J.; le Bras, Y.; Guinet, C. Measuring the Marine Soundscape of the Indian Ocean with Southern Elephant Seals Used as Acoustic Gliders of Opportunity. *J. Atmos. Ocean. Technol.* **2017**, *34*, 207–223. [CrossRef]
66. Franco, B.C. Procesos Acoplados Bento-Pelágicos Relacionados con el Establecimiento y Deriva Larval de la Vieira Patagónica (*Zygochlamys patagónica*) en el Océano Atlántico Sudoeste. Ph.D. Thesis, Universidad de Buenos Aires, Facultad de Ciencias Exactas y Naturales, Buenos Aires, Argentina, 2013. Available online: https://hdl.handle.net/20.500.12110/tesis_n5466_Franco (accessed on 15 March 2023).
67. McMahon, C.R.; Hindell, M.A.; Charrassin, J.B.; Coleman, R.; Guinet, C.; Harcourt, R.; Labrousse, S.; Raymond, B.; Sumner, M.; Ribeiro, N. Southern Ocean pinnipeds provide bathymetric insights on the East Antarctic continental shelf. *Commun. Earth Environ.* **2023**, *4*, 266. [CrossRef]
68. Franco, B.C.; Ruiz-Etcheverry, L.A.; Marrari, M.; Piola, A.R.; Matano, R.P. Climate Change Impacts on the Patagonian Shelf Break Front. *Geophys. Res. Lett.* **2022**, *49*, e2021GL096513. [CrossRef]

remote sensing

MDPI

Article

Near-Real-Time Turbidity Monitoring at Global Scale Using Sentinel-2 Data and Machine Learning Techniques

Masuma Chowdhury [1,2,*], Ignacio de la Calle [1], Irene Laiz [2] and Ana B. Ruescas [3]

1 Quasar Science Resources, S. L. Camino de las Ceudas 2, 28232 Las Rozas de Madrid, Madrid, Spain; ignaciodelacalle@quasarsr.com
2 Department of Applied Physics, University Marine Research Institute (INMAR), Campus of International Excellence of the Sea (CEI MAR), University of Cádiz, Puerto Real, Cadiz, Spain; irene.laiz@uca.es
3 Image Processing Laboratory, University of Valencia, 46980 Paterna, Valencia, Spain; ana.b.ruescas@uv.es
* Correspondence: masuma.chowdhury@quasarsr.com or masuma.chowdhury@uca.es

Highlights

What are the main findings?

- We have developed a machine-learning algorithm able to quantify high-turbid environments.
- We use open-source and free databases to train the model and open-source and free tools to develop it, which makes it easily replicable and transferable.

What are the implications of the main findings?

- The social impact of our work implies improved monitoring of areas with high turbidity, which can lead to a better understanding and use of forecasting.
- Ocean color and water quality communities can take advantage of the lessons learned for developing new products or services.

Abstract: Reliable global turbidity monitoring is crucial for water resource management, yet existing satellite-based methods face limitations in accuracy, generalization, and scalability across diverse aquatic environments. This study presents a robust, globally applicable turbidity estimation model using Sentinel-2 imagery and a machine-learning approach, developed based on harmonized global open-source datasets (GLORIA and MAGEST; turbidity range: 0–2200 FNU) encompassing 68 lakes, 2 rivers, 2 estuaries, and 11 coastal oceans across 17 countries. Among the evaluated machine-learning models, gradient boosting regression demonstrated the best performance, achieving a high correlation (r: 0.95) with minimal bias (1.32 FNU) and robust generalization across all water types, outperforming existing turbidity models when evaluated on the same test dataset. Shapley Additive exPlanations-based model interpretability identified the Rrs865/Rrs560 ratio as the dominant predictor, with critical contributions from Rrs783, Rrs665, and Rrs865. The model's performance is evaluated across various optical water types and aquatic systems in diverse geographical settings, showcasing its robustness in sediment-rich and highly turbid environments that underscores its suitability for reliable turbidity monitoring after severe storms or extreme precipitation. Additionally, innovative automated pipelines integrated within a scientific exploitation platform facilitate scalable and near-real-time operational monitoring. This methodological integration provides a significant advancement in satellite-based turbidity monitoring, enabling informed water quality management under diverse environmental and climatic conditions.

Keywords: water quality; spectral convolution; data harmonization; machine-learning; gradient boosting; SHAP; optical water types; uncertainty analysis; operational monitoring; automation and scalability

1. Introduction

Turbidity, the cloudiness of water caused by suspended particles, serves as a critical indicator of aquatic ecosystem health and water resource sustainability. This optical property reflects both natural processes (river discharge, wave-induced resuspension, erosion, phytoplankton blooms) and anthropogenic pressures (dredging, construction, agricultural runoff), making it indispensable for water resource management across the globe. Whether caused by natural or anthropogenic disturbances, turbidity profoundly influences aquatic life by altering light penetration, thermal stratification, and nutrient cycling. Elevated turbidity levels trigger cascading ecological effects: reduced light penetration impairs photosynthesis and primary productivity, disrupted thermal stratification alters oxygen distribution, and increased nutrient availability may initiate harmful algal blooms, ultimately degrading ecosystem resilience and water security [1–5]. Climate change further exacerbates these impacts through intensified precipitation patterns, increased storm frequency, and accelerated sea-level rise, driving unprecedented sediment mobilization and turbidity episodes across coastal and inland waters [6–10]. As anthropogenic pressures and climate variability intensify, robust turbidity monitoring at relevant spatial and temporal scales has become essential for predicting ecosystem responses and implementing effective mitigation strategies.

Traditional in situ turbidity measurements, while accurate, cannot meet the spatiotemporal demands of modern water resource management. These point-based observations are resource-intensive, logistically challenging in remote regions, and inadequate for capturing episodic events or large-scale disturbances. Satellite remote sensing emerged as a compelling alternative, offering synoptic, repeated observations across vast areas. Early satellite missions, such as Sea-Viewing Wide Field-of-view Sensor (SeaWiFS), Moderate Resolution Imaging Spectroradiometer (MODIS), and Medium Resolution Imaging Spectrometer (MERIS), demonstrated turbidity monitoring potential [11–14]; however, their coarse spatial resolution (250–1000 m) rendered them inadequate for smaller water bodies and complex coastal zones where fine spatial details are necessary.

The Sentinel-2 MultiSpectral Instrument (MSI) transformed this landscape with 10 m spatial resolution and 5-day revisit frequency, enabling detailed assessments at management-relevant scales [15–20]. Yet despite these technological advances, a critical gap persists between algorithm innovation and operational implementation. Current turbidity algorithms remain largely confined to specific regions, sensors, or turbidity ranges, lacking the versatility required for global deployment (Table 1). Empirical and semi-empirical models, while successful within their design parameters, fail when confronted with optical conditions beyond their training domains. Even the most widely adopted global turbidity model [13], though more robust within its design parameters, cannot address extreme turbidity conditions common in highly dynamic riverine-estuarine environments [17].

Table 1. Summary of existing satellite-based turbidity algorithms showing their regional scope, sensor specifications, and operational turbidity ranges.

Algorithms	Study Region	Satellites/Sensors (Operational Period)	Turbidity Levels
Semi-empirical single-band algorithm using Rrs859 for high turbidity, and Rrs645 for medium to low turbidity [13]	Southern North Sea, French Guyana, Scheldt, Gironde, Rio Plata	MERIS (2002–2012), MODIS (1999–Present), SeaWiFS (1997–2010)	1–1000 FNU *
Single band algorithm using Rrs842 and Rrs665 [21]	Gironde Estuary, France	Pléiades (2011–Present)	200–900 FNU
Multi-conditional algorithm using Rrs665 and Rrs704 [17]	Guadalquivir Estuary, Spain	Sentinel-2 (2015–Present)	0–600 FNU
Normalized difference turbidity index [22]	Panchet Hill Dam, India	Landsat 5 (1984–2013), Landsat 8 (2013–Present)	<700 FNU
Machine learning [15,23]	Taihu Lake, China; North Tyrrhenian Sea, Italy	Sentinel-2 (2015–Present), Landsat 8/9 (2013–Present)	0–200; 0–30 FNU
Generalized additive models [24]	Doñana Marshes, Spain	Landsat 5 (1984–2013), Landsat 7 (1999–Present)	1–500 FNU
CMEMS [25]	European Seas	Sentinel-2 (2015–Present)	-

* FNU denotes Formazin Nephelometric Units.

The limitations of existing approaches (Table 1) stem from their reliance on simplified spectral relationships that cannot capture the full complexity of turbidity's optical signatures across diverse aquatic environments. Machine-learning (ML) techniques offer a promising alternative, with their ability to model complex, nonlinear relationships and adapt to varied optical conditions [15,23,26,27]. However, leveraging this potential requires comprehensive training datasets that span the global aquatic conditions, which have historically been unattainable.

Recent community datasets—GLORIA (GLObal Reflectance community dataset for Imaging and optical sensing of Aquatic environments) and MAGEST (Monitoring the water quality of the Gironde Estuary)—provide unprecedented coverage of in situ radiometry and co-located water quality measurements (i.e., turbidity, among others) across diverse geographic regions and water types [28,29]. While these datasets offer valuable resources, they also present significant challenges due to measurement inconsistencies, methodological variations, and gaps in turbidity observations that require careful harmonization and quality control to ensure reliable model development.

This study presents a comprehensive solution to these challenges by developing a globally applicable turbidity monitoring system that bridges the gap between methodological innovation and operational implementation. Our approach integrates Sentinel-2 observations with advanced ML techniques, trained on rigorously harmonized global datasets to achieve unprecedented monitoring capabilities across a wide turbidity range (0–2200 FNU). We address the following critical challenges: (1) establishing robust data curation approach for merging heterogeneous in situ measurements with satellite observations; (2) developing ML models capable of accurate predictions across diverse optical environments, including extreme turbidity conditions previously beyond algorithmic reach; (3) ensuring model transparency and interpretability through explainable Artificial Intelligence (AI) techniques, i.e., SHapley Additive exPlanations (SHAP) [30] that reveal model's physical basis of spectral-turbidity relationships; and (4) implementing comprehensive uncertainty quantification. Furthermore, recognizing the operational demand for timely environmental insights, this study presents an innovative automated processing framework that transforms our methodological innovations into practical monitoring tools. Leveraging containerized computing architecture (Docker, v4.44.2 (202017)) and orchestration technologies (Kubernetes, v1.24.3), our system enables near-real-time turbidity monitoring at global scales while maintaining the flexibility to incorporate future improvements and additional data sources. Thus, by effectively advancing the current capabilities of satellite-based extreme turbidity monitoring, as well as bridging methodological innovation with practical applicability, this study presents a high-resolution (10 m) satellite-based turbidity monitoring system, essential for managing aquatic resources under accelerating environmental change.

2. Materials and Methods

2.1. Dataset Integration and Quality Control

This study integrated two complementary datasets to achieve comprehensive global turbidity coverage. The GLORIA dataset (https://doi.org/10.1594/PANGAEA.948492, accessed 22 March 2023) provided 7572 hyperspectral remote sensing reflectance (Rrs, sr^{-1}) measurements spanning 350–900 nm at 1 nm resolution, with co-located turbidity, total suspended solids (TSS), Secchi-depth transparency (SDT), and chlorophyll-*a* (Chl-*a*) measurements from lakes, rivers, estuaries, and coastal oceans worldwide [28]. The MAGEST dataset (https://magest.oasu.u-bordeaux.fr/index.php, accessed 19 August 2024) contained 2581 direct turbidity measurements from the Gironde River Estuary, France, a highly turbid and sediment-rich system, providing critical training data for extreme conditions [29]. Together, these datasets spanned turbidity levels from 0 to 2200 FNU across diverse aquatic environments.

Quality control employed the Quality Water Index Polynomial (QWIP) metric [31] to identify and remove optically inconsistent spectra from GLORIA, reducing the datasets to 5471 valid observations. For observations lacking direct turbidity measurements, we developed predictive models using co-located parameters. Linear regression converted TSS to turbidity ($y = 0.86x + 0.48$; root mean squared error (RMSE) = 11.94 FNU, correlation coefficient (r) = 0.75) [32], while power regression transformed SDT values ($y = \exp(2.19 - 1.02\ln(x))$; RMSE = 10.90 FNU, r = 0.73) [33]. These models, developed using 80% of paired observations and validated on the remaining 20%, demonstrated acceptable predictive performance as shown in Table 2 and Figure A1. Therefore, these in situ proxy-derived turbidity values were combined with direct in situ available measurements to maximize data coverage across the turbidity spectrum. The MAGEST dataset, consisting entirely of field turbidity measurements, required no proxy estimation, thus providing direct data for model training in the extreme turbidity range.

Table 2. Performance metrics for turbidity estimation from TSS and SDT using the test dataset (n = 206).

Model	Equation	Root Mean Squared Error (RMSE), FNU	Mean Absolute Error (MAE), FNU	Bias FNU	Correlation Coefficient (r)
$Turbidity_{TSS}$	$y = 0.86x + 0.48$	11.94	4.48	0.63	0.75
$Turbidity_{SDT}$	$y = \exp(2.19 - 1.02\ln(x))$	10.90	5.51	1.23	0.73

2.2. Hyperspectral to Multispectral Conversion

Sentinel-2 equivalent reflectances were derived from GLORIA in situ hyperspectral measurements through the spectral response function (SRF) [34] convolution, enabling comparison between in situ-convolved and satellite-derived reflectances. The Sentinel-2 SRF characterizes the wavelength-dependent sensitivity of each MSI band, accounting for the detector's spectral response across its operational range. The convolution process transformed continuous hyperspectral Rrs measurements (sr^{-1}) at 1 nm resolution into discrete multispectral bands corresponding to Sentinel-2A's spectral configuration. For each band i, normalized SRF weights were first computed as:

$$\hat{S}i(\lambda j) = Si(\lambda j)/\Sigma_j\, Si(\lambda j) \quad (1)$$

where $Si(\lambda j)$ represents the SRF value at wavelength λj within the band's range [λi, *min*, λi, *max*], and the summation extends over all wavelengths within this range. The convolved Sentinel-2 reflectance for band i was then calculated as:

$$Ri = \Sigma_j\, \hat{S}i(\lambda j).\ \mathrm{Rrs}(\lambda j) \quad (2)$$

The resulting Sentinel-2 equivalent reflectances maintained spectral fidelity while accounting for the instrument's specific optical characteristics, enabling robust algorithm development using the GLORIA dataset. Only bands within GLORIA's spectral range (350–900 nm) were retained for analysis, ensuring that all the convolved values were based on measured data.

2.3. Satellite Data Processing

Sentinel-2 Level-1C imagery was acquired through the Copernicus Data Space Ecosystem (https://dataspace.copernicus.eu/, accessed 1 March 2023 onwards) API (Application Programming Interface) for locations and times matching in situ observations. A ±1 h temporal window relative to in situ sampling was applied to minimize temporal discrepancies between satellite and field measurements. Initial acquisition yielded 259 scenes (139 GLORIA, 120 MAGEST), which underwent visual screening using true color composites to identify and exclude images affected by clouds, haze, fog, or strong sun-glint, retaining 199 scenes (101 GLORIA, 98 MAGEST).

Atmospheric correction was performed using the dark spectrum fitting (DSF) algorithm with sun-glint correction implemented in ACOLITE (v20231023) [35], generating 10 m resolution Rrs products. At each sampling location, 3×3 pixel windows were extracted and subjected to quality filtering: z-score outlier removal eliminated anomalous pixels, while coefficient of variation (CV) screening retained only homogeneous windows (CV < 20%) [36] across all bands to ensure spatial consistency. For GLORIA locations, mean reflectances were computed from valid pixels within each window and used for validating the convolved spectra derived in Section 2.2. Strong correlations were observed across all bands (r = 0.52–0.93), with the highest agreement for Rrs704, Rrs665, and Rrs560 (Figure 1). RMSE values remained consistently low (0.004–0.005 sr^{-1}) with minimal positive biases (0.0008–0.0029 sr^{-1}), indicating strong agreement and validating the conversion approach. In contrast, MAGEST extraction meeting homogeneity criteria retained all valid pixels, maximizing turbidity-reflectance pairs for model training. In this case, given the homogeneity threshold (CV < 20%), pixel-level variation was considered negligible for this dataset.

Therefore, the final dataset preparation merged convolved GLORIA Rrs and the corresponding combined turbidity measurements with the Sentinel-2 MAGEST Rrs and the corresponding in situ turbidity measurements. Quality screening removed observations with missing values or extreme reflectances (Rrs < 0.001 or >0.1), yielding 1373 high-quality observation pairs spanning diverse aquatic environments, i.e., lakes (55%), rivers (24%), coastal oceans (18%), and estuaries (2.5%) across 17 countries. Following GLORIA and MAGEST dataset, this final dataset represents sediment-dominated (61%), Chl-*a*-dominated (32.4%), colored dissolved organic matter (CDOM)-dominated (0.4%), both Chl-*a* and CDOM-dominated (3.5%), clear waters (0.5%), and other water types (2.2%) (Figures 2 and A2, Table A1).

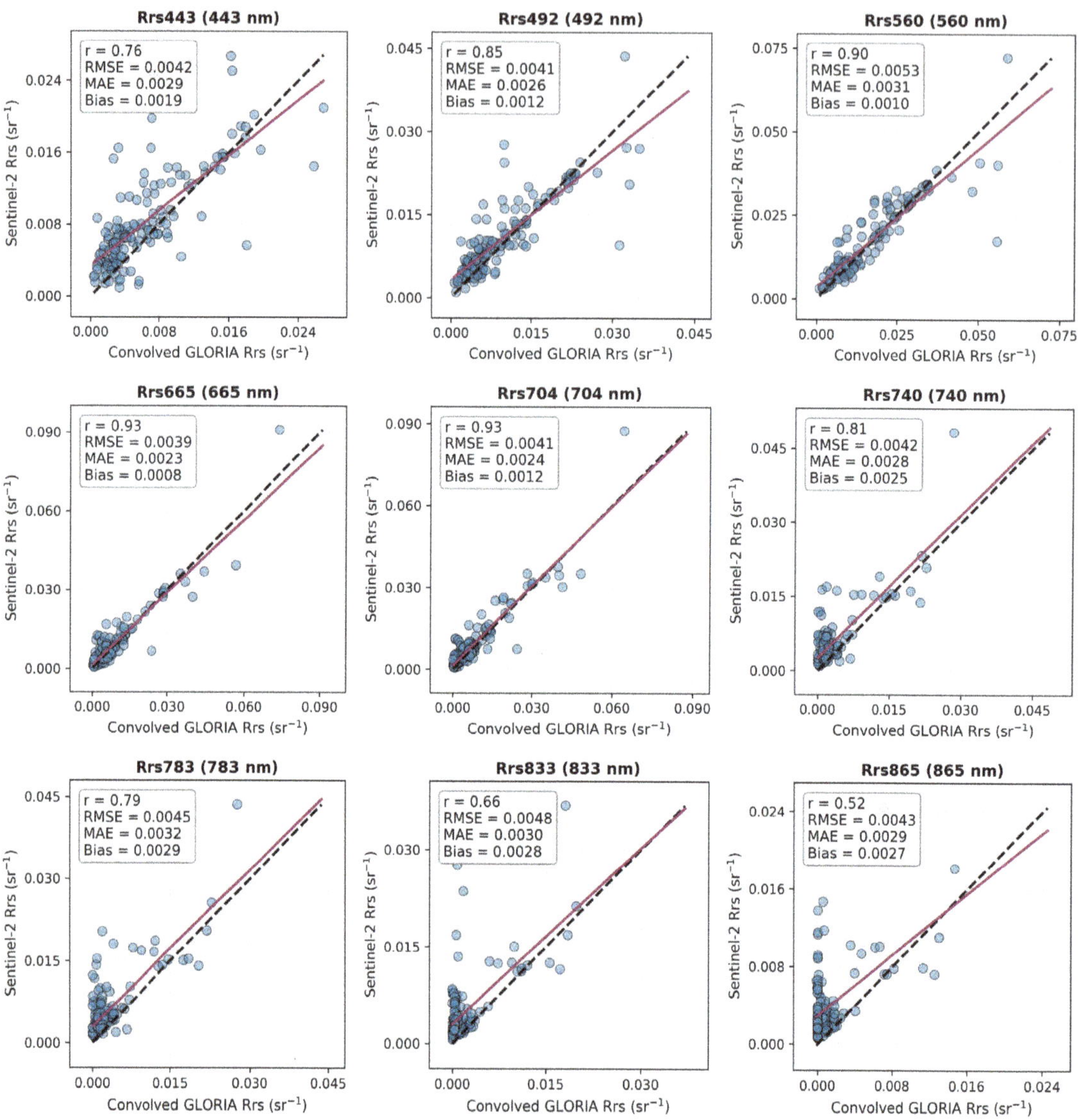

Figure 1. Comparison between ACOLITE-derived Sentinel-2 Rrs and in situ convolved GLORIA Rrs.

2.4. Machine Learning Framework

2.4.1. Algorithm Selection and Configuration

Four regression algorithms were evaluated: Elastic Net (ENR), Random Forest (RFR), Gradient Boosting (GBR), and Extreme Gradient Boosting (XGBR). These span from linear (ENR) to complex ensemble methods, providing a comprehensive performance assessment. ENR [37] combines Ridge [38] and Lasso [39] regressions to handle multicollinearity while performing feature selection. It is quite flexible, allowing tuning between both models, enabling shrinkage and selection of correlated feature groups. RFR, an ensemble method, aggregates multiple decision trees for robust nonlinear modeling with minimal tuning requirements [40]. However, it may be biased toward dominant feature groups and can be slow to train. GBR is another ensemble method that sequentially builds trees to correct predecessor errors, offering high accuracy through iterative refinement [41]. It is flexible and often accurate but sensitive to noise, slower to train, and more tuning-intensive than RFR. XGBR extends GBR with regularization and computational optimizations for enhanced efficiency [42].

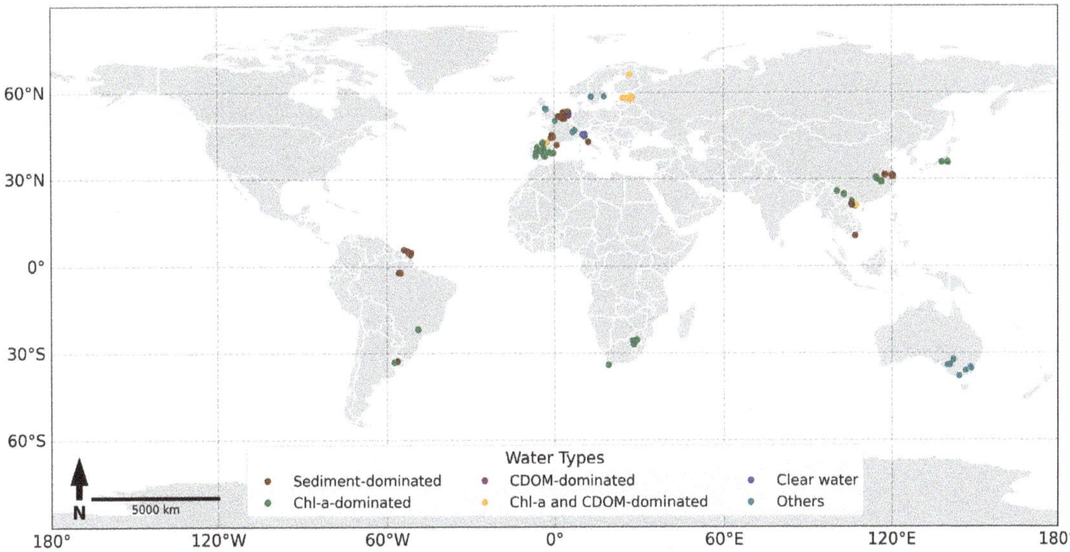

Figure 2. Study area map. Water types are defined according to GLORIA and MAGEST dataset.

Input feature comprised all nine Sentinel-2 visible and near-infrared (VNIR) bands (443–865 nm) to leverage the full spectral information available for turbidity characterization. This comprehensive approach allowed the model to capture subtle spectral variations across different water types and turbidity ranges, as turbidity affects reflectance differently across the VNIR spectrum [17]. The Rrs865/Rrs560 ratio was included as an additional feature based on its strong correlation with turbidity (r = 0.81) among all the tested spectral ratios during the exploratory analysis (Figures A3 and A4). All features underwent min-max normalization to the [0,1] range, ensuring comparable scales and preventing bias toward high-magnitude features during model training. Turbidity values were mean-centered to enhance model sensitivity to relative variations rather than absolute magnitudes.

Data partitioning employed a 75:25 train–test split using a fixed random state (random state = 42) to ensure reproducibility across experiments. This approach randomly assigned observations to training (75%, n = 1029) or testing (25%, n = 344) datasets while maintaining the overall turbidity distribution. In addition, this fixed random state eliminated variability between model runs, enabling consistent performance comparisons across different algorithms and hyperparameter configurations.

To minimize overfitting and ensure robust hyperparameter selection, 5-fold cross-validation was implemented during training. This approach evaluated model performance across five independent data splits within the training set, with each fold serving as validation once, while the remaining four provided training data. Cross-validation provides more reliable performance estimation than single train-validation splits by averaging results across multiple data configurations, reducing the influence of any particular data partition [43]. Hyperparameters were optimized through grid search across the cross-validation folds, tuning tree count, maximum depth, learning rate, and regularization parameters for ensemble models; alpha and L1-ratio for ENR.

2.4.2. Model Evaluation and Interpretability

After training, the model's performance was assessed on the unseen test dataset and on an independent dataset from the Albufera Lake, Valencia, Spain (2023–2025) [44], with predictions reconstructed by adding the training mean to the model outputs. To provide a comprehensive analysis, the trained models were compared against two established empirical models from Table 1 [13,17] using the identical test dataset. Comparative performance of the six models was assessed using RMSE, r, mean absolute error (MAE), and bias.

Model interpretability employed SHAP analysis to quantify feature contributions through Shapley values derived from cooperative game theory [30,45]. SHAP decomposes each prediction into additive feature contributions, providing both global feature importance rankings and local explanations for individual predictions. Summary plots visualized feature importance with directional effects, while dependence plots revealed nonlinear relationships between features and turbidity across different optical regimes. Pairwise interaction effects among the most influential features generated unique visualizations, identifying synergistic or antagonistic relationships critical for understanding the model behavior. Decision plots illustrated cumulative feature contributions along prediction paths, revealing the model's internal logic from baseline to final prediction.

2.4.3. Uncertainty Quantification

The comprehensive uncertainty analysis quantified prediction reliability across the turbidity spectrum. The residual analysis examined the prediction errors (differences between observed and predicted turbidity) to identify systematic biases and heteroscedasticity patterns [46,47]. The confidence intervals (CI) for the regression line and prediction intervals (PI) for individual observations were computed following standard regression theory [48,49].

The 95% CI, representing uncertainty in the mean prediction, was computed as:

$$\mathrm{CI} = \bar{y} \pm t_{0.025,n-2} \times \sigma_{res} \times \sqrt{\left(1/n + (x - \bar{x})^2/\Sigma(xi - \bar{x})^2\right)} \tag{3}$$

where $\bar{y}$ represents the predicted value, σ_{res} is the residual standard error, n is the sample size, x is the prediction point, $\bar{x}$ is the mean of observations, and $t_{0.025,n-2}$ is the critical t-value at 95% confidence with $n - 2$ degrees of freedom.

The 95% PI, capturing uncertainty for individual predictions including natural variability, incorporated an additional unit term:

$$\mathrm{PI} = \bar{y} \pm t_{0.025,n-2} \times \sigma_{res} \times \sqrt{\left(1 + 1/n + (x - \bar{x})^2/\Sigma(xi - \bar{x})^2\right)} \tag{4}$$

These intervals provided operational guidance for end-users, with CI indicating the reliability of the regression line (mean predictions) and PI bounding the range within which 95% of the individual observations are expected to fall [48,50].

2.5. Automated Processing Pipelines

Two processing pipelines were developed to support both algorithm development and operational deployment: a semi-automated scientific pipeline for scientific experimentation, model development, and validation, and a fully automated operational pipeline for routine turbidity monitoring (Figure 3). Both pipelines were implemented within the scientific exploitation platform (SEP) under the SIMBAD (Sentinel Imagery Multiband Analysis and Dissemination) project (https://simbad.quasarsr.com/) (accessed on 7 November 2025), designed for Earth observation data processing in distributed computing environments [51].

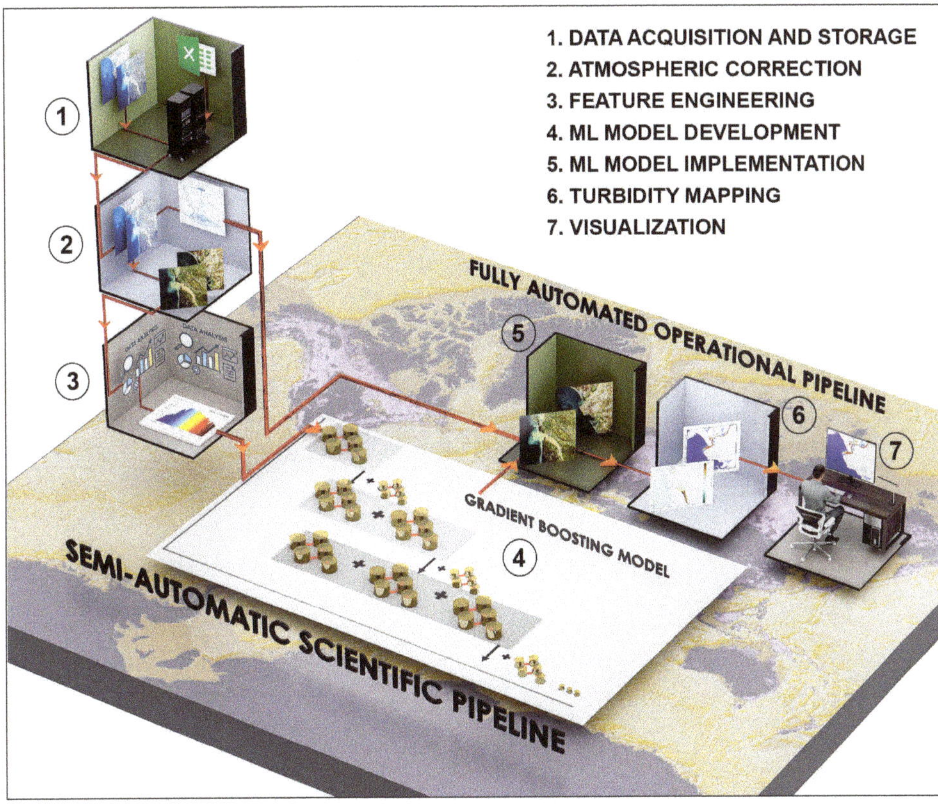

Figure 3. Schematic pipelines: The semi-automatic scientific pipeline begins with satellite and in situ data acquisition (**1**), followed by atmospheric correction of Sentinel-2 Level-1C data (**2**), feature selection through feature engineering (**3**), ML model development (**4**), turbidity prediction via model implementation on satellite data (**5**), generation of georeferenced turbidity maps (**6**), and visualization (**7**). The fully automated operational pipeline includes steps **1**, **2**, **5**, **6**, and **7**.

The pipelines utilize Docker containerization to ensure consistent execution across different computing environments by bundling application code, dependencies, and runtime configurations into portable containers. Kubernetes orchestration manages computational resources and enables parallel processing through dynamic workload distribution across available nodes. Data storage and accessibility within SEP are handled by a centralized Network File System, facilitating seamless data access across the distributed infrastructure. Pipeline configurations are externalized into environment-specific files, allowing modification of processing parameters (e.g., file paths, satellite identifiers, temporal windows, etc.) without code changes.

The semi-automated scientific pipeline supports iterative model development through modular components for satellite data download, atmospheric correction and matchup extraction, feature engineering and data preprocessing, model training with cross-validation, and model implementation, validation, performance evaluation and visualization. Manual intervention points allow quality control and parameter adjustment during the model development phases. In contrast, the operational pipeline automates end-to-end processing for routine monitoring: downloading Sentinel-2 imagery, applying atmospheric corrections, executing the pre-trained model, and generating georeferenced turbidity products in GeoTIFF format. The modular architecture facilitates maintenance and updates, while containerization and Kubernetes orchestration ensure reproducibility across parallel deployments.

3. Results

3.1. Model's Performance Evaluation

The comprehensive analysis of six turbidity prediction models on the test dataset (n = 344) revealed distinct performance characteristics, as evidenced by both the regression analysis and comprehensive statistical metrics (Table 3 and Figure 4). The four ML models (ENR, RFR, GBR, and XGBR) demonstrated substantially better performance compared to the two empirical algorithms, with the ensemble methods (RFR, GBR, and XGBR) achieving r-values of 0.95 and R^2 values of 0.89–0.90, indicating that these models explain approximately 90% of the measured turbidity variance. Among these, GBR emerged as the optimal model, combining the highest slope (0.887)—the closest to the ideal value of 1.0, the lowest bias (1.32 FNU), and the second-best MAE (43.24 FNU), demonstrating both accuracy and precision across the entire measurement range.

Table 3. Performance metrics for turbidity models on test dataset (n = 344).

Model	r	R^2	RMSE (FNU)	MAE (FNU)	Bias (FNU)
ENR	0.83	0.69	226.56	127.11	21.19
RFR	0.95	0.89	117.12	47.10	1.86
GBR	0.95	0.90	116.62	43.24	1.32
XGBR	0.95	0.90	114.21	41.56	−3.29
Chowdhury et al., 2023 [17]	0.05	−0.13	369.92	152.61	−123.81
Dogliotti et al., 2015 [13]	0.06	−0.18	377.29	151.79	−149.11

The GBR model's performance was further validated by its comprehensive statistical metrics. With an R^2 value of 0.90 and RMSE of 116.62 FNU, GBR demonstrated robust predictive capability while maintaining minimal systematic error. The remarkably low bias of 1.32 FNU indicated that the model's predictions are nearly unbiased on average, a critical characteristic for operational water quality monitoring. While XGBR showed marginally better RMSE (114.21 FNU) and MAE (41.56 FNU) values, its negative bias (−3.29 FNU) and lower slope (0.849) suggested a tendency toward systematic underestimation, which is particularly problematic for high turbidity detection. RFR, despite sharing the same correlation coefficient (0.95) and similar R^2 (0.89), exhibited slightly higher errors (RMSE: 117.12 FNU, MAE: 47.10 FNU) and a lower slope (0.886) compared to GBR, confirming GBR's superiority in balancing multiple performance criteria. ENR, while demonstrating reasonable correlation (r = 0.83, R^2 = 0.69), showed substantially higher error metrics with an RMSE of 226.56 FNU and MAE of 127.11 FNU. The model's slope of 0.646 indicated systematic underestimation by approximately 35%, and despite a moderate bias of 21.19 FNU, the large standard error (±175.03 FNU) limited its utility for precise turbidity quantification.

The empirical algorithms revealed fundamental limitations. Both models demonstrated severe underestimation, with near-zero correlation coefficients (0.05 and 0.06) and negative R^2 values (−0.13 and −0.18). The extreme negative biases (−123.81 and −149.11 FNU) and high error metrics (RMSE > 369 FNU, and MAE > 151 FNU) confirmed severe systematic underestimation when applied to the same test dataset. The models' near-zero slopes (0.006 and 0.003) indicated a lack of sensitivity to turbidity variations, with predictions remaining clustered below 500 FNU regardless of the actual turbidity levels exceeding 2000 FNU.

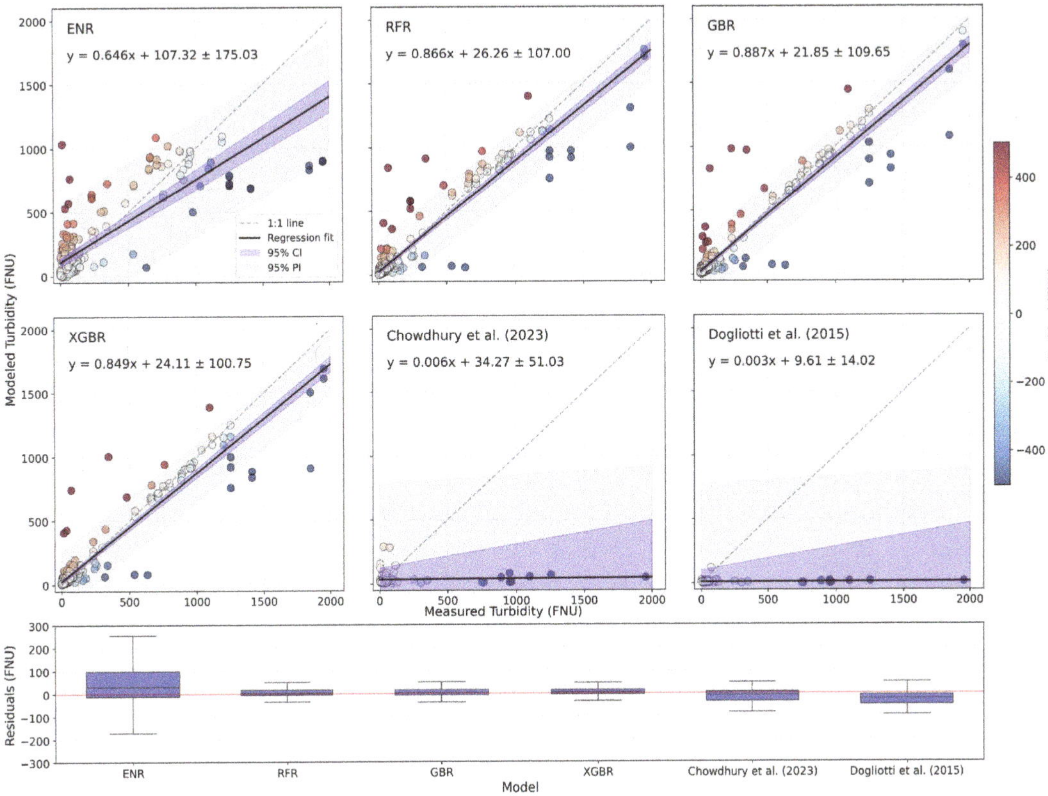

Figure 4. Scatter plots comparing measured against predicted turbidity for six models, i.e., ENR, RFR, GBR, XGBR, Chowdhury et al., (2023) [17], and Dogliotti et al., (2015) [13] (the top and middle panels) applied to the test dataset (n = 344), with color gradient indicating residual magnitude. Grey dashed lines represent 1:1 perfect prediction, black lines show regression fits. Grey and purple shaded areas indicate 95% prediction (PI) and confidence (CI) intervals, respectively. The bottom panel shows residual distributions across all models.

In addition, the uncertainty analysis provided essential insights into model reliability and uncertainty quantification that extended beyond simple error metrics. While CI represented uncertainty in the mean prediction (the regression line itself), the PI captured the total prediction uncertainty, including model variance and measurement noise. The 95% PI represented the range within which 95% of future observations are expected to fall, given the model's inherent uncertainty. Though both GBR and XGBR showed similar PI and CI width, unlike XGBR, the 1:1 line fell entirely within GBR's 95% PI across the full turbidity range, indicating that the model is well calibrated and the uncertainty estimates are statistically consistent, which means users can trust that these uncertainty bounds are neither overly optimistic nor unnecessarily conservative. The residual distribution analysis also showed that GBR maintained consistent prediction variance across turbidity ranges, whereas XGBR exhibited heteroscedastic behavior with increasing underestimation at high turbidity levels (>1000 FNU). Based on this comprehensive evaluation, GBR was selected as the optimal model for operational turbidity monitoring.

3.2. Model Interpretation

SHAP analysis revealed the physical basis underlying GBR predictions by quantifying feature contributions across the turbidity range. The Rrs865/Rrs560 ratio emerged as the dominant predictor, with SHAP values extending up to 800 units, exceeding all other features (Figure 5). This ratio, comparing NIR backscattering (865 nm) to visible reflectance (560 nm), effectively discriminated turbidity through its sensitivity to particle-induced spectral changes. Low ratio values (blue points) consistently produced negative SHAP contributions, suppressing turbidity predictions, while high values (pink points) generated strong positive contributions, amplifying predictions.

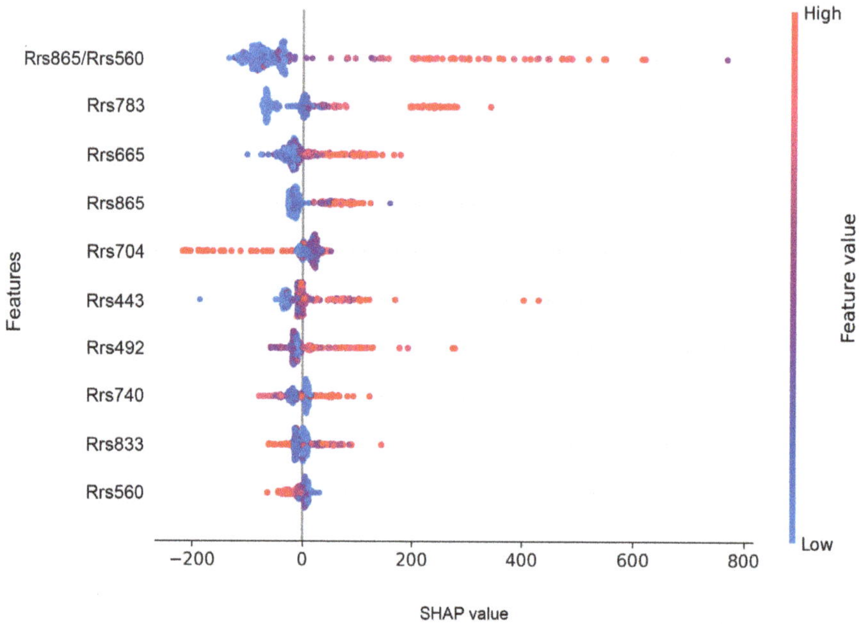

Figure 5. SHAP summary plot showing feature importance ranking and impact direction on turbidity predictions. Points represent individual predictions, with color indicating feature values (blue: low, pink: high) and horizontal position showing SHAP contribution magnitude. Overlapping points are jittered along the y-axis to visualize the distribution of Shapley values for each feature.

Individual bands showed hierarchical importance, with Rrs783, Rrs665, and Rrs865 providing the next tier of contributions (Figure 5). These red and NIR bands, sensitive to suspended sediment backscattering, showed clear positive relationships—higher reflectance values (pink) increased the predicted turbidity while lower values (blue) decreased it. The remaining features (Rrs443, Rrs492, Rrs740, Rrs833, and Rrs560) contributed minimally, with SHAP values clustering near zero, suggesting they provided minor refinements rather than primary predictive power.

Examining SHAP values across the turbidity spectrum revealed distinct threshold behaviors and saturation patterns (Figure 6). The Rrs865/Rrs560 ratio exhibited a sharp nonlinear response: SHAP values remained near zero for turbidity below 30 FNU and ratio values below 0.75, then increased dramatically as the ratio exceeded 1.0, particularly between 100–1000 FNU. This threshold at the ratio ≈ 1.0 represents a physical transition where NIR backscattering begins to dominate over visible reflectance, signaling high particle concentrations. The contribution plateaued around 1500 FNU, suggesting saturation in the model's reliance on this feature.

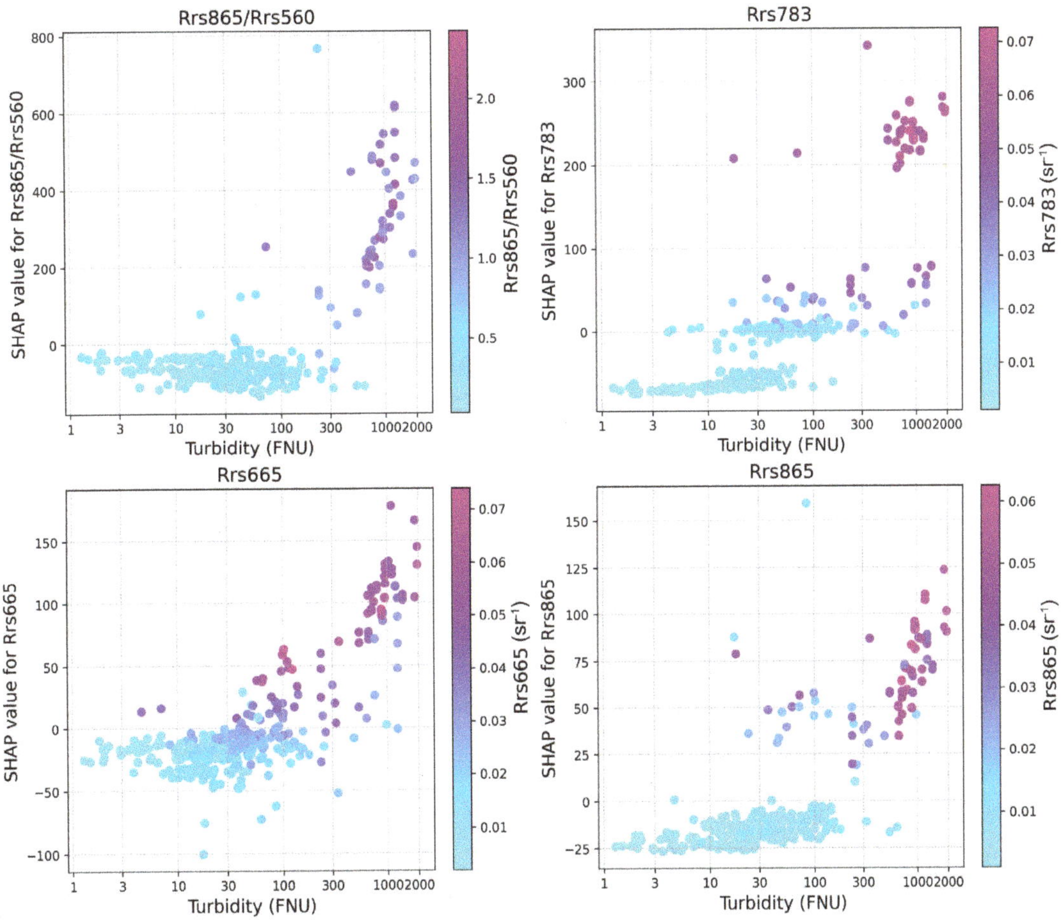

Figure 6. SHAP values versus turbidity levels, showing feature contributions across the 0–2200 FNU range. Colors indicate feature values, with a logarithmic x-axis emphasizing behavior at both low and high turbidity.

Rrs783 showed a similar but less pronounced pattern, with SHAP values remaining minimal below 30 FNU, then increasing steadily between 30–500 FNU before plateauing. Rrs665 demonstrated relatively linear behavior up to 500 FNU before saturating beyond ~750 FNU. In high turbid waters, the reflectance in Rrs665 often reaches an optical maximum, where additional suspended sediments do not further increase reflectance due to multiple scattering and absorption [17]. The model appears to have internalized this physical behavior, reducing the marginal weight of Rrs665 in the upper turbidity range.

Rrs865 provided gradual, consistent contributions across the entire range without sharp thresholds or clear saturation (Figure 6). This suggested that while Rrs865 was not the primary trigger for high turbidity predictions, it provided steady contextual information that reinforced trends identified by other features. Unlike Rrs865/Rrs560 or Rrs665, it did not dominate the model's output in any specific regime. Instead, it appeared to function as a stabilizing feature, contributing modestly but consistently to turbidity prediction across the measurement range.

The SHAP interaction plots revealed how secondary features modulated the primary features' contributions to turbidity predictions (Figure 7). The Rrs865/Rrs560 ratio showed consistent threshold behavior across all interactions (top row). When Rrs865/Rrs560 values remained below 0.5, SHAP contributions clustered near zero regardless of secondary feature values. However, once the ratio exceeded 1.0, SHAP values increased dramatically to 400–800 units, with the steepest increase occurring between ratios of 1.0–1.5. Furthermore, the threshold effects of Rrs865/Rrs560 reinforced when Rrs783, Rrs665, and Rrs865 reflectance values were also elevated, revealing a synergistic pattern, where multiple indicators of particle scattering aligned, strengthening the model's output. In contrast, the interactions between Rrs783 and Rrs665 were more additive and linear, suggesting that the model relied on both reflectance values independently to characterize turbidity. Notably, the SHAP contribution of Rrs665 saturated at high reflectance, which the model appeared to account for by tapering its influence in highly turbid waters.

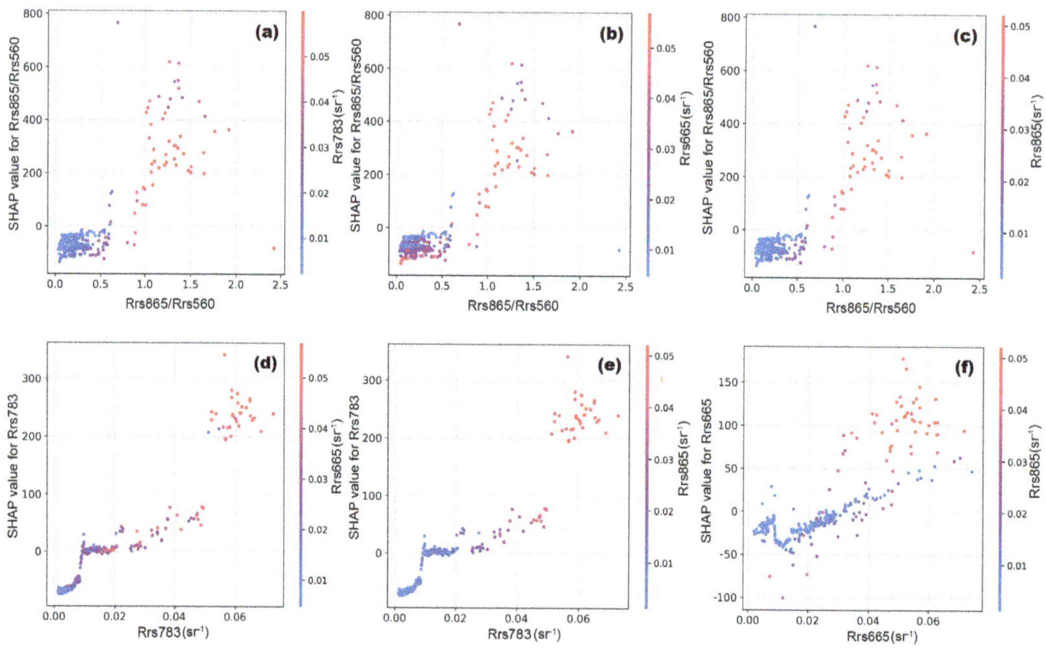

Figure 7. SHAP interaction plots showing pairwise feature dependencies, i.e., interactions between Rrs865/Rrs560 and Rrs783 (**a**), Rrs865/Rrs560 and Rrs665 (**b**), Rrs865/Rrs560 and Rrs865 (**c**), Rrs783 and Rrs665 (**d**), Rrs783 and Rrs865 (**e**), and Rrs665 and Rrs865 (**f**). Primary feature values on the x-axis, SHAP contributions on the y-axis, with color indicating secondary feature values that modulated the primary feature's impact.

Finally, the SHAP decision plot illustrated how the model sequentially accumulated evidence from individual features to reach final turbidity predictions (Figure 8). The x-axis at the top shows cumulative SHAP values that track the running sum of feature contributions, while the bottom x-axis displays the final model output values after adding these contributions to the baseline. Each line represents a single prediction's path, starting from the baseline (0, representing mean training turbidity) and sequentially incorporating feature contributions as the path moves upward through the feature stack.

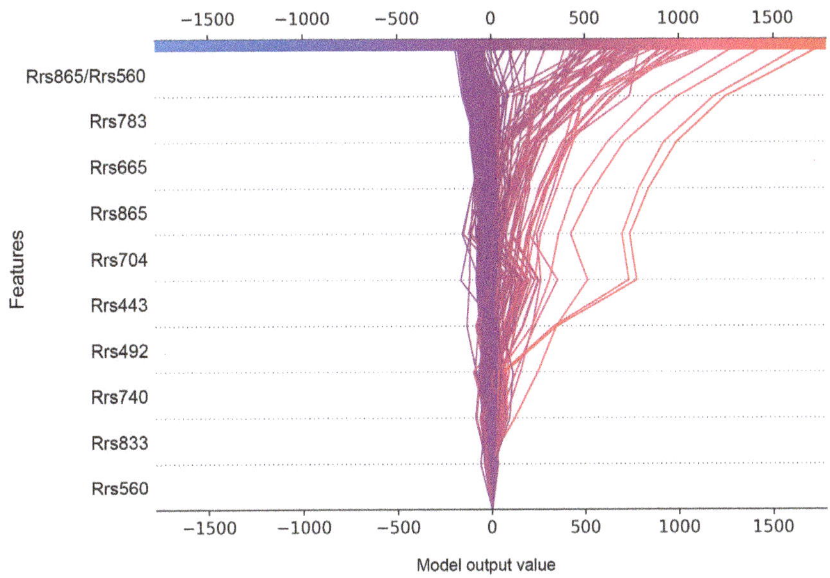

Figure 8. SHAP decision plot showing cumulative feature contributions from baseline (mean training turbidity) to final predictions. Each line represents one observation's prediction path, with colors indicating the final predicted values.

The plot revealed distinct prediction regimes. For example, low turbidity predictions (0–100 FNU) remained tightly clustered near zero, with minimal deviation even after incorporating all features. These paths showed that the low Rrs865/Rrs560 ratio values immediately pulled predictions leftward by −300 SHAP units, and subsequent features made only minor adjustments, keeping final predictions below 100 FNU. Moderate turbidity predictions between 100–1000 FNU showed gradual divergence from the baseline. The Rrs865/Rrs560 ratio provided positive contributions of SHAP units, followed by incremental additions from Rrs783 and Rrs665. These paths exhibited more variation as they progressed through the feature stack, reflecting the model's integration of multiple spectral signals to refine predictions within this intermediate range. At high turbidity, predictions (>1000 FNU) demonstrated dramatic rightward trajectories. The Rrs865/Rrs560 ratio alone often contributed to high SHAP units, immediately pushing predictions toward extreme values. Subsequent features (Rrs783 and Rrs665) provided secondary adjustments, while Rrs865 and Rrs704 offered minor refinements. The remaining features (Rrs443, Rrs492, Rrs740, Rrs833, and Rrs560) contributed negligibly, with paths showing no horizontal movement at these levels.

3.3. Model Performance Across Optical Water Types

The GBR model maintained robust performance across optically distinct water types (Figure 9), classified following the established spectral criteria [31]. In this case, the combined dataset containing both training and test observations was first classified as brown waters when Rrs665 > Rrs560 or Rrs665 > 0.025. If neither condition was met, but Rrs560 < Rrs490, the water was categorized as blue-green waters. Otherwise, the water was classified as green water.

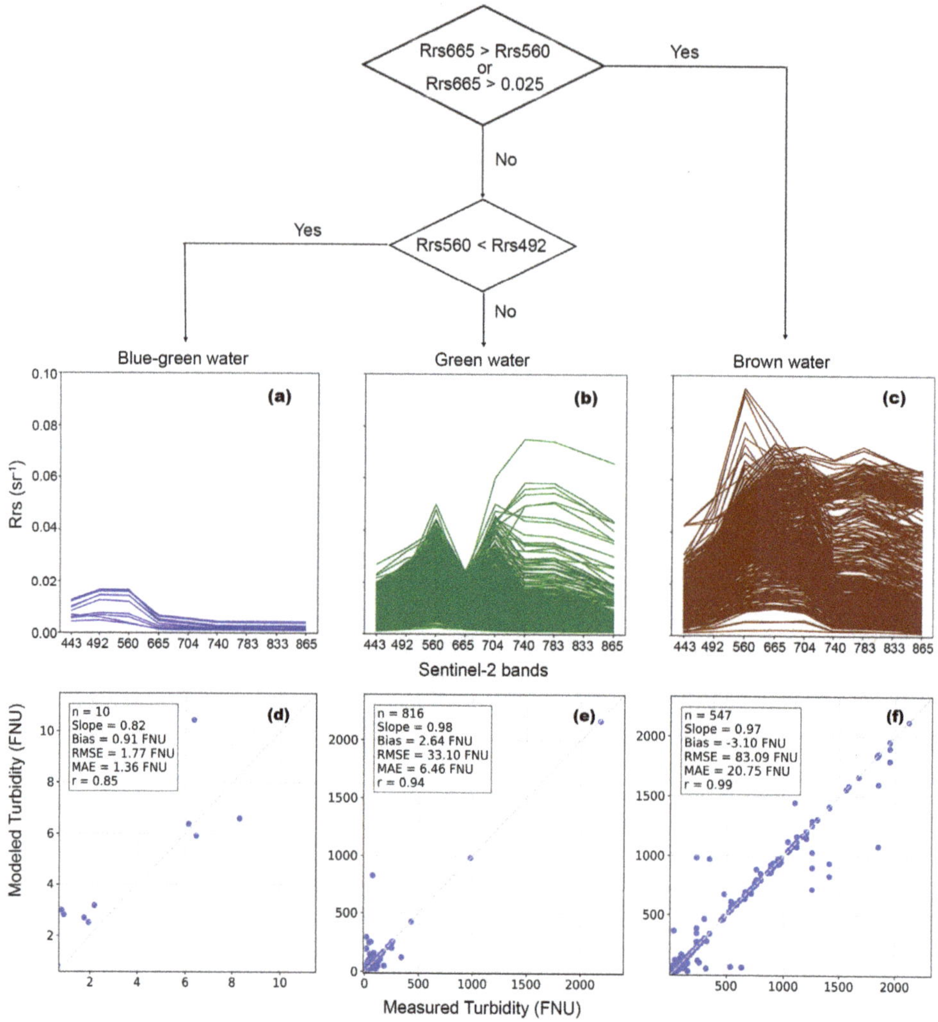

Figure 9. Performance of the proposed turbidity model across optically distinct water types classified using spectral criteria. Mean spectral reflectance signatures for blue-green (**a**), green (**b**), and brown (**c**) waters, showing distinct optical characteristics across Sentinel-2 bands. Scatterplots comparing model-derived and measured turbidity (in FNU) for the corresponding water types: blue-green (**d**), green (**e**), and brown (**f**) waters, indicating model's performance metrices across water types.

In oligotrophic blue-green waters (n = 10), the model achieved a strong correlation (r = 0.85) with minimal bias (0.91 FNU) and low errors (RMSE: 1.77 FNU, MAE: 1.36 FNU), though slight underestimation was observed (slope: 0.82). Green waters dominated by Chl-*a* and suspended matter (n = 816) showed excellent agreement (r: 0.94, slope: 0.98) with moderate errors (RMSE: 33.10 FNU, MAE: 6.46 FNU). Sediment-dominated brown waters (n = 547) demonstrated exceptional predictive accuracy (r: 0.99, slope: 0.97) despite the high absolute errors (RMSE = 83.09 FNU, MAE = 20.75 FNU), which are attributable to the wide turbidity range of up to 2200 FNU.

3.4. Model Application Across Diverse Geographic Settings

Model applications across four distinct aquatic systems demonstrated consistent performance independent of geographic location (Figure 10). Lake Taihu, China's third-largest freshwater lake experiencing severe eutrophication and cyanobacterial blooms [52], showed strong agreement (r: 0.91) across turbidity ranges from 0 to 250 FNU using data from 2008–2011. The model's application on the Red River (Vietnam), a major sediment-laden river system with an annual suspended sediment flux exceeding 100 million tons [53], exhibited a correlation of 0.80 for similar turbidity ranges (0 to 250 FNU) using field measurements from the year of 2017.

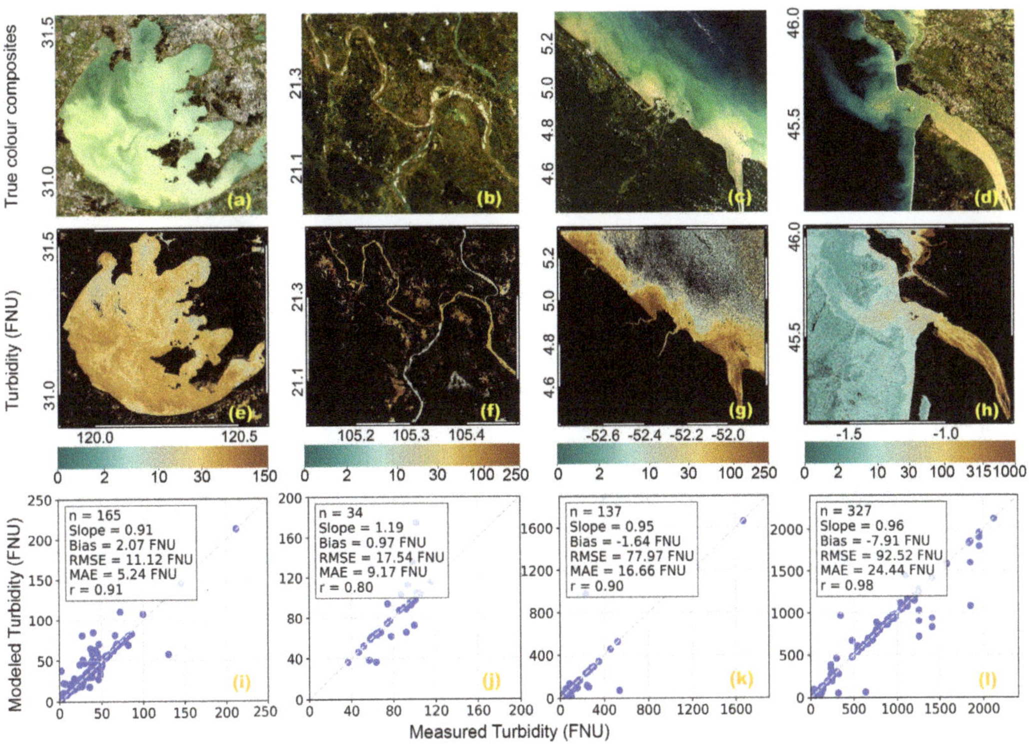

Figure 10. Model performance across diverse aquatic environments. Lake Taihu: (**a**,**e**,**i**), Red River: (**b**,**f**,**j**), French Guiana: (**c**,**g**,**k**), and Gironde Estuary: (**d**,**h**,**l**). Top and middle panels: example Sentinel-2 true-color composites and turbidity maps (FNU, land in black) from 3 March 2024, 14 November 2020, 7 September 2024, and 10 October 2021, respectively; bottom panel: scatterplots comparing model-derived and measured turbidity (in FNU): Lake Taihu (2008–2011), Red River (2017), French Guiana (2009–2017), and Gironde Estuary (2012–2023). Note: For Lake Taihu, Red River, and French Guiana, the GLORIA convolved Rrs were utilized to retrieve model-derived turbidity measurements and compared with co-located in situ turbidity measurements, whereas for Gironde Estuary, actual Sentinel-2 imagery was utilized.

Coastal waters of French Guiana (North Atlantic coast), characterized by extensive Amazon River sediment plumes and complex optical properties from CDOM and suspended sediments [54], showed strong model performance (r: 0.9) despite these challenging conditions. The Gironde Estuary (France), Europe's largest estuary with a well-documented maximum turbidity zone, where suspended sediment concentrations can exceed 1000 mg/L [55], maintained high correlation (r: 0.98) even for extreme turbidity values of up to 2000 FNU.

Turbidity maps generated for each system revealed characteristic spatial patterns consistent with documented hydrodynamic processes: wind-driven resuspension gradients in shallow Lake Taihu [56], downstream sediment transport patterns in the Red River delta, estuarine circulation in the Gironde Estuary [57], and northwest-flowing coastal plumes along French Guiana driven by the North Brazil Current [58]. These spatially coherent patterns, aligning with known regional oceanographic and limnological understanding, confirm the model's ability to capture physical phenomena across diverse environmental settings.

3.5. Model Validation in an Independent Site

The independent validation results (Figure 11) demonstrated the GBR model's exceptional generalization capability when applied to Albufera Lake, Spain. The model achieved noteworthy performance metrics with an RMSE of 19.39 FNU and an MAE of 15.29 FNU, representing an 83% reduction in RMSE compared to the test dataset (116.62 FNU). The high correlation (r = 0.81) and minimal bias (9.46 FNU) confirmed the model's ability to accurately predict turbidity in this independent water body without site-specific calibration.

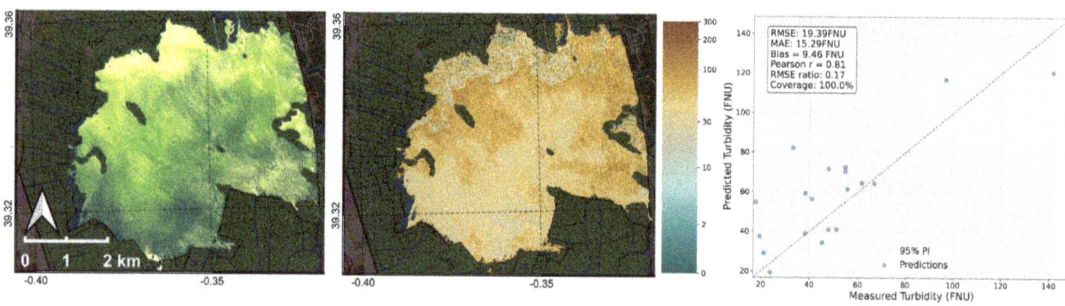

Figure 11. GBR model validation at Albufera Lake, Spain. (**Left**) Sentinel-2 RGB composite on 15 May 2023, (**center**): corresponding spatial distribution map of predicted turbidity (in FNU), and (**right**): measured versus predicted turbidity during 2023–2025 with 95% PI (n = 20).

The uncertainty quantification analysis revealed a particularly robust model behavior at Albufera Lake. All validation samples (100% coverage) fell within the 95% prediction intervals established from the test dataset, indicating that the model's uncertainty bounds remain reliable when extrapolated to new independent data. The RMSE ratio of 0.17 (validation/test) demonstrated that the model performed approximately six times better at Albufera than on the heterogeneous test dataset, likely due to the lake's more homogeneous optical properties compared to the diverse water bodies in the training and test datasets. This performance improvement, rather than degradation, suggested that the model successfully learned generalizable relationships between spectral signatures and turbidity that transferred well to lacustrine environments.

3.6. Performance of Automated Pipelines

The dockerized workflows demonstrated operational readiness through sustained performance metrics (Table 4). For turbidity monitoring, the pipeline achieved near-real-time capability through several optimizations. Processing time from satellite overpass to product delivery consistently remained under 3 h, with a typical runtime of less than 3 min per Sentinel-2 tile, excluding atmospheric correction. Leveraging SEP's Kubernetes parallel cluster processing capabilities, turbidity mapping across the entire Spanish–Portuguese coast (38 Sentinel-2 tiles) can be completed in under 3 min without atmospheric correction, and within 30 min to 1 h with atmospheric correction.

Table 4. Average runtime per step for the operational turbidity monitoring system.

Processing Steps	Description	Avg. Runtime
Satellite data download	Fetching Sentinel-2 imagery from defined sources	~30–45 min/100 images
Atmospheric correction	DSF-based correction using ACOLITE	~10–45 min/image
Feature extraction, and model implementation	Preparing input feature variables, and applying the pre-trained model to produce turbidity maps	Few min to <1 h
Validation and diagnostics	Comparing predictions with observations, computing accuracy metrics, etc.	Few min/image
Output logging	Saving trained model, metadata, and logs	Few s/image

4. Discussion

In this study, a globally applicable turbidity monitoring system has been developed using a Gradient Boosting Regression (GBR) model. GBR is a data-driven machine-learning (ML) technique, capable of capturing complex optical signatures across diverse aquatic environments at 10 m spatial resolution. The model's superior performance (r = 0.95, bias = 1.32 FNU) (Figure 4) across a 0–2200 FNU range substantially exceeded existing empirical algorithms [13,17], which failed at high turbidity levels with near-zero correlations and systematic underestimation. This performance gain stems from GBR's capacity to learn nonlinear spectral-turbidity relationships through iterative error correction [41,59–61], rather than relying on predetermined mathematical functions that cannot accommodate global optical variability.

Typically, turbidity increases reflectance in the visible and near-infrared (VNIR) wavelengths, particularly in green (500–600 nm), red (600–700 nm), and NIR (~800 nm) bands [62–66]. However, its spectral responses vary with sediment composition, phytoplankton abundance, and dissolved organic matter [67]. The physical interpretability revealed through SHAP analysis (Figures 5–8) demonstrated that the model has internalized these fundamental optical principles. The dominance of the Rrs865/Rrs560 ratio, with its sharp transition at the ratio $\approx$ 1.0 indicated the physical shift from absorption-dominated to scattering-dominated optical regimes as particle concentrations increase. Similarly, the model automatically accounts for the saturation of Rrs665 reflectance at high turbidity levels (a known optical limit where multiple scattering prevents further reflectance increases [17,65]), by reducing band weighting at extreme turbidity levels. This learned representation of optical physics provides confidence in the model's extrapolation capabilities beyond its training domain.

The model's validation at the Albufera Lake (Figure 11) and its performance across different optical water types, i.e., blue-green, green, and sediment-rich brown waters, as well as across different geographical settings (Figures 9 and 10), demonstrated a consistently high predictive reliability. Although a slight underestimation in low-turbidity blue-green waters was observed, exceptional performance was achieved in sediment-dominated waters. The model's performance from the eutrophic lakes Albufera [68] and Taihu [52] to the hyper-turbid Gironde Estuary [55] confirmed that a single global model can capture site-specific hydrodynamic processes without regional tuning. The spatial representation of turbidity plumes (Figures 10 and 11) indicated the model's sensitivity to underlying physical drivers.

While region-specific models might achieve marginally higher local accuracy through reduced complexity [26,69], the operational advantages of a unified global approach—consistency, interoperability, and simplified maintenance outweigh minor performance trade-offs.

The successful integration of heterogeneous datasets (GLORIA and MAGEST) through rigorous quality control and spectral harmonization addressed a fundamental challenge in the global algorithm development: the lack of consistent, quality-assured training data spanning diverse optical conditions. Though the proxy-derived GLORIA turbidity values (Table 2, Figure A1) introduced quantifiable uncertainty (i.e., 37–44% relative MAE), they enabled critical coverage of under-sampled environments. This pragmatic approach balanced data quality with representativeness, a trade-off inherent in transitioning from local to global monitoring capabilities.

Atmospheric correction remained another source of uncertainty, particularly in NIR bands (Figure 1), where ACOLITE's performance degrades at high turbidity levels [35]. This uncertainty in Rrs783 and Rrs865 (key bands for the GBR model) within the MAGEST dataset propagated through turbidity estimates, suggesting that further improvements in atmospheric correction algorithms would yield improved accuracy. Future investigations should systematically compare existing correction processors across diverse atmospheric and aquatic conditions to quantify and potentially correct these uncertainties. Additionally, it is crucial to assess the impact of adjacency effects and bottom reflectance contributions, both of which are prevalent yet challenging to quantify accurately. These factors can significantly influence the retrieved water properties and, if not properly accounted for, may introduce biases in the analysis. A comprehensive evaluation of these elements in future analysis, i.e., evaluating and utilizing the RAdCor algorithm within the ACOLITE processor, will further enhance the reliability of remote sensing observation and improve the accuracy of the turbidity assessment.

The operational pipeline's achievement—3 h processing time from satellite overpass to turbidity product—represents a critical advancement for environmental monitoring applications requiring timely information. The containerized architecture ensured reproducibility while enabling seamless integration of future improvements, including new training data, alternative sensors, or enhanced atmospheric correction algorithms. This technical infrastructure transformed the algorithm from a research tool to an operational capability suitable for routine monitoring programs [44].

However, data limitations at extreme turbidity ranges highlight priorities for future development. The underrepresentation of oligotrophic waters (n = 10) (Figure 9) explains the observed underestimation in clear conditions, while sparse sampling above 1500 FNU limits confidence in hyperturbid environments (Figure 4). Recognizing that data sparsity, particularly at extreme turbidity levels, poses a significant challenge, alternative solutions should be explored. Approaches such as data simulations, data augmentation strategies, or advanced ML techniques can help expand the dataset, ensuring a more comprehensive and representative training set. Additionally, incorporating sensors with different spectral or spatial characteristics can address current limitations in extending the observable turbidity range.

5. Conclusions

This study presents a globally applicable turbidity retrieval model developed from harmonized global datasets (GLORIA and MAGEST) and Sentinel-2 observations using a Gradient Boosting Regression model. The model successfully extends the measurable turbidity range up to 2200 FNU, surpassing existing algorithms, while maintaining strong predictive accuracy (r = 0.95, bias = 1.32 FNU). Its interpretability through SHAP analysis confirms a physically consistent representation of optical behavior, with the Rrs865/Rrs560 ratio emerging as the dominant predictor across diverse aquatic environments. The model demonstrates robust transferability across optical water types and geographic regions, establishing its readiness for operational monitoring.

By combining open-source datasets, machine-learning transparency, and automated processing pipelines, this research bridges the gap between algorithm development and real-time implementation. The proposed system can enhance near-real-time global turbidity monitoring, supporting evidence-based water management and environmental response to climate change impacts and anthropogenic pressures. Future work should focus on refining atmospheric corrections (e.g., adjacency effect adjustments), expanding training datasets to oligotrophic and hyperturbid conditions, and integrating multisensory data to further improve reliability and scalability. As environmental pressures intensify and monitoring demands expand, the capacity for consistent, automated, and physically interpretable turbidity assessment becomes increasingly critical for understanding and mitigating impacts on aquatic ecosystems globally.

Author Contributions: Conceptualization, M.C., A.B.R., I.L. and I.d.l.C.; methodology, M.C. and A.B.R.; software, M.C.; validation, M.C., A.B.R. and I.L.; formal analysis, M.C.; investigation, M.C. and A.B.R.; resources, M.C., A.B.R., I.d.l.C. and I.L.; data curation, M.C.; writing—original draft preparation, M.C.; writing—review and editing, M.C., A.B.R., I.d.l.C. and I.L.; visualization, M.C., A.B.R., I.L. and I.d.l.C.; supervision, A.B.R., I.d.l.C. and I.L.; project administration, I.L. and I.d.l.C.; funding acquisition, I.L. and I.d.l.C. All authors have read and agreed to the published version of the manuscript.

Funding: M.C. is a student at the University of Cadiz and is currently employed by the company Quasar Science Resources S.L. Consequently, M.C.'s research is funded 50% by Quasar S.R. and 50% by the Industrial Doctorate Program of the Spanish Ministerio de Ciencia e Innovación (ref. DIN2020-010979/AEI/10.13039/501100011033). This work forms part of M.C.'s PhD research within the SIMBAD project (ref. QSR-ESABIC-2018-001), incubated by ESA-BIC Madrid Region, and includes a research collaboration with the University of Valencia. A.B.R. is involved in the AI4CS–GVA PROMETEO project, titled "Artificial Intelligence for complex systems: Brain, Earth, Climate, Society", funded by Conselleria de Innovación, Universidades, Ciencia y Sociedad Digital, CIPROM/2021/56. I.L. is part of the Projects ref. PID2020-112488RB-I00 and ref. TED2021-132439B-I00, funded by the Spanish Ministerio de Ciencia e Innovación, and project ref. HORIZON-CL5-2022-D1-02, funded by the EU HORIZON-CL5-2022-D1-02-05 Programme. M.C. and I.L. are members of the project ref. PID2023-146617OB-I00, funded by the Spanish Ministerio de Ciencia, Innovación y Universidades.

Data Availability Statement: The necessary procedures to generate the maps and reproduce the methodology have been outlined in the manuscript. In situ GLORIA and MAGEST datasets used in this study were collected from https://doi.org/10.1594/PANGAEA.948492 [28] and https://magest.oasu.u-bordeaux.fr/index.php [29] (accessed on 7 November 2025), respectively. Sentinel-2 data were downloaded from https://dataspace.copernicus.eu/ (accessed on 7 November 2025). The data that supports the findings of this study are available in Table A1. This study primarily employed established workflows using widely available open-source tools, including ACOLITE (v20231023) for atmospheric correction, and Python libraries (e.g., numpy (v1.26.4), pandas (v2.2.2), sklearn (v1.5.1), matplotlib (v3.9.1), seaborn (v0.13.2)) for model training, statistical analysis, and visualization. Additional open-source tools, such as Dockers (v4.44.2 (202017)), were used for pipeline development. We would like to respectfully note that this research was conducted as part of an Industrial Doctoral Programme in collaboration with a private company. As such, while we have provided complete details of all analytical procedures in Section 2 to ensure scientific reproducibility, we are unable to share internal scripts or components that may be subject to commercial confidentiality. We trust that the level of methodological detail provided in the manuscript meets the Journal's standards, and we remain fully committed to supporting reproducibility to the extent possible within the constraints of an industrial research setting.

Acknowledgments: The authors are grateful to Lehmann et al. (2022) [28] and MAGEST Network on Monitoring the water quality of the Gironde Estuary [29] for freely distributing the GLORIA and MAGEST data, and to the European Space Agency, the European Commission, and the Copernicus Programme for freely distributing the Sentinel-2 imagery. The authors are also thankful to David Doxaran, Julia Amorós-López, Jorge García-Jiménez, and Patricia Urrego for providing data or assistance to improve the research methodology. Additionally, the authors would like to thank A. Ashiqul Mursalin Chy for preparing the 3D methodological diagram presented in Figure 3.

Conflicts of Interest: The authors declare no conflicts of interest. The funders had no role in the design of the study, in the collection, analyses, or interpretation of data, in the writing of the manuscript, or in the decision to publish the results.

Abbreviations

The following abbreviations are used in this manuscript:

SeaWiFS	Sea-Viewing Wide Field-of-view Sensor
MODIS	Moderate Resolution Imaging Spectroradiometer
MERIS	Medium Resolution Imaging Spectrometer
MSI	MultiSpectral Instrument
ML	Machine-learning
FNU	Formazin Nephelometric Units
GLORIA	GLObal Reflectance community dataset for Imaging and optical sensing of Aquatic environments
MAGEST	Monitoring the water quality of the Gironde Estuary
SHAP	SHapley Additive exPlanations
AI	Artificial Intelligence
Chl-a	Chlorophyll-a
Rrs	Remote Sensing Reflectance
TSS	Total Suspended Solids
SDT	Secchi-depth Transparency
QWIP	Quality Water Index Polynomial
RMSE	Root Mean Squared Error
MAE	Mean Absolute Error
r	Correlation Coefficient
SRF	Spectral Response Function
API	Application Programming Interface
DSF	Dark Spectrum Fitting
CV	Coefficient of Variation
CDOM	Colored Dissolved Organic Matter
ENR	Elastic Net
RFR	Random Forest
GBR	Gradient Boosting
XGBR	Extreme Gradient Boosting
VNIR	Visible and Near-Infrared
CI	Confidence Intervals
PI	Prediction Intervals
SEP	Scientific Exploitation Platform
SIMBAD	Sentinel Imagery Multiband Analysis and Dissemination

Appendix A

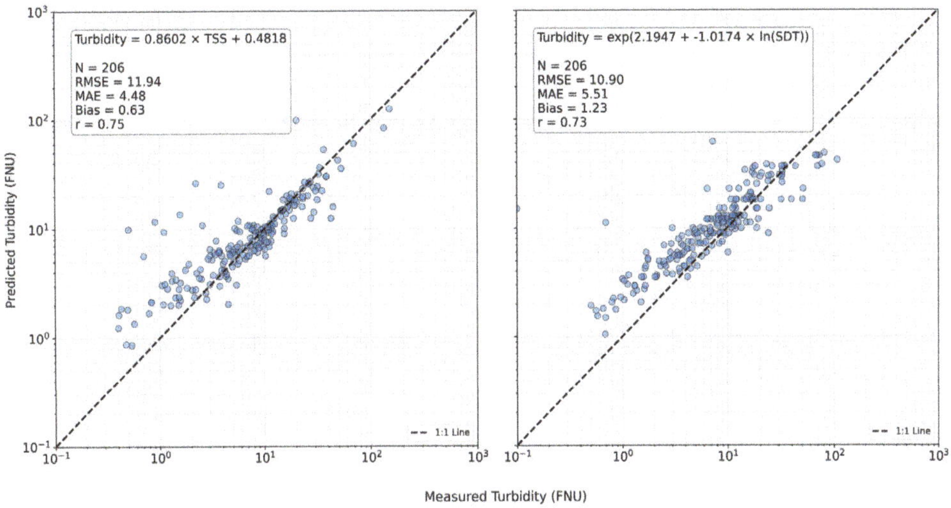

Figure A1. Turbidity estimation from TSS (left panel) and SDT (right panel) for the GLORIA dataset.

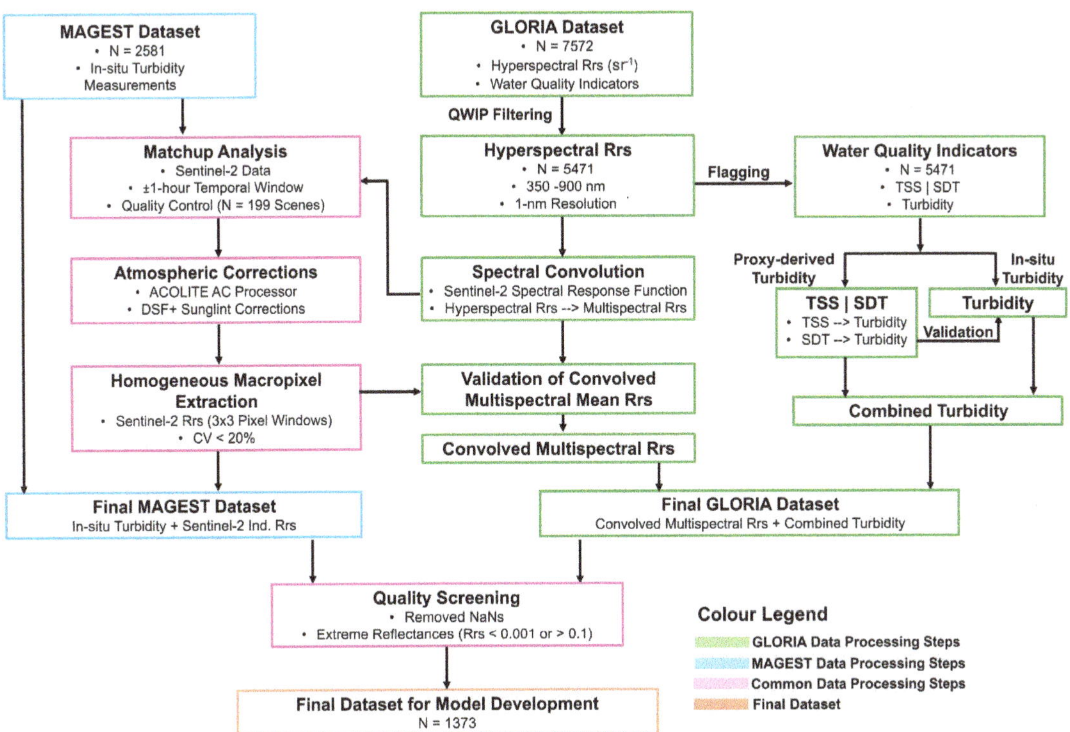

Figure A2. GLORIA and MAGEST data pre-processing workflow.

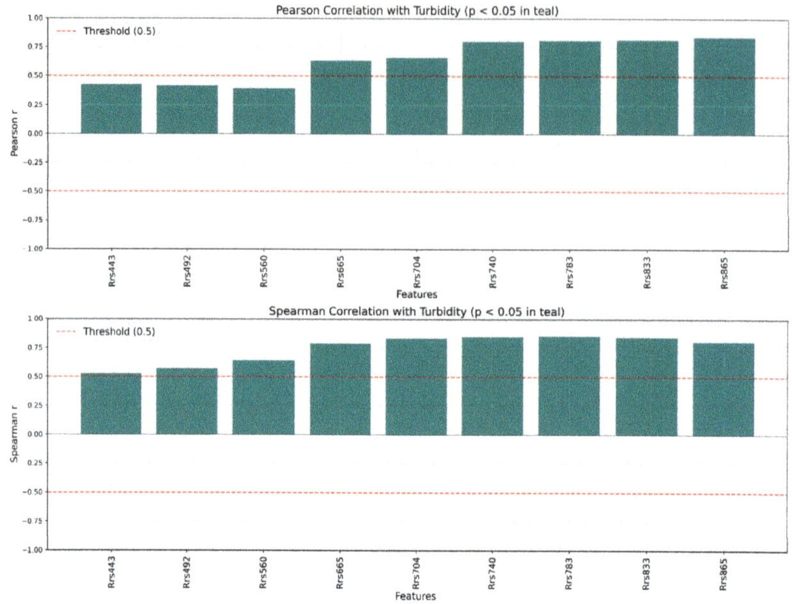

Figure A3. Linear and Spearman correlation analysis between in situ turbidity and Sentinel-2 bands. Statistically significant correlations ($p < 0.05$) are highlighted in teal color.

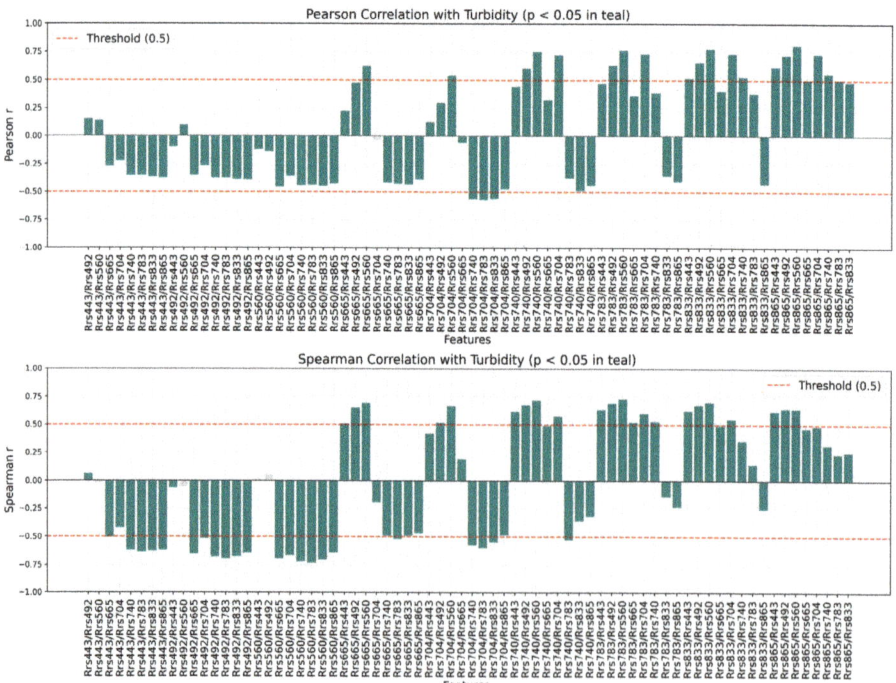

Figure A4. Linear and Spearman correlation analysis between in situ turbidity and Sentinel-2 band ratios. Statistically significant correlations ($p < 0.05$) are highlighted in teal color.

Table A1. Data used for model training and performance evaluation to support the findings of this study.

Country	Site Name	Water Body Type	Water Type	No. of Samples	Turbidity Range (FNU)
Australia					
Australia	Burrinjuck Dam	Lake	Others	1	11.49–11.49
Australia	Lake Burley Griffin	Lake	Others	2	10.23–13.67
Australia	Lake Hume	Lake	Others	1	21.92–21.92
Australia	Lake Pamamaroo	Lake	Others	7	102.56–151.02
Australia	Lake Victoria	Lake	Others	2	17.49–22.85
Australia	Wachtels Lagoon	Lake	Others	3	32.40–36.76
Australia	Western Treatment Plant	Others	Others	1	51.52–51.52
Belgium					
Belgium	English Channel	Coastal ocean	Sediment-dominated	12	4.58–46.74
Brazil					
Brazil	Curuai Lake	Lake	Sediment-dominated	14	3.92–37.90
Brazil	Ibitinga Reservoir	Lake	Chl-*a*-dominated	15	3.23–50.60
China					
China	Chaohu	Lake	Sediment-dominated	40	12.52–339.40
China	Dianchi	Lake	Chl-*a*-dominated	63	19.84–158.76
China	Erhai	Lake	Chl-*a*-dominated	3	6.00–8.30
China	Hou Lake	Lake	Chl-*a*-dominated	22	18.40–102.50
China	Liangzi Lake	Lake	Chl-*a*-dominated	33	9.14–108.00
China	Poyang Lake	Lake	Chl-*a*-dominated	12	7.18–30.60
China	Taihu	Lake	Sediment-dominated	165	1.89–212.09
China	Wuhan East Lake	Lake	Chl-*a*-dominated	20	3.13–37.55
Estonia					
Estonia	Lake Holstre	Lake	Chl-*a* and CDOM-dominated	1	9.73–9.73
Estonia	Lake Kaiavere	Lake	Chl-*a* and CDOM-dominated	1	9.66–9.66
Estonia	Lake Kooraste Linajarv	Lake	Chl-*a* and CDOM-dominated	1	16.65–16.65
Estonia	Lake Nohipalu Valgjarv	Lake	Chl-*a* and CDOM-dominated	1	2.20–2.20
Estonia	Lake Pangodi	Lake	Chl-*a* and CDOM-dominated	1	4.07–4.07
Estonia	Lake Peipsi	Lake	Chl-*a* and CDOM-dominated	21	1.13–22.85
Estonia	Lake Rouge Suurjarv	Lake	Chl-*a* and CDOM-dominated	2	2.57–3.92

Table A1. Cont.

Country	Site Name	Water Body Type	Water Type	No. of Samples	Turbidity Range (FNU)
Estonia	Lake Vortsjarv	Lake	Chl-a and CDOM-dominated	7	11.09–17.69
Estonia	Parnu Bay	Estuary	Chl-a and CDOM-dominated	4	12.87–17.34
Finland					
Finland	Lake Vanttausjarvi	Lake	Chl-a and CDOM-dominated	1	2.37–2.37
France					
France	Arcachon Bay	Coastal ocean	Sediment-dominated	17	1.64–1562.04
France	Arcachon Bay	Coastal ocean	Others	1	1.42–1.42
France	English Channel	Coastal ocean	Others	1	0.84–0.84
France	Gironde River	Estuary	Sediment-dominated	31	7.48–2124.59
France	Gironde River	River	Sediment-dominated	296	21.70–1953.84
France	Guiana	Coastal ocean	Sediment-dominated	126	2.70–1673.86
France	Guiana	Coastal ocean	Others	2	30.12–32.21
France	Guiana	Coastal ocean	Others	2	32.31–69.19
France	Guiana	Coastal ocean	Chl-a-dominated	7	10.28–258.20
Italy					
Italy	Garda	Lake	Clear water	2	0.57–5.00
Italy	Iseo	Lake	Clear water	4	2.04–3.49
Italy	Mantova	Lake	Chl-a-dominated	15	4.97–15.11
Italy	Trasimeno	Lake	Sediment-dominated	7	3.92–24.57
Japan					
Japan	Lake Kasumigaura	Lake	Chl-a-dominated	84	8.87–42.67
Japan	Shirakaba	Lake	Chl-a-dominated	1	8.28–8.28
Japan	Suwa	Lake	Chl-a-dominated	3	7.56–8.75
Netherlands					
Netherlands (the)	English Channel	Coastal ocean	Sediment-dominated	12	1.92–58.40
Netherlands (the)	Ijsselmeer De Oude Zeug	Lake	Chl-a-dominated	1	21.07–21.07
Netherlands (the)	Loosdtrechtse plassen nr5	Lake	CDOM-dominated	1	14.68–14.68
Netherlands (the)	North Sea	Coastal ocean	Sediment-dominated	24	1.45–21.09
South Africa					
South Africa	Bronkhorstspruit	Lake	Chl-a-dominated	4	7.02–11.66
South Africa	Hartbeespoort	Lake	Chl-a-dominated	11	4.78–2179.62
South Africa	Loskop	Lake	Chl-a-dominated	8	15.02–23.62

Table A1. *Cont.*

Country	Site Name	Water Body Type	Water Type	No. of Samples	Turbidity Range (FNU)
South Africa	Roodeplaat	Lake	Chl-*a*-dominated	9	3.92–50.80
South Africa	Theewaterskloof	Lake	Chl-*a*-dominated	10	8.90–25.53
South Africa	Vaal	Lake	Chl-*a*-dominated	5	4.61–72.74
Spain					
Spain	Aguilar	Lake	Chl-*a*-dominated	1	8.54–8.54
Spain	Alarcón	Lake	Chl-*a*-dominated	1	5.57–5.57
Spain	Albufera	Lake	Chl-*a*-dominated	19	26.91–93.45
Spain	Alcántara	Lake	Chl-*a*-dominated	6	4.22–18.06
Spain	Almendra	Lake	Chl-*a*-dominated	1	4.80–4.80
Spain	Brovales	Lake	Chl-*a*-dominated	1	14.37–14.37
Spain	Contreras	Lake	Chl-*a*-dominated	1	3.88–3.88
Spain	Cortes	Lake	Chl-*a*-dominated	1	2.19–2.19
Spain	Ebro	Lake	Chl-*a*-dominated	3	4.94–17.46
Spain	Giribaile	Lake	Chl-*a*-dominated	1	3.86–3.86
Spain	Guadalén	Lake	Chl-*a*-dominated	1	8.71–8.71
Spain	Navalcán	Lake	Chl-*a*-dominated	3	12.54–18.17
Spain	Pinilla	Lake	Chl-*a*-dominated	3	4.44–8.15
Spain	Rosarito	Lake	Chl-*a*-dominated	30	1.82–25.39
Spain	Santa Teresa	Lake	Chl-*a*-dominated	1	3.85–3.85
Spain	Santillana	Lake	Chl-*a*-dominated	1	8.98–8.98
Spain	Terradets	Lake	Sediment-dominated	2	10.47–29.56
Spain	Ullívarri	Lake	Chl-*a* and CDOM-dominated	1	1.90–1.90
Spain	Valuengo	Lake	Chl-*a*-dominated	2	12.37–20.70
Spain	Vega de Jabalón	Lake	Chl-*a*-dominated	1	13.92–13.92
Sweden					
Sweden	Baltic Sea	Coastal ocean	Others	1	1.06–1.06
Sweden	Lake Vänern	Lake	Others	2	1.25–1.29
Sweden	Lake Vanern	Lake	Chl-*a*-dominated	1	2.31–2.31
Switzerland					
Switzerland	Lake Biel	Lake	Others	3	2.19–2.95
Switzerland	Lake Geneva	Lake	Others	3	0.74–2.89
United Kingdom					
United Kingdom of Great Britain and Northern Ireland (the)	Bassenthwaite Lake	Lake	Others	1	1.49–1.49
United Kingdom of Great Britain and Northern Ireland (the)	English Channel	Coastal ocean	Sediment-dominated	9	1.74–102.27

Table A1. *Cont.*

Country	Site Name	Water Body Type	Water Type	No. of Samples	Turbidity Range (FNU)
United Kingdom of Great Britain and Northern Ireland (the)	Windermere	Lake	Others	2	1.98–5.51
Uruguay					
Uruguay	Embalse de Paso del Palmar	Lake	Chl-*a*-dominated	24	12.70–100.00
Uruguay	Embalse de Paso del Palmar	Lake	Sediment-dominated	9	12.20–19.00
Uruguay	Lago Rincon del Bonete	Lake	Sediment-dominated	10	1.90–15.00
Uruguay	Lago Rincon del Bonete	Lake	Clear water	1	14.20–14.20
Viet Nam					
Viet Nam	Ba Be Lake	Lake	Chl-*a*-dominated	2	13.38–15.19
Viet Nam	Gulf of Tonkin	Coastal ocean	Sediment-dominated	23	21.71–127.52
Viet Nam	Ha Long Bay, Quang Ninh Province	Coastal ocean	Chl-*a* and CDOM-dominated	7	12.01–14.29
Viet Nam	Red River	River	Sediment-dominated	34	36.95–115.40
Viet Nam	Soai Rap River	Coastal ocean	Sediment-dominated	6	35.97–207.05
Viet Nam	West Lake, Hanoi	Lake	Chl-*a*-dominated	15	20.95–70.37

References

1. Cloern, J.E. Turbidity as a Control on Phytoplankton Biomass and Productivity in Estuaries. *Cont. Shelf Res.* **1987**, *7*, 1367–1381. [CrossRef]
2. Doan, P.T.K.; Némery, J.; Schmid, M.; Gratiot, N. Eutrophication of Turbid Tropical Reservoirs: Scenarios of Evolution of the Reservoir of Cointzio, Mexico. *Ecol. Inform.* **2015**, *29*, 192–205. [CrossRef]
3. Neukermans, G.; Ruddick, K.G.; Greenwood, N. Diurnal Variability of Turbidity and Light Attenuation in the Southern North Sea from the SEVIRI Geostationary Sensor. *Remote Sens. Environ.* **2012**, *124*, 564–580. [CrossRef]
4. Potes, M.; Costa, M.J.; Salgado, R. Satellite Remote Sensing of Water Turbidity in Alqueva Reservoir and Implications on Lake Modelling. *Hydrol. Earth Syst. Sci. Discuss.* **2012**, *16*, 1623–1633. [CrossRef]
5. Wang, Y.; Wu, H.; Lin, J.; Zhu, J.; Zhang, W.; Li, C. Phytoplankton Blooms off a High Turbidity Estuary: A Case Study in the Changjiang River Estuary. *J. Geophys. Res. Ocean.* **2019**, *124*, 8036–8059. [CrossRef]
6. Lee, H.W.; Kim, E.J.; Park, S.S.; Choi, J.H. Effects of Climate Change on the Movement of Turbidity Flow in a Stratified Reservoir. *Water Resour Manag.* **2015**, *29*, 4095–4110. [CrossRef]
7. Mi, H.; Fagherazzi, S.; Qiao, G.; Hong, Y.; Fichot, C.G. Climate Change Leads to a Doubling of Turbidity in a Rapidly Expanding Tibetan Lake. *Sci. Total Environ.* **2019**, *688*, 952–959. [CrossRef]
8. U.S. Environmental Protection Agency. Climate Adaptation and Erosion & Sedimentation. Available online: https://www.epa.gov/arc-x/climate-adaptation-and-erosion-sedimentation (accessed on 2 May 2025).
9. Vergara, I.; Garreaud, R.; Ayala, Á. Sharp Increase of Extreme Turbidity Events Due to Deglaciation in the Subtropical Andes. *J. Geophys. Res. Earth Surf.* **2022**, *127*, e2021JF006584. [CrossRef]
10. León-Muñoz, J.; Aguayo, R.; Corredor-Acosta, A.; Tapia, F.J.; Iriarte, J.L.; Reid, B.; Soto, D. Hydrographic Shifts in Coastal Waters Reflect Climate-Driven Changes in Hydrological Regimes across Northwestern Patagonia. *Sci. Rep.* **2024**, *14*, 20632. [CrossRef]
11. Hou, X.; Feng, L.; Duan, H.; Chen, X.; Sun, D.; Shi, K. Fifteen-Year Monitoring of the Turbidity Dynamics in Large Lakes and Reservoirs in the Middle and Lower Basin of the Yangtze River, China. *Remote Sens. Environ.* **2017**, *190*, 107–121. [CrossRef]

12. Nechad, B.; Ruddick, K.G.; Park, Y. Calibration and Validation of a Generic Multisensor Algorithm for Mapping of Total Suspended Matter in Turbid Waters. *Remote Sens. Environ.* **2010**, *114*, 854–866. [CrossRef]
13. Dogliotti, A.I.; Ruddick, K.G.; Nechad, B.; Doxaran, D.; Knaeps, E. A Single Algorithm to Retrieve Turbidity from Remotely-Sensed Data in All Coastal and Estuarine Waters. *Remote Sens. Environ.* **2015**, *156*, 157–168. [CrossRef]
14. van der Woerd, H.; Pasterkamp, R. Mapping of the North Sea Turbid Coastal Waters Using SeaWiFS Data. *Can. J. Remote Sens.* **2004**, *30*, 44–53. [CrossRef]
15. Magrì, S.; Ottaviani, E.; Prampolini, E.; Besio, G.; Fabiano, B.; Federici, B. Application of Machine Learning Techniques to Derive Sea Water Turbidity from Sentinel-2 Imagery. *Remote Sens. Appl. Soc. Environ.* **2023**, *30*, 100951. [CrossRef]
16. Chen, C.; Zhang, C.; Tian, B.; Wu, W.; Zhou, Y. Mapping Intertidal Topographic Changes in a Highly Turbid Estuary Using Dense Sentinel-2 Time Series with Deep Learning. *ISPRS J. Photogramm. Remote Sens.* **2023**, *205*, 1–16. [CrossRef]
17. Chowdhury, M.; Vilas, C.; VanBergeijk, S.; Navarro, G.; Laiz, I.; Caballero, I. Monitoring Turbidity in a Highly Variable Estuary Using Sentinel 2-A/B for Ecosystem Management Applications. *Front. Mar. Sci.* **2023**, *10*, 1186441. [CrossRef]
18. Jiang, D.; Matsushita, B.; Pahlevan, N.; Gurlin, D.; Fichot, C.G.; Harringmeyer, J.; Sent, G.; Brito, A.C.; Brotas, V.; Werther, M.; et al. Estimating the Concentration of Total Suspended Solids in Inland and Coastal Waters from Sentinel-2 MSI: A Semi-Analytical Approach. *ISPRS J. Photogramm. Remote Sens.* **2023**, *204*, 362–377. [CrossRef]
19. Sebastiá-Frasquet, M.-T.; Aguilar-Maldonado, J.A.; Santamaría-Del-Ángel, E.; Estornell, J. Sentinel 2 Analysis of Turbidity Patterns in a Coastal Lagoon. *Remote Sens.* **2019**, *11*, 2926. [CrossRef]
20. Zeng, F.; Song, C.; Cao, Z.; Xue, K.; Lu, S.; Chen, T.; Liu, K. Monitoring Inland Water via Sentinel Satellite Constellation: A Review and Perspective. *ISPRS J. Photogramm. Remote Sens.* **2023**, *204*, 340–361. [CrossRef]
21. Luo, Y.; Doxaran, D.; Vanhellemont, Q. Retrieval and Validation of Water Turbidity at Metre-Scale Using Pléiades Satellite Data: A Case Study in the Gironde Estuary. *Remote Sens.* **2020**, *12*, 946. [CrossRef]
22. Bid, S.; Siddique, G. Identification of Seasonal Variation of Water Turbidity Using NDTI Method in Panchet Hill Dam, India. *Model. Earth Syst. Environ.* **2019**, *5*, 1179–1200. [CrossRef]
23. Yang, Z.; Gong, C.; Lu, Z.; Wu, E.; Huai, H.; Hu, Y.; Li, L.; Dong, L. Combined Retrievals of Turbidity from Sentinel-2A/B and Landsat-8/9 in the Taihu Lake through Machine Learning. *Remote Sens.* **2023**, *15*, 4333. [CrossRef]
24. Bustamante, J.; Pacios, F.; Díaz-Delgado, R.; Aragonés, D. Predictive Models of Turbidity and Water Depth in the Doñana Marshes Using Landsat TM and ETM+ Images. *J. Environ. Manag.* **2009**, *90*, 2219–2225. [CrossRef] [PubMed]
25. CMEMS. CMEMS Copernicus Marine Service to Deliver High-Resolution Ocean Colour Products Using Sentinel-2. Available online: https://marine.copernicus.eu/news/copernicus-marine-service-deliver-high-resolution-ocean-colour-products-using-sentinel-2 (accessed on 27 June 2025).
26. Cao, Z.; Ma, R.; Duan, H.; Pahlevan, N.; Melack, J.; Shen, M.; Xue, K. A Machine Learning Approach to Estimate Chlorophyll-*a* from Landsat-8 Measurements in Inland Lakes. *Remote Sens. Environ.* **2020**, *248*, 111974. [CrossRef]
27. Ruescas, A.B.; Hieronymi, M.; Mateo-Garcia, G.; Koponen, S.; Kallio, K.; Camps-Valls, G. Machine Learning Regression Approaches for Colored Dissolved Organic Matter (CDOM) Retrieval with S2-MSI and S3-OLCI Simulated Data. *Remote Sens.* **2018**, *10*, 786. [CrossRef]
28. Lehmann, M.K.; Gurlin, D.; Pahlevan, N.; Alikas, K.; Anstee, J.M.; Balasubramanian, S.V.; Barbosa, C.C.F.; Binding, C.; Bracher, A.; Bresciani, M.; et al. GLORIA—A Global Dataset of Remote Sensing Reflectance and Water Quality from Inland and Coastal Waters [dataset]. *PANGAEA* **2022**. [CrossRef]
29. MAGEST. MAGEST Network. Available online: https://magest.oasu.u-bordeaux.fr/index.php (accessed on 19 August 2024).
30. Lundberg, S.M.; Erion, G.; Chen, H.; DeGrave, A.; Prutkin, J.M.; Nair, B.; Katz, R.; Himmelfarb, J.; Bansal, N.; Lee, S.-I. From Local Explanations to Global Understanding with Explainable AI for Trees. *Nat. Mach. Intell.* **2020**, *2*, 56–67. [CrossRef]
31. Dierssen, H.M.; Vandermeulen, R.A.; Barnes, B.B.; Castagna, A.; Knaeps, E.; Vanhellemont, Q. QWIP: A Quantitative Metric for Quality Control of Aquatic Reflectance Spectral Shape Using the Apparent Visible Wavelength. *Front. Remote Sens.* **2022**, *3*, 869611. [CrossRef]
32. Hern, T.; Lai, S.; Ibrahim, S.; Nik Sulaiman, N.M.; Sharifi, M.; Abe, S. Impact of Fine Sediment on TSS and Turbidity in Retention Structure. *J. Geosci. Environ. Prot.* **2014**, *2*, 1–8. [CrossRef]
33. Baughman, C.; Jones, B.; Bartz, K.; Young, D.; Zimmerman, C. Reconstructing Turbidity in a Glacially Influenced Lake Using the Landsat TM and ETM+ Surface Reflectance Climate Data Record Archive, Lake Clark, Alaska. *Remote Sens.* **2015**, *7*, 13692–13710. [CrossRef]
34. ESA. Sentinel-2 Spectral Response Functions (S2-SRF). Available online: https://sentinels.copernicus.eu/-/copernicus-sentinel-2c-spectral-response-functions (accessed on 18 March 2024).
35. Vanhellemont, Q. Adaptation of the Dark Spectrum Fitting Atmospheric Correction for Aquatic Applications of the Landsat and Sentinel-2 Archives. *Remote Sens. Environ.* **2019**, *225*, 175–192. [CrossRef]

36. EUMETSAT. *Algorithm Theoretical Baseline Document for Matchup Generation*; EUMETSAT: Darmstadt, Germany, 2021.
37. Zou, H.; Hastie, T. Regularization and Variable Selection Via the Elastic Net. *J. R. Stat. Soc. Ser. B Stat. Methodol.* **2005**, *67*, 301–320 [CrossRef]
38. Hoerl, A.E.; Kennard, R.W. Ridge Regression: Biased Estimation for Nonorthogonal Problems. *Technometrics* **1970**, *12*, 55–67 [CrossRef]
39. Tibshirani, R. Regression Shrinkage and Selection via the Lasso. *J. R. Stat. Soc.* **1996**, *58*, 267–288. [CrossRef]
40. Breiman, L. Random Forests. *Mach. Learn.* **2001**, *45*, 5–32. [CrossRef]
41. Friedman, J.H. Greedy Function Approximation: A Gradient Boosting Machine. *Ann. Stat.* **2001**, *29*, 1189–1232. [CrossRef]
42. Chen, T.; Guestrin, C. XGBoost: A Scalable Tree Boosting System. In Proceedings of the 22nd ACM SIGKDD International Conference on Knowledge Discovery and Data Mining, San Francisco, CA, USA, 13–17 August 2016; Association for Computing Machinery: New York, NY, USA, 2016; pp. 785–794.
43. Qiu, J. An Analysis of Model Evaluation with Cross-Validation: Techniques, Applications, and Recent Advances. *Adv. Econ. Manag. Political Sci.* **2024**, *99*, 69–72. [CrossRef]
44. Chowdhury, M. Mapping and Monitoring Valencia Flood 2024 Using Sentinel-2 MSI Data and Machine-Learning-Based Turbidity Models. In Proceedings of the ESA Living Planet Symosium 2025, Vienna, Austria, 23–27 June 2025.
45. Shapley, L.S. A Value for n-Person Games. In *Contributions to the Theory of Games*; Kuhn, H.W., Tucker, A.W., Eds.; Princeton University Press: Princeton, NJ, USA, 2016; Volume 2, pp. 307–318. ISBN 978-1-4008-8197-0.
46. Cook, R.D.; Weisberg, S. *Residuals and Influence in Regression*; Chapman and Hall: New York, NY, USA, 1982; ISBN 978-0-412-24280-9.
47. Kutner, M.H. *Applied Linear Statistical Models*; McGraw-Hill Irwin: Boston, MA, USA, 2005; ISBN 978-0-07-238688-2.
48. Draper, N.R.; Smith, H. *Applied Regression Analysis*; John Wiley & Sons: Hoboken, NJ, USA, 1998; ISBN 978-0-471-17082-2.
49. Montgomery, D.C.; Peck, E.A.; Vining, G.G. *Introduction to Linear Regression Analysis*; John Wiley & Sons: Hoboken, NJ, USA, 2012; ISBN 978-0-470-54281-1.
50. Chatterjee, S.; Hadi, A.S. *Regression Analysis by Example*; John Wiley & Sons: Hoboken, NJ, USA, 2013; ISBN 978-1-118-45624-8.
51. Chowdhury, M.; Martínez-Sansigre, A.; Mole, M.; Alonso-Peleato, E.; Basos, N.; Blanco, J.M.; Ramirez-Nicolas, M.; Caballero, I.; de la Calle, I. AI-Driven Remote Sensing Enhances Mediterranean Seagrass Monitoring and Conservation to Combat Climate Change and Anthropogenic Impacts. *Sci. Rep.* **2024**, *14*, 8360. [CrossRef]
52. Qin, B.; Zhu, G.; Gao, G.; Zhang, Y.; Li, W.; Paerl, H.W.; Carmichael, W.W. A Drinking Water Crisis in Lake Taihu, China: Linkage to Climatic Variability and Lake Management. *Environ. Manag.* **2010**, *45*, 105–112. [CrossRef]
53. Le, T.P.Q.; Garnier, J.; Gilles, B.; Sylvain, T.; Van Minh, C. The Changing Flow Regime and Sediment Load of the Red River, Viet Nam. *J. Hydrol.* **2007**, *334*, 199–214. [CrossRef]
54. Martinez, J.M.; Guyot, J.L.; Filizola, N.; Sondag, F. Increase in Suspended Sediment Discharge of the Amazon River Assessed by Monitoring Network and Satellite Data. *CATENA* **2009**, *79*, 257–264. [CrossRef]
55. Sottolichio, A.; Castaing, P. A Synthesis on Seasonal Dynamics of Highly-Concentrated Structures in the Gironde Estuary. *Comptes Rendus L'académie Sci.-Ser. IIA-Earth Planet. Sci.* **1999**, *329*, 795–800. [CrossRef]
56. Zhang, Y.L.; Qin, B.Q.; Liu, M.L. Temporal–Spatial Variations of Chlorophyll a and Primary Production in Meiliang Bay, Lake Taihu, China from 1995 to 2003. *J. Plankton Res.* **2007**, *29*, 707–719. [CrossRef]
57. Allen, G.P.; Salomon, J.C.; Bassoullet, P.; Du Penhoat, Y.; de Grandpré, C. Effects of Tides on Mixing and Suspended Sediment Transport in Macrotidal Estuaries. *Sediment. Geol.* **1980**, *26*, 69–90. [CrossRef]
58. Froidefond, J.-M.; Gardel, L.; Guiral, D.; Parra, M.; Ternon, J.-F. Spectral Remote Sensing Reflectances of Coastal Waters in French Guiana under the Amazon Influence. *Remote Sens. Environ.* **2002**, *80*, 225–232. [CrossRef]
59. Woo Kim, Y.; Kim, T.; Shin, J.; Lee, D.-S.; Park, Y.-S.; Kim, Y.; Cha, Y. Validity Evaluation of a Machine-Learning Model for Chlorophyll a Retrieval Using Sentinel-2 from Inland and Coastal Waters. *Ecol. Indic.* **2022**, *137*, 108737. [CrossRef]
60. Emami, S.; Martínez-Muñoz, G. Condensed-Gradient Boosting. *Int. J. Mach. Learn. Cyber.* **2025**, *16*, 687–701. [CrossRef]
61. Maciel, D.A.; Barbosa, C.C.F.; Novo, E.M.L.d.M.; Flores Júnior, R.; Begliomini, F.N. Water Clarity in Brazilian Water Assessed Using Sentinel-2 and Machine Learning Methods. *ISPRS J. Photogramm. Remote Sens.* **2021**, *182*, 134–152. [CrossRef]
62. Chen, J.; Cui, T.; Qiu, Z.; Lin, C. A Three-Band Semi-Analytical Model for Deriving Total Suspended Sediment Concentration from HJ-1A/CCD Data in Turbid Coastal Waters. *ISPRS J. Photogramm. Remote Sens.* **2014**, *93*, 1–13. [CrossRef]
63. Chen, S.; Han, L.; Chen, X.; Li, D.; Sun, L.; Li, Y. Estimating Wide Range Total Suspended Solids Concentrations from MODIS 250-m Imageries: An Improved Method. *ISPRS J. Photogramm. Remote Sens.* **2015**, *99*, 58–69. [CrossRef]
64. Doxaran, D.; Froidefond, J.-M.; Castaing, P. Remote-Sensing Reflectance of Turbid Sediment-Dominated Waters. Reduction of Sediment Type Variations and Changing Illumination Conditions Effects by Use of Reflectance Ratios. *Appl. Opt.* **2003**, *42*, 2623–2634. [CrossRef] [PubMed]

5. Ouillon, S.; Douillet, P.; Petrenko, A.; Neveux, J.; Dupouy, C.; Froidefond, J.-M.; Andréfouët, S.; Muñoz-Caravaca, A. Optical Algorithms at Satellite Wavelengths for Total Suspended Matter in Tropical Coastal Waters. *Sensors* **2008**, *8*, 4165–4185. [CrossRef] [PubMed]
6. Wang, C.; Zhou, C.; Zhou, X.; Duan, M.; Yan, Y.; Wang, J.; Wang, L.; Jia, K.; Sun, Y.; Wang, D.; et al. A Universal Method to Recognize Global Big Rivers Estuarine Turbidity Maximum from Remote Sensing. *ISPRS J. Photogramm. Remote Sens.* **2025**, *220*, 509–523. [CrossRef]
7. Güttler, F.N.; Niculescu, S.; Gohin, F. Turbidity Retrieval and Monitoring of Danube Delta Waters Using Multi-Sensor Optical Remote Sensing Data: An Integrated View from the Delta Plain Lakes to the Western–Northwestern Black Sea Coastal Zone. *Remote Sens. Environ.* **2013**, *132*, 86–101. [CrossRef]
8. Martín, M.; Hernández-Crespo, C.; Andrés-Doménech, I.; Benedito-Durá, V. Fifty Years of Eutrophication in the Albufera Lake (Valencia, Spain): Causes, Evolution and Remediation Strategies. *Ecol. Eng.* **2020**, *155*, 105932. [CrossRef]
9. Xue, K.; Ma, R.; Duan, H.; Shen, M.; Boss, E.; Cao, Z. Inversion of Inherent Optical Properties in Optically Complex Waters Using Sentinel-3A/OLCI Images: A Case Study Using China's Three Largest Freshwater Lakes. *Remote Sens. Environ.* **2019**, *225*, 328–346. [CrossRef]

remote sensing

Article

Links between Land Cover and In-Water Optical Properties in Four Optically Contrasting Swedish Bays

Susanne Kratzer [1,*] and Martin Allart [1,2]

[1] Department of Ecology, Environment and Plant Sciences (DEEP), Stockholm University, 10691 Stockholm, Sweden
[2] National Institute of Applied Sciences (INSA), 20 Av. Albert Einstein, 69100 Villeurbanne, France
* Correspondence: susanne.kratzer@su.se

Abstract: The optical complexity of coastal waters is mostly caused by the water discharged from land carrying optical components (such as dissolved and particulate matter) into coastal bays and estuaries, and increasing the attenuation of light. This paper aims to investigate the links between in-water optical properties in four Swedish bays (from the northern Baltic proper up to the Bothnian bay) and the land use and land cover (LULC) in the respective catchment of each bay. The optical properties were measured in situ over the last decade by various research and monitoring groups while the LULC in each bay was classified using the Copernicus Land Monitoring Service based on Landsat 8/OLI data. The absorption coefficient of colored dissolve organic matter (CDOM) at 440 nm, a_{CDOM} (440), was significantly correlated to *Wetlands* which may act as sources of CDOM, while *Developed* areas (*Agricultural* and *Urban* classes) were negatively correlated. The *Agriculture* class was also negatively related to suspended particulate organic matter (SPOM), whilst *Coniferous Forests* and *Mixed Forests* as well as *Meadows* were positively correlated. SPOM seems thus to mostly originate from *Natural* classes, possibly due to the release of pollen and other organic matter. Overall, the methods applied here allow for a better understanding of effects of land use and land cover on the bio-optical properties, and thus coastal water quality, on a macroscopic scale.

Keywords: land use and land cover (LULC); in-water optical properties; bio-optics; suspended particulate matter; colored dissolved organic matter; chlorophyll-a; catchment area; water discharge; land–sea interactions

1. Introduction

Seen from space, or from a plane, coastal waters are usually not blue but often yellow or brownish, or even green or red during certain phytoplankton blooms. The yellow-brown shades are mostly due to one of the main optical components often found in coastal waters: CDOM, also referred to as "yellow substance". It consist of humic and fulvic substances [1] which have a complex chemical nature, consisting mostly of diverse polymers with aromatic rings and long chains of alkyls [2]. Humic substances are derived from the decomposition of plants and are ultimately drained into coasts by rivers and by runoff from land [1]. CDOM has a high absorption in the UV-blue spectrum which decreases with increasing wavelength towards the red in a logarithmic manner. Thus, CDOM appears yellow to brown. In addition to particle scatter and chlorophyll-a absorption, CDOM absorption is one of the key inherent optical properties (IOPs) of seawater. Preisendorfer [3] defined the IOPs as those optical properties that are independent of the radiance distribution, while apparent optical properties (AOPs)—such as diffuse attenuation or Secchi depth—are also influenced by the sun angle, and thus the radiance distribution.

Carder et al. [4] studied the nature of CDOM in the Gulf of Mexico and determined that they could either be of a more humic character (i.e., larger molecules and a slope factor S_{CDOM} closer to -0.0011) when related to bogs, or a more fulvic character (i.e., smaller, i.e., more degraded molecules, and a slope factor closer to −0.0022). The Baltic Sea is rich in dissolved organic carbon (DOC), the light-absorbing fraction of which is referred to as CDOM. Harvey et al. [5] found that the slope factor S_{CDOM} varies over the year, both in the Gulf of Bothnia (with a high terrestrial input) as well as in the northwestern Baltic proper (with a relatively low terrestrial input).

CDOM absorption is particularly high in the Baltic Sea compared to other seas and oceans around the world [1]. This is due to the land-locked situation of the sea and the low water exchange with the North Sea. Thus, CDOM is the main optical constituent determining the light attenuation [6,7]. Another important optical component in coastal waters is SPM, consisting of solid particles made up by both inorganic sediments (sand and silt), by organic material (usually in minor proportions) and by phytoplankton. The absorption characteristics of SPM are usually difficult to measure because it also scatters light, especially its inorganic fraction [8,9]. Light scattering by particles depends on particle composition, size and shape as well as the refraction index of the particles [10]. Because of their high water fraction, phytoplankton have a relatively low refractive index of 1.02–1.07 while inorganic particles have a high refractive index ranging from 1.1 to 1.26 [11–14]. Inorganic SPM is usually found in coastal areas and indicates coastal influence [7,15].

SPM is often found in large proportions in coastal waters and may indicate physical forcing such as wind-wave action, eddies and tidal currents, causing the resuspension of sediments derived from discharge of particle-laden rivers [16,17]. The absorption properties of non-algal organic SPM are similar to those of CDOM but with a different spectral slope [18].

The study by Le et al. [19] evaluated the effect of LULC derived from remote sensing data on the inherent optical properties of several estuaries in the northern Gulf of Mexico. Initially, the authors derived eight main classes: *Urban*, *Agriculture*, *Evergreen Forest*, *Deciduous Forest*, *Water*, *Wetland*, *Pastures*, and *Barren Land*. Next, they compiled the *Urban* and *Agriculture* class into a *'Developed'*, and *Evergreen Forest* and *Wetlands* into a *'Natural'* class. They could then show that there were significant correlation between the optical properties and the ratio of *Developed* to *Natural* LULC classification across six bays.

Our main hypothesis is that the dominant LULC characteristics also influence coastal water quality in Swedish bays. The aim of this paper is thus to describe the optical characteristics of four Swedish bays in relation to the LULC characteristics of the respective catchment of each bay. First, the catchment area of each bay is defined based on the hydrology. The LULC characteristics are then derived from remote sensing data, and are further analyzed using an Open Source Geographic Information System (GIS). Subsequently, the relationship between LULC cover and optical properties is evaluated via correlation analysis. Furthermore, we hypothesize that catchments with a relative large percentage of bogs are characterized by higher CDOM absorption.

The objectives of this study thus can be summarized in the following way: Firstly, we will assess the nature of the CDOM absorption at 440 nm and the CDOM slope factor of each bay in relation to the proportion of different natural land categories. Secondly, we will evaluate if different land cover categories may have an effect on the SPM load, or on the nature of the SPM found in the four coastal bays. Thirdly, we will investigate if LULC has an effect on the levels of chlorophyll-a (Chl-a) concentration found in the bays, and, finally, we will discuss if there are any relationships between LULC cover and bio-optical water quality that are specific for Scandinavia and the high latitudes.

2. Materials and Methods

2.1. Site Descriptions

For the analysis, we selected four Swedish bays where measurement transects of optical in-water variables (e.g., Chl-a, CDOM, SPM and turbidity) were performed over the last decade by the Marine Remote Sensing Group (MRSG) and the pelagic Monitoring Group (MG) at the Department of Ecology, Environment and Plant Sciences (DEEP),

Stockholm University, and also during 2018 through a collaboration between MRSG and the monitoring group from the Umeå Marine Sciences Centre (UMF), Umeå University.

Figure 1 indicates the two zones of this study: Zone I comprising the northern Baltic proper and the Åland Sea, and Zone II in the northwestern Bothnian Bay. Optical measurements were performed along transects through the different bays (see Section 2.2 below). A numerical analysis of the land cover in the catchments surrounding those bays was carried out using the Geographic Information System (GIS) open source software, QGIS. A catchment is here defined as the natural drainage area, i.e., a zone where rain water flows into streams, rivers, lakes and ground waters, and eventually ends up in certain water bodies (basins) or bays. Section 2.1.1 aims at describing the selected bays and the main surrounding elements (both natural and artificial) that affect these coastal water bodies while Section 2.1.2 explains how the catchments were defined using hydrological data.

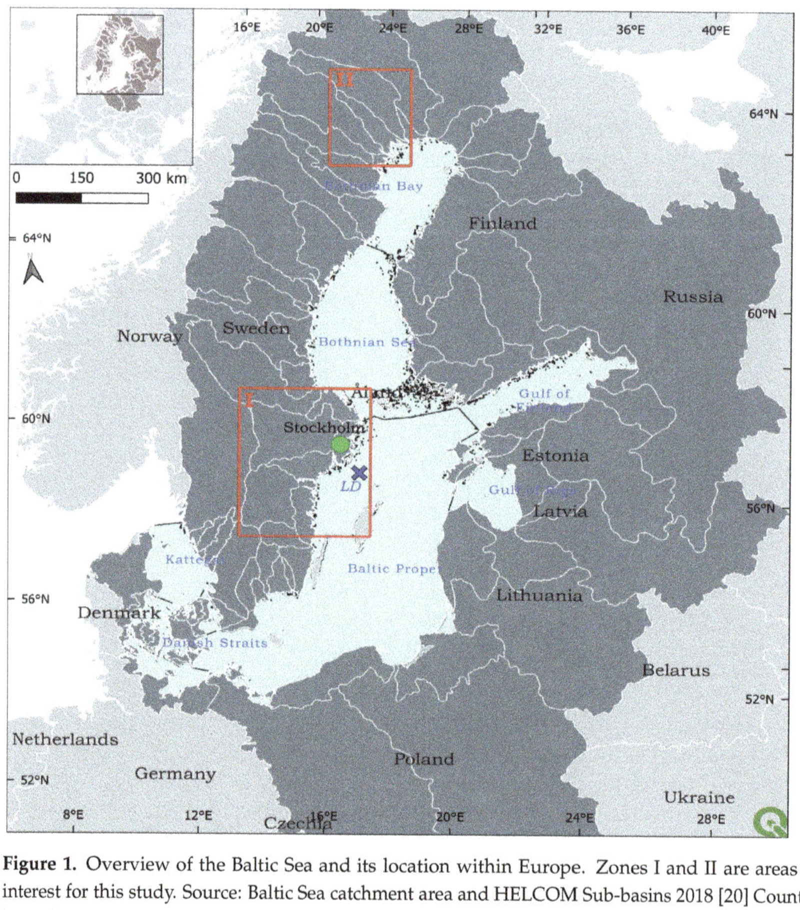

Figure 1. Overview of the Baltic Sea and its location within Europe. Zones I and II are areas of interest for this study. Source: Baltic Sea catchment area and HELCOM Sub-basins 2018 [20] Country boundaries: Natural Earth [21], European coastline shapefile: EEA [22], SMHI's sub-basin division (havs-områden_SVAR_2016 [23]). LD refers to Landsort Deep, the deepest part in the Baltic Sea (459 m).

2.1.1. Description of Each Site (Bay)

Bråviken bay (Figure 2a) is a large east-west facing bay. Its topography is particularly shallow with a depth gradient from its southern bank (mean depth: 10 m) to its northern bank (mean depth: 40 m). The Swedish Water Archive divides the bay into 5 sub-basins: Pampusfjärden (A), Inner Bråviken (B), Mid-Bråviken (C), Outer Bråviken (D), Bråviken's coastal waters (E). Several sampling stations (BR1-BR9) of the MRSG (SU) are located along a salinity gradient from the inner bay towards (A) the open sea (E). Many industrial sites are established in the bay and several pulp industry premises are located on the shores of Pampusfjärden and the bay also hosts Europe's largest grain facility port for export which leads to a lot of ship and boat traffic inside the bay.

Himmerfjärden bay (Figure 2b), is a narrow, elongated bay facing from north to south and it is deeper than Bråviken (40 m mean depth). In the inner bay, there is a sewage treatment plant, which is the third largest in Stockholm County. Himmerfjärden bay consists of several basins that are separated by sills, and is also fragmented by many islands of various size which makes the water circulations more complex than in Bråviken. The water basins in HFj are called Näslandfjärden (W), Himmerfjärden (X), Svärdsfjärden (Y) and Krabbfjärden (Z). Himmerfjärden bay surrounds a central island called Mörkö which hosts recreational sites and natural reserves as well as summer houses. There is a salinity gradient from the inner bay (W) out into the open sea (X), as well as a gradient of optical properties [7]. The optical transects considered in this study starts at station H6 Näslandfjärden (W) going out to H2 in Svärdsfjärden (Y). Station B1 is located in Krabbfjärden in the SW of the island Askö (which is a nature reserve and home of Askö Laboratory, sustained by Stockholm University). Even though B1 is shielded by the island and not part of the chosen optical transect, it has shown to have similar optical characteristics as station H2 just inside Himmerfjärden bay [7]. This is why the optical variable of station B1 and H2 were combined and averaged in the analysis detailed below.

Östhammar bay (Figure 2c) is very shallow and lies in the Åland Sea which is adjacent to the most southern part of the Bothnian Sea. It can be regarded as a bay even though its shape is more complex than those of the other bays of Zone I. Its main branch stretches from SW to NE. Its sub-basins are: Östhammarfjärden (Q), Hargsviken (R), Galtfjärden (S) and Singöfjärden (T). The transect of measurement originates at station U22 towards the outermost station U15, along an increasing salinity gradient. Station U18, U14, U13 and U11 lie in between. At the junction of the sub-basins Q, R and S there is a large island called Värlingsö. The water flows from the inner Östhammar bay (Q) into R where the municipality of Hargshamn is located with its large, industrial port. This part of the bay is deeper than the passage flowing from Q to S-basin, which allows the passage of cargo boats. The water flow is thus more important in the R-S than in Q-S passage. The second inlet branch of the bay is called Edeboviken, where station U12 is usually sampled by the pelagic Monitoring Group (MG) from Stockholm University, but it was rejected from the analysis here because its optical properties have shown to be very different from U13. The presence of Väddöviken bay at the end of the transect might also influence the optical characteristics of the last sampling station U15. The remainder of the bay, i.e., denoted basins S and T (Figure 2c), are fragmented by many islands which makes the circulation of water rather complex.

Råneå bay (Figure 2d) is located in the western Gulf of Bothnia (Zone II, Figure 1), where the waters in the surface layers are frozen during winter, and usually have a very high content of CDOM from riverine sources [24]. The coastal waters are characterized by a very strong coastal influence both after the ice thawing in spring and all during summer. The sampling station RA1, RA2, RA6, RA7 and RA8 are located in 4 sub-basins, stretching from the inner bay towards the outer bay: Rånefjärden (L), Gussöfjärden (M), Tistersöfjärden (N) and Fjusköfjärden (O).

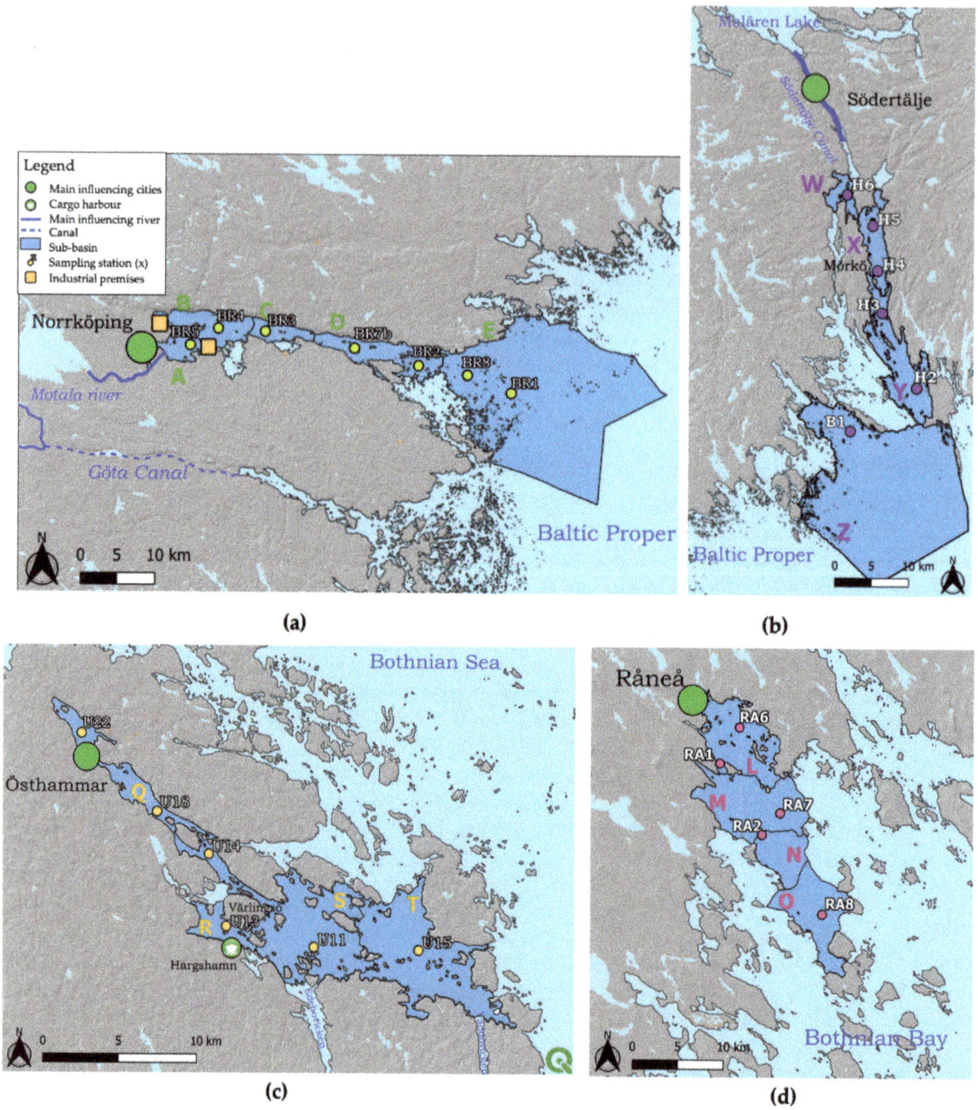

Figure 2. The four Swedish bays, their sub-basins and sampling stations (different coloured circles in different bays) (**a**) Bråviken, (**b**) Himmerfjärden, (**c**), Östhammar (**d**) Råneå. Note (**a**–**c**) are situated in Zone I (see also Figures 1 and 3a) and (**d**) is situated in Zone II (Figures 1 and 3b). Maps generated in QGIS using predefined shapefiles (European coastline shapefile [22]), SMHI's sub-basin division (havs-områden_SVAR_2016 [23]). Recurrence layer—Global Surface Water, Surface elevation—Copernicus EU-DEM- v1.1; Sentinel Hub [25]).

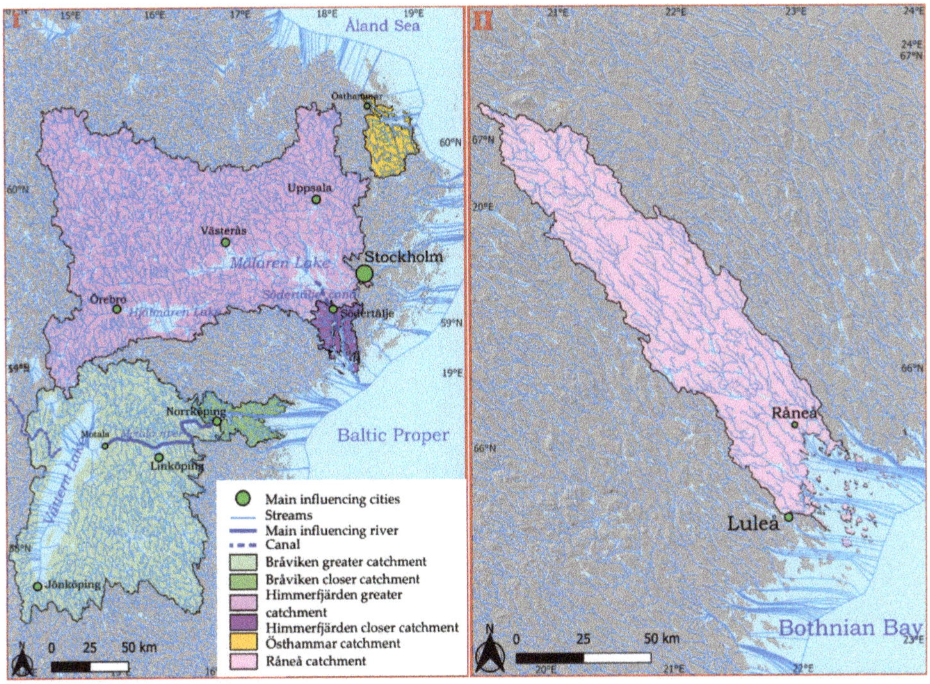

(a) (b)

Figure 3. Water course network and catchments of (**a**) Zone I: Bråviken bay (light and dark green), Himmerfjärden bay (light and dark purple), Östhammar bay (orange); (**b**) Råneå bay, located in Zone II in Swedish Lapland (see also Figure 1 for situation of Zones I and II). Maps generated in QGIS using predefined shapefiles (European coastline shapefile [22]). Recurrence layer—Global Surface Water, Surface elevation—Copernicus EU-DEM- v1.1; Sentinel Hub [25]), Water course network [26].

2.1.2. Selection of Catchments

The catchments surrounding the bays were selected based on the Vattenweb (in English: water web) of the Swedish Meteorological and Hydrological Institute (SMHI, Norrköping, Sweden), that can be used to assess the Current Hydrological Status in a certain catchment area [26], as mentioned above, a catchment is defined as an area from which rain and river water flows into other downstream waterbodies. We initially selected both the greater catchment zone for each bay (shapefile haro_y_2016_3) as well as the sub-catchments directly surrounding the bays (shapefile aro_y_2016_3). Eventually, these areas were merged as to constitute the greater catchment zone of interest for each of the 4 bays, respectively (see Figure 3a,b). The average discharge value for each month over the period 1991–2021 and the size of each catchment was derived for each discharge basin, and summarized in an Excel spreadsheet for further analysis.

2.2. Optical Transects

The MRSG performed two satellite validation campaigns in Bråviken bay on board R/V Electra during 1–2 July 2021 and 28–29 April 2022. Optical measurements were taken along transects with a salinity gradient (Figure 2a) from the inner bay (A) out to the open Baltic Sea (E). The data from the field campaigns in 2021 and 2022 were merged with other transects from past expeditions performed over the last decade in the four bays (Figure 2, Table 1). The MRSG has trained the pelagic monitoring groups at Stockholm and Umeå Universities on how to perform optical measurements (e.g., CDOM, SPM and turbidity) and the groups share their optical protocols and also intercompare their methods [27]). Table 1 provides a summary of the measurements performed, including dates, operating group and number of samples gathered. Altogether, n = 116 stations were sampled for this study, including measurements of Chl-a, SPM, SPIM, SPOM concentrations, turbidity, Secchi Depth, CDOM absorption, a_{CDOM}, and the CDOM slope factor, S_{CDOM}. The largest collection of data comes from the Himmerfjärden area.

Table 1. Overview table of the number of stations (n) used in the final analysis. The full dataset can be found in Supplementary Table S2.

Bay							
Bråviken		**Himmerfjärden**		**Östhammar**		**Råneå**	
April 2022	n = 5 (MRSG)	April 2018	n = 4 (MRSG) SPM, SPIM, SPOM missing	August 2021	n = 4 (MG) SPM, SPIM, SPOM missing	July 2018	n = 4 (MRSG and MG UMF) SPM, SPIM, SPOM not valid
July 2021	n = 6 (MRSG) SPM, SPIM, SPOM missing	August 2017	n = 5 (MG)	Jul 2021	n = 4 (MG) SPM, SPIM, SPOM missing	June 2018	n = 4 (MRSG and UMF) SPM, SPIM, SPOM not valid
April 2018	n = 8 (MRSG) S_{CDOM} missing	July 2017	n = 4 (MRSG)	August 2020	n = 4 (MG) SPM, SPIM, SPOM missing	May 2018	n = 4 (MRSG and UMF) SPM, SPIM, SPOM not valid
August 2013	n = 2 (MG) turbidity, S_{CDOM} missing	May 2012	n = 4 (MRSG)	July 2020	n = 4 (MG) SPM, SPIM, SPOM missing		
July 2013	n = 2 (MG) turbidity, S_{CDOM} missing	April 2012	n = 3 (MRSG)	August 2019	n = 4 (MG) SPM, SPIM, SPOM missing		
June 2013	n = 2 (MG) turbidity, S_{CDOM} missing	August 2010	n = 8 (MRSG)	July 2019	n = 4 (MG) S_{CDOM}, a_{CDOM}, SPM, SPIM, SPOM missing		
August 2012	n = 2 (MG) S_{CDOM} missing	July 2007	n = 13 (MRSG)	August 2010	n = 4 (MG) turbidity missing		
June 2012	n = 2 (MG) S_{CDOM} missing			July 2010	n = 6 (MG) turbidity missing		
Total stations sampled: n = 29		Total stations sampled: n = 41		Total stations sampled: n = 34		Total stations sampled: n = 12	

2.2.1. Water Sampling and Measurement Protocols

Note that the procedures below follow the Optical Measurement Protocols of the Marine Remote Sensing Group. During the field work, the samples were gathered 20–30 cm below the water surface with a sturdy sampling bucket with a pip. The bucket was rinsed twice with sea water before sampling as to avoid contamination. The sampling bottles were also rinsed twice with sea water and then filled with sea water from the bucket. The same standard protocol was applied for all other field campaigns by the MRSG but other laboratories sometimes have slightly different protocols. Nonetheless, intercomparison workshops are held regularly to make sure that the measurements between groups are comparable.

SPM, SPIM and SPOM Measurements

Depending on the Secchi depth reading, 1–2 L of sea water were sampled for SPM measurements. The measurements were performed using the gravimetric method [7,28]. This method consists of filtering a certain quantity of natural water through microfibers glass filters (GF/F filters) of 47 mm diameter and with a nominal pore size of 0.7 μm. Before filtration, the filters were pre-rinsed with ultra-pure water (UPW), combusted and pre-weighed (tare weight) with a high-precision scale (Satorius MP3 microbalance—±1 μg). Each sampling bottle was gently mixed before filtration to ensure even distribution of suspended material. All filters were rinsed at the end of the filtering process with 100 mL

UPW for removal of salt residuals. After filtration, each filter was placed in a numbered (by etching), clean square aluminum foil (10 × 10 cm). Then, the filters were dried overnight at 60 °C. The filters were subsequently stored in a desiccator in order to keep them dry until weighing. The dry weight of SPM was derived by subtracting the tare weight from the dry weight. In order to derive SPIM, the filters were combusted for 5 h at 480 °C, and were weighed again. The organic fraction was then derived from the weight difference between the total SPM (dry weight) and the SPIM (combusted weight), assuming all organic matter has been combusted. The handling of the filters sometimes leads to the loss of filter bits. Therefore, triplicate filters were taken for every station as quality control of the measurements. Outliers were removed if the standard deviation was more than 20%. For every field campaign between 5 and 10 blank filters were processed in the same way except that ultra-pure water (UPW) was used (0.5 L) instead of sea water, and their average weight was then used to correct for handling errors. In the Bothnian bay, the rinsing with UPW was sometimes omitted by mistake by the local monitoring group, and during the data quality control (i.e., comparison against the measured turbidity values), the samples were found to be invalid and therefore had to be omitted from the final analysis.

Turbidity Measurements

Turbidity was measured during field campaigns with a portable turbidity meter (Hach Lange 2100Qis, HACH LANGE GmbH, Düsseldorf, Germany). The calibration of the device is performed using standard formazin solutions of known turbidity (10, 20, 100, 800 FNU). Formazin Nephelometric Unit (FNU) is a unit equivalent to NTU but is measured in the near infrared (NIR) rather than using white light. The standard calibration solutions are provided by the manufacturer. Before performing a measurement, one gently mixes a sample [16]. It is important to wait for 10 s before starting the measurement in order to remove air bubbles that could cause extra scattering, and thus wrongly increase the turbidity value. Each turbidity sample was measured 5 times and the values were averaged and corrected for the average turbidity value of ultrapure water (UPW) which was also measured 5 times.

CDOM Measurements

The CDOM samples were collected in 250 mL amber glass bottles and the samples were filtered through 0.22 µm membrane filters using glass filtration unit fitted with a metal mesh to avoid clogging of the filtering unit. The filtrate was then transferred into 100 mL amber glass bottles and stored in a refrigerator at 4 °C for up to 3 months. The absorption spectra of CDOM was measured by scanning the sample in a Shimadzu UVPV-2401 dual beam spectrophotometer (Shimadzu Corporation, Kyoto, Japan). Before scanning, the samples were removed from the refrigerator and allowed to reach room temperature (overnight). The absorption spectrum was measured in the range from 350 to 800 nm in cylindrical quartz cuvettes (10 cm). First, a baseline was performed, using both cuvettes filled with UPW. The cuvette in the back was then used as a reference (UPW) while the front cuvette was filled with the respective water sample and scanned. The absorbance was measured from 190 to 850 nm (in 1 nm steps). The absorbance at 700 nm was then used to correct for measuring errors, and the spectral absorption for CDOM was derived from the following relation, according to Kirk [1]:

$$a_{CDOM}(\lambda) = \ln(10)\mathrm{OD}(\lambda)\mathrm{L}^{-1},\ [\mathrm{m}^{-1}] \tag{1}$$

where OD(λ) being optical density (absorbance), L being the optical path length of the cuvette in meter, here 0.1 m. Note that the OD is measured in logarithmic values (Ln10). The CDOM absorption at 440, a_{CDOM}, is then derived per meter after log transformation. Then, the whole absorption spectrum is log transformed and the slope of the derived line (in the range of 350–500 nm) corresponds to the slope factor of CDOM, S_{CDOM}.

Chlorophyll-a Measurements

0.2–0.5 L of sea water were filtered onto 25 mm diameter GF/F filters. Filtering was performed on board ship and the filters were flash-frozen in liquid nitrogen and stored for up to 2 months. Each sample was filtered and measured in triplicates. Before the spectral analysis, the samples were placed in 5 mL of 90% acetone solution and sonicated for 30 s for destruction of the cell walls and extraction in acetone. The samples were then centrifuged for 10 min at 3000 rpm so that all particles could settle in the bottom of the polypropylene test tube and thus avoid the effect of scattering on the absorption spectrum. After 30 min of extraction, the samples were transferred into 1 cm quartz cuvettes and scanned against 90% acetone (350–850 nm) in a UVPC-2401 dual beam spectrophotometer (Shimadzu Corporation, Kyoto, Japan). The spectra were then processed in Excel (2019) with the algorithms and coefficients provided by Strickland and Parsons (trichromatic method) [29] in order to derive the Chl-*a* concentration in μgL^{-1}. This method has shown to be within 10% error during an inter-calibration performed by the ESA MERIS Validation Team [30] and within 2–10% when compared to the Swedish monitoring groups [27]. The method has also been intercompared regularly with HPLC measurements [27,30].

2.3. Land Use and Land Cover Analysis

The land use and land cover (LULC) of each bay was analyzed in QGIS (version 3.26.2 'Buenos Aires'), an open-source Geographic Information System (GIS). Data from two sources were used to define the extent of the catchment of each bay. Firstly, shapefiles from the Swedish Hydrological Meteorological Institute (SMHI), provided the polygons shapefiles of the Swedish catchments (Haro) and sub watersheds (aro) around the different bays of the study using SVAR version 2016_3 [23]. Maps of the different catchment areas are shown in the results section in Figure 3. The final catchment area selection was not trivial and was performed with regard to modelled hydrological data from the SMHI's Water Web [26].

Secondly, we used the CORINE Land Cover 2018 (CLC2018) dataset derived from Landsat-8 satellites images [25] for the LULC classification. These products are based on data from 2017 to 2018 and have a 100 m resolution with a thematic accuracy of ≥85%. They are generated regularly in 6-yearly cycles. The CLC2018 dataset was published in 2020 and provided initially 44 land cover categories (see Supplementary Table S1) which were further aggregated into 10 classes for this study. The following operations were performed to process the data in QGIS:

- Reprojection of CORINE into the same geographical projection as for the Catchment shapefile EPSG3006 SWEREF99 TM used by the Swedish Meteorological and Hydrological Institute, SMHI, Norrköping, Sweden [23].
- Fixing geometries of CORINE data via the vector operation "fix geometries".
- Reducing the number of Level 1 attributes from 44 to 10 categories (so called Code 18).
- Aggregating the original 44 to 10 polygons of the same Level 1 class.
- Intersection of the dissolved LULC polygons with the catchment boundaries in order to exclude information outside the areas of interests.
- Eventually, computation of percentage area per category of Level 1.

All 11 urban classes from the original 44 classes were thus aggregated under the class *Urban*. All arable land, permanent crops and heterogeneous agricultural areas (consisting of 10 classes) were aggregate into *Agriculture*. The class *Pastures* was left as designated by CORINE. It may differ from the *Agriculture* class due to the manure from cattle, and also from the class *Meadow* which refers to land that is covered by different natural grass types and bushes. All open spaces with little or no vegetation were assembled into one class: *Barren Land*. The four shrub/herbaceous vegetation classes were aggregated into one *Meadow* class. All 5 wetland categories were aggregated into a *Wetland* class. Similarly, the 5 water subcategories into one *Water* class. *Deciduous Forest*, *Coniferous Forest* and *Mixed Forest* were left unchanged.

2.4. Combining Optical and LULC Data

All data were collected and processed in Excel for further analysis. A first analysis was performed in order to operate a data quality check. From the geographical coordinates of a given station the horizontal distance from an outlet was derived (in km) with the aim to make the transects within the 4 bays comparable. The outlet was chosen to be either a defined river mouth (cf. Östhammar and Råneå), or the innermost station usually sampled by the respective monitoring or research group (cf. station H5 and BR5 in Himmerfjärden and Bråviken). A new data sheet was then compiled for each optical parameter, and various plots were made to visualize the data, either season-wise or regardless of the time of the year. It allowed to identify outliers and observe variability between the bays as well as seasonal variability, and to choose which variable to focus on.

Finally, the stations were aggregated based on their horizontal distance from the outlet. The mean horizontal distance from the outlet ranged from approximately 15 $\pm$ 5 km in the different areas. Any innermost or outermost station making the mean distance deviating from this range were rejected from the study in order to make the datasets comparable. The optical properties in each transect were then averaged by taking the mean values from geographic middle stations, the rationale being that this allows for a comparison of the different bays of different sizes, and across different times of year. Note that the derived optical products are thus averaged over space and time for every transect. Eventually, the optical variables were plotted each against a LULC information, such as a percentage of a given land cover, or a ratio of several land covers which helped to identify the main trends.

Additionally, a correlation matrix was prepared to identify possible links between LULC class and optical variables. The correlations in the matrix were then tested using a significance test based on the p-value test ($\alpha = 0.05$). The workflow of the analysis is summarized in Figure 4.

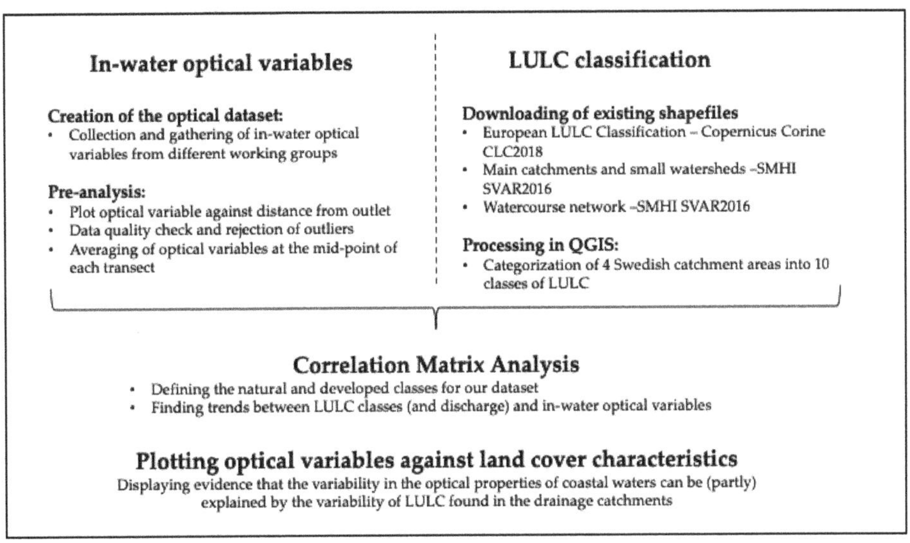

Figure 4. Workflow diagram of the data processing method.

Le et al. [19] had observed in their study that groups of LULC categories could be assembled in order to relate them to optical properties on a macroscopic scale. They derived a *Natural* category (*Coniferous Forest* + *Wetland*) as well as a *Developed* category (*Urban* + *Agriculture*). Eventually, they inferred that the *Ratio Developed/Natural* category showed interesting results when comparing to the optical properties of various estuaries in their study. As a first step, such group categories and ratios were also derived and tested for this study but were also adapted with regard to our correlation matrix and typical trends observed in Swedish bays (see results of the correlation analyses and trend analysis below). The natural class used here, denoted *Natural**, was thus merged to include *Coniferous Forest + Wetland + Meadows* while the developed class, denoted *Developed**, included *Urban + Agriculture + Pastures* (although the latter was so small in % area that it did not show any effect on the results; see below). Finally, the ratio of *Developed** over *Natural** was denoted *Ratio**. Correlations were applied between optical properties and the final LULC categories as well as the grouped categories *Natural**, *Developed** and *Ratio** as defined above, and also between optical properties and the discharge values derived for the different bays from the Current Hydrological Status on the SMHI Water Web [26].

3. Results

3.1. Derived Catchment Areas and Description of Their Hydrology

The derived catchment areas varied from only 993 km^2 for Östhammar bay to 23,368 km^2 for Himmerfjärden bay (Figure 3). Due to the large differences in surface area and hydrology, the monthly and yearly discharge varied considerably in and between bays (see Figure 5 and Table 2). Himmerfjärden has the lowest yearly discharge with approximately 49 m^{3}s^{-1}, followed by Östhammar bay with approximately 95 m^{3}s^{-1}. Bråviken bay and Råneå bay show a substantially higher yearly discharge of approximately 766 m^{3}s^{-1} and 725 m^{3}s^{-1}, respectively. It must be noted that in Råneå the monthly discharge is much lower from January to March (approximately 20 m^{3}s^{-1}), while there is a very high peak stretching over April to June due to the ice melt in spring.

The Bråviken catchment (Figure 3a) is composed of two parts—its greater catchment (light green) and its surrounding catchment (dark green). The Motala river is connected to Lake Vättern and leads into Pampusfjärden (area A, Figure 2a) passing through Motala, Linköping, and eventually Norrköping. Lake Vättern, with a total surface area of 1912 km^2 is the second largest lake of Sweden and the sixth largest lake in Europe.

It thus generates a considerable monthly freshwater flow of approximately 50 m^{3}s^{-1} at its source in Motala, gathering even more discharge water on its way to Norrköping where the water flow reaches approximately 100 m^{3}s^{-1} in winter and spring, and approximately 50 m^{3}s^{-1} in summer and autumn (Figure 5). Vättern also exhibits downstream connections with the Göta Canal, while one section flows down into Lake Vänern and another into the Baltic Sea, but the water flows are negligible when compared to Motala river (less than 1 m^{3}s^{-1}).

The water in Himmerfjärden bay originates by approximately 50% from the streams of its surrounding catchment (dark purple) and ca. 50% from its greater catchment (light purple), collecting all the discharge water into Lake Mälaren (Figure 3a). Unlike the greater catchment of Bråviken, it is not connected by a natural river but with the Södertälje Canal, named after the town it passes through. This canal is regulated by locks and has a very low and almost constant flow of freshwater of approximately 5 m^{3}s^{-1}. During events of floods and heavy rainfall, the locks can be opened up to alleviate the flooding risk in Stockholm City and the discharge jumps then suddenly up to 100 m^{3}s^{-1}, but these kinds of events are quite rare and did not occur during any field campaign considered in this study.

The Östhammar bay receives its waters from various streams and small rivers (Figure 3a). The catchment includes many small lakes and minor bogs which explains why the water outflow is only approximately 4 m^{3}s^{-1} on a yearly average.

The landscape in the Råneå catchment is highly influenced by a myriad of various streams, converging into the Råneå river (Figure 3a). The shape of the monthly discharge curve along the year is remarkably different from the catchments of Zone I (Figure 5b). A major peak in mid-May can be observed induced by the thawing of sea ice and snow. It gives rise to a very large monthly discharge of approximately 320 m^{3}s^{-1}. The inflow of freshwater decreases to approximately 50 m^{3}s^{-1} during July, and then progressively further down to ca. 20 m^{3}s^{-1} during the winter months.

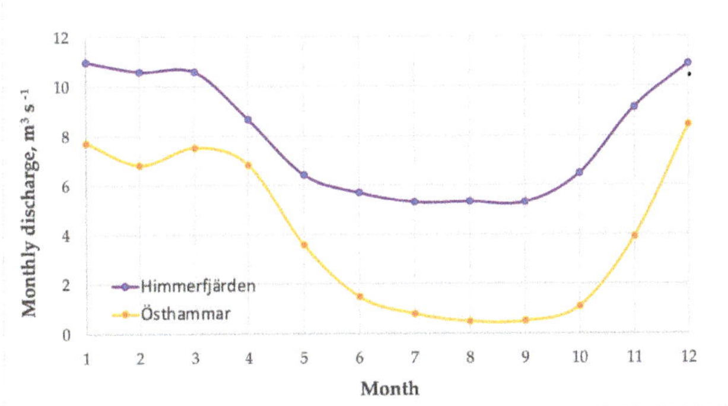

(a)

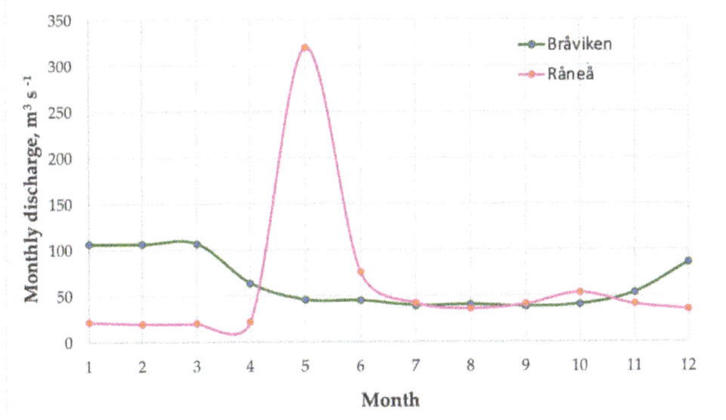

(b)

Figure 5. Changes in the monthly discharge averaged over the years 1991–2021 for (**a**) Himmerfärden and Östhammar bays, and (**b**) for Råneå and Bråviken bay. The values were derived from the SMHI interactive map (SMHI, Hydrologiskt Nuläge, Vattenwebb—in English: Current Hydrological Status, Water Web), available online [26] by summing up all the contributions of the smaller watersheds and greater catchments contributing to the discharge in the considered bay.

Table 2. Overview of the average monthly and cumulative yearly discharge (1991–2021).

Bay	Surface Area	Average Monthly Discharge	Standard Deviation	Min	Max	Yearly Discharge
Himmerfjärden	23,370 km^2	7.9	±2.41	5.3	11	95.4
Östhammar	993 km^2	4.1	±3.18	0.5	8.4	49.1
Bråviken	16,400 km^2	63.9	±28.60	38.1	106.2	766.4
Råneå	5670 km^2	60.4	±83.16	19.7	319.5	725.1

3.2. Results of the LULC Analysis

Figure 6 shows the map of Himmerfjärden bay and its closer catchment area before and after the LULC analysis was applied.

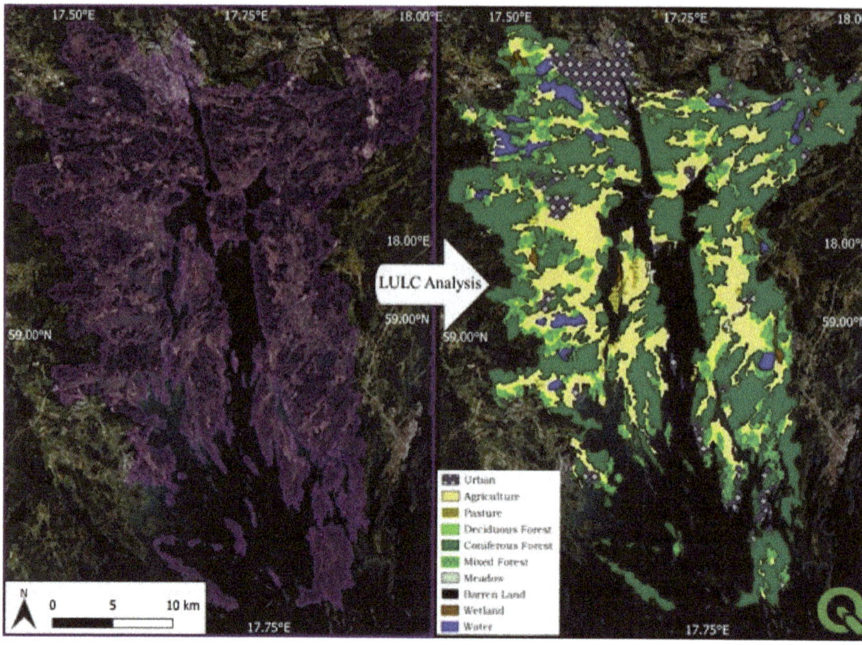

Figure 6. Image of the closer Himmerfjärden catchment area before and after LULC analysis. Maps generated in QGIS using predefined shapefiles (European coastline shapefile [22]. Recurrence layer—Global Surface Water, Surface elevation—Copernicus EU-DEM- v1.1, downloaded from the Sentinel Hub [25]; Water course network (vattendragslinjer nätverk): SVAR (Svenskt Vat-tenARkiv [26]).

The results of the LULC classification for all four bays are listed (as % coverage) in Table 3 and shown in Figure 7 (see also enlarged images in Supplementary Figure S2c,d). The catchment areas of Himmerfjärden bay (Figure 7c) and Bråviken bay (Figure 7d) exhibit quite similar land cover profiles, except that Himmerfjärden is more artificially developed due to its slightly more important *Agriculture* and *Urban* classes The Råneå estuary stands out with a much more *Natural* profile (Figure 7a). In this part of Swedish Lapland, the flora is not only dominated by *Coniferous Forest* but also by vast *Meadows* (13.2%) and *Wetlands* (19.1%). Östhammar, located slightly to the North of Stockholm archipelago, displays an intermediate profile between that of Råneå, Himmerfjärden bay and Bråviken bay. On the one hand, *Agriculture* (11.3%) is the third most important class, but to a lesser extent than for Bråviken bay (20.0%) or Himmerfjärden bay (23.2%). On the other hand, its *Forest* coverage is more similar to that of Råneå bay.

Table 3. The variability of LULC in the wider catchment areas of four Swedish bays. Values were obtained after processing of Copernicus CORINE_2018 data in QGIS. The main LULC classes are shown in bold. Classes for which the surface area was 0–3% across all bays were rejected from further analysis (marked in light grey) as they did not have a visual impact on the correlation analysis (see below), and therefore were assumed not to have a significant effect on the optical properties in the coastal bays.

	Urban	*Agriculture*	*Coniferous Forest*	*Deciduous Forest*	*Mixed Forest*	*Meadow*	*Pasture*	*Wetland*	*Water*	*Barren Land*
Bråviken	2.6%	**20.0%**	**47.3%**	2.6%	4.3%	2.5%	1.4%	1.0%	**18.4%**	0.0%
Himmerfjärden	4.4%	**23.2%**	**47.4%**	1.2%	6.5%	4.0%	0.7%	1.5%	**10.6%**	0.5%
Östhammar	2.9%	**11.3%**	**56.2%**	1.8%	**14.3%**	8.1%	1.1%	0.7%	3.7%	0.0%
Råneå	0.3%	1.5%	**52.4%**	1.1%	8.6%	**13.2%**	0.2%	**19.1%**	3.8%	0.0%

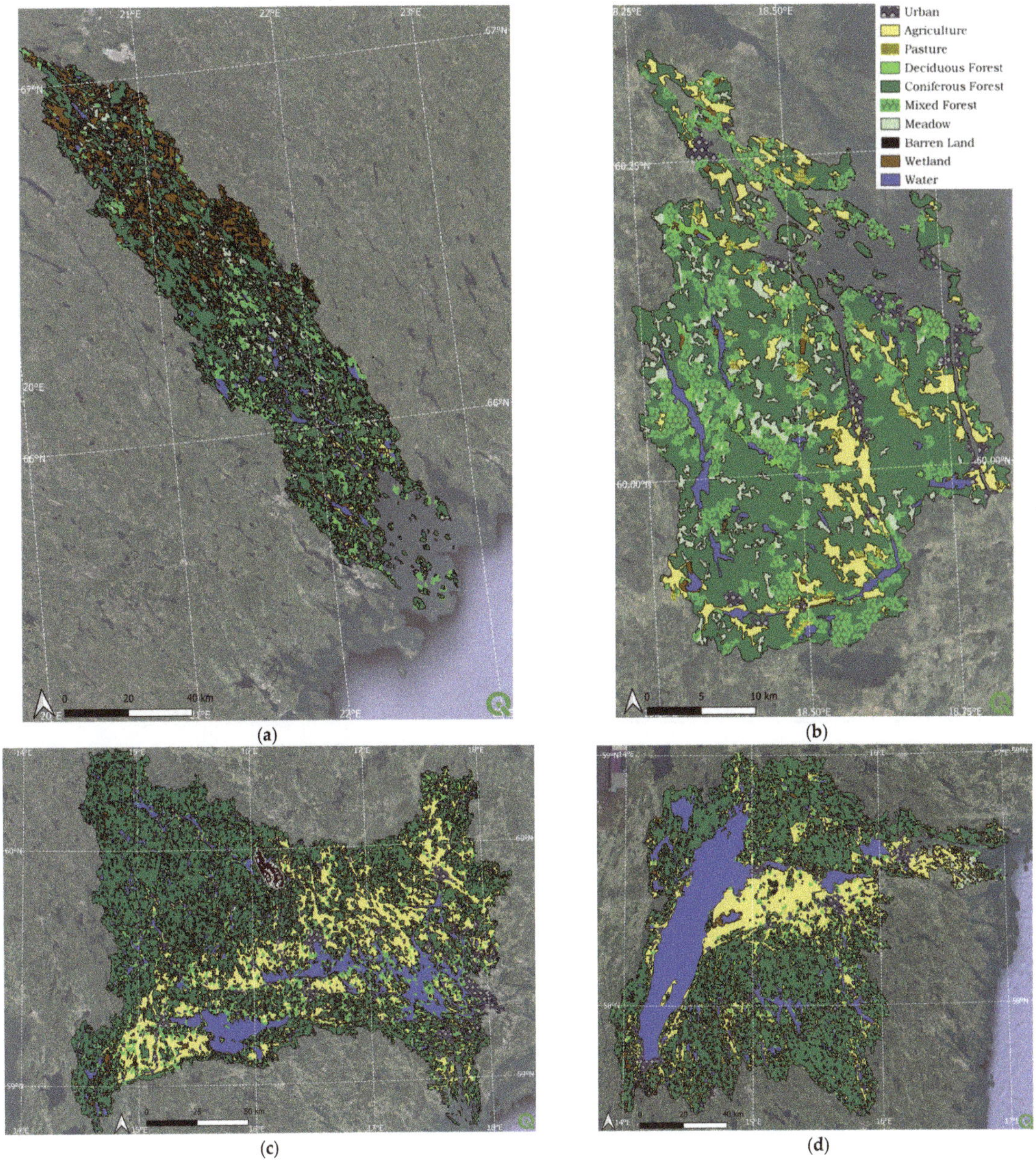

Figure 7. Landcover maps for the greater catchment areas of (**a**) the Råneå bay, (**b**) the Östhammar bay, (**c**) the Himmerfjärden bay, and for (**d**) the Bråviken bay. Note that the same figures are shown enlarged in the Supplementary Figure S2a–d (1 map per page). This allows for a more detailed assessment of the LULC in the greater catchment of each bay.

Table 3 also shows that the leading LULC class in each catchment area is the class *Coniferous Forest*. This is very typical for Swedish forests which are dominated by spruce and pine which together make up approximately 80% of tree coverage [31]. Overall, *Agriculture* was the second most important land cover type except in the more pristine area Råneå in the northwestern Bothnian Bay. Furthermore, *Mixed Forest* made up 4.3% to 14.3% across all catchments, while the *Water* class was also important in the catchments of Bråviken (18.4%) and Himmerfjärden (10.6%) due to the influence of Lake Vättern (Bråviken) and the combined influence from Mälaren and Hjälmaren (Himmerfjärden).

3.3. Ranges of Optical Properties in Each Bay

Table 4 gives an overview of the minimum and maximum values of the optical properties in each bay, as well as the median and the number of observations (n).

Table 4. Ranges of optical variables (shown in bold) in the four different bays. The median values of each optical variable in each bay are highlighted in bold.

		Chl-a	SPM	SPIM	SPOM	Turbidity	a_{CDOM}	S_{CDOM}
		$\mu g\ L^{-1}$	$g\ m^{-3}$	$g\ m^{-3}$	$g\ m^{-3}$	FNU	m^{-1}	
Himmerfjärden	Min	1.32	0.48	0.18	0.28	0.58	0.30	−0.021
	Max	13.70	2.69	1.36	1.59	1.96	0.80	−0.014
	Median	**4.17**	**1.65**	**0.61**	**0.92**	**1.29**	**0.46**	**−0.018**
	N	36	31	31	31	20	37	37
Bråviken	Min	2.90	2.12	1.41	0.27	1.24	0.42	−0.019
	Max	25.05	6.77	4.90	2.55	7.48	1.62	−0.017
	Median	**6.90**	**4.00**	**2.86**	**0.96**	**3.21**	**0.92**	**−0.018**
	N	29	23	23	23	23	29	19
Östhammar	Min	2.30	1.75	0.26	1.49	0.93	0.60	−0.018
	Max	89.88	14.54	3.89	13.59	7.25	2.46	−0.010
	Median	**7.97**	**6.24**	**0.68**	**4.29**	**3.16**	**0.91**	**−0.017**
	N	34	6	6	6	24	27	27
Råneå	Min	0.55				0.64	1.15	−0.018
	Max	5.69				8.90	5.18	−0.016
	Median	**2.21**				**1.34**	**1.71**	**−0.017**
	N	12				12	12	12

3.4. Investigating the Nature of CDOM Due to LULC

Figure 8 displays CDOM absorption at 440 nm, a_{CDOM} (440) against the distance from the outlet for each respective bay (data from all seasons and years included). The maximum CDOM values were found in Råneå (up to approximately 6 m^{-1}) during the transect in May 2018, which corresponds to the snow and ice thawing period. At that time of the year the average water catchment discharge, Q, is approximately 300 m^3s^{-1} (Figure 5b). The results from the correlation analysis of LULC categories vs. a_{CDOM} and S_{CDOM}, respectively are shown in Table 5.

Table 5. Correlation matrix for CDOM absorption at 440 nm, a_{CDOM}, and spectral slope coefficient. **Significant correlations are marked in bold characters** after the p-value test (with significance level α = 0.05). Natural classes: *Natural** = *Coniferous Forest* + *Wetland* + *Meadows*; Developed classes: *Developed* = Urban + Agriculture; *Ratio** (*Dev/Nat*): the ratio of developed to natural classes (*Dev/Nat*).

	Water	*Coniferous Forest*	*Mixed Forest*	*Meadow*	*Wetland*	*Agriculture*
S_{CDOM}	−0.744	0.585	0.387	**0.984**	**0.916**	**−0.980**
a_{CDOM}	−0.510	0.421	0.166	0.882	**0.933**	**−0.944**
	Urban	*Pasture*	*Discharge*	*Dev**	*Natural**	*Ratio**
S_{CDOM}	−0.878	−0.767	0.829	**−0.985**	**0.998**	**−0.975**
a_{CDOM}	**−0.980**	−0.648	0.815	**−0.964**	**0.931**	**−0.926**

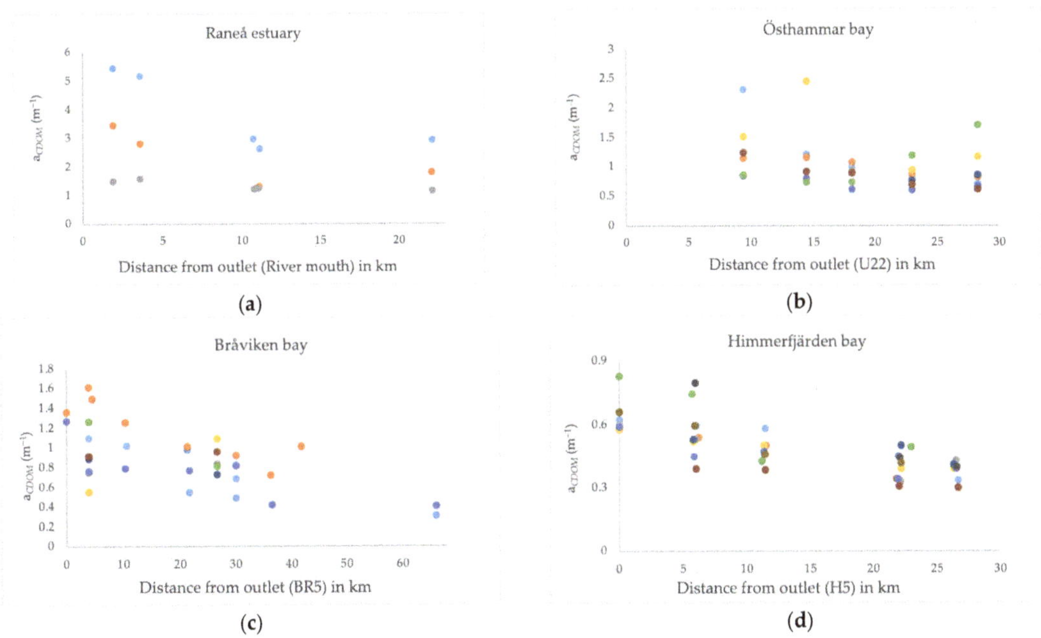

Figure 8. CDOM absorption at 440 nm, a_{CDOM}, in four Swedish bays: (**a**) Råneå bay, (**b**) Östhammar bay (**c**) Bråviken bay and (**d**) Himmerfärden bay. Values are plotted against horizontal distance from the outlet (in km). All transects were plotted within each bay regardless of the season. The different colors in each plot refer to different days for the respective in situ transects.

In Figure 9a one can see how a_{CDOM} is related to the ratio of *Developed* to *Natural* classes in the wider catchment areas. The lower the proportion of developed areas, i.e., *Ratio** (*Dev/Nat*), the higher the CDOM absorption. The results show that Råneå bay has the most natural catchment with a *Ratio** (*Dev/Nat*) of approximately 0.023, followed by Öshammar bay (0.235), Bråviken (0.47) and Himmerfjärden bays (0.535). Figure 9b shows that the average CDOM slope factor per transects is also dependent on the ratio of developed to natural areas, and there is a negative trend over the four bays, although Himmerfjärden shows a large spread of data. Seasonal variability can be here rejected as explanation as both the highest (August 2010) and lowest value (August 2017) were both obtained during measurements in summer. The large variability in Himmerfjärden bay is likely to be due to the influence of Himmerfjärden Sewage Treatment Plant situated in the inner bay, where nitrate is treated by bacterial breakdown [31]. In addition to breaking down nitrates, the bacteria may also use CDOM as a substrate, and break down the larger humic acids to fulvic acids. On the other hand during periods of flooding, untreated stormwater may enter the bay, and thus giving rise to larger humic acids. The decreasing trend from Råneå to Bråviken in Figure 9b indicates that more natural LULC types will generate more humic acids, whereas the more developed bays tend to have a lower spectral slope factor, and thus contain a higher proportion of fulvic acids.

Figure 10 shows that there is a clear difference in the spatial trend of a_{CDOM} across the bays between spring and summer. a_{CDOM} is especially high in the Bothnian Sea during spring presumably due to the high coastal runoff after the spring melt, bringing in high concentrations of larger humic acids [5].

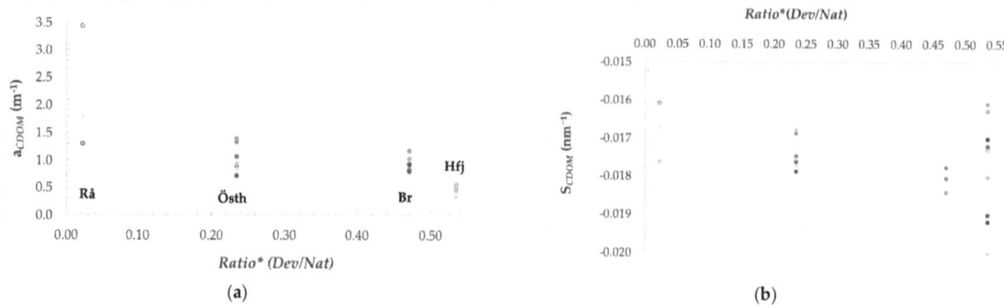

Figure 9. (**a**) CDOM absorption at 440 nm, a_{CDOM}, against the LULC category *Ratio** (*Dev/Nat)*, i.e., *the* ratio of developed to natural classes; all transects for all seasons. (**b**) Averaged slope factor per transects against *Ratio** (*Dev/Nat*) per bay; Råneå (Rå), Östhammar (Östh), Bråviken (Br) and Himmerfjärden (Hfj)—all transects for all seasons.

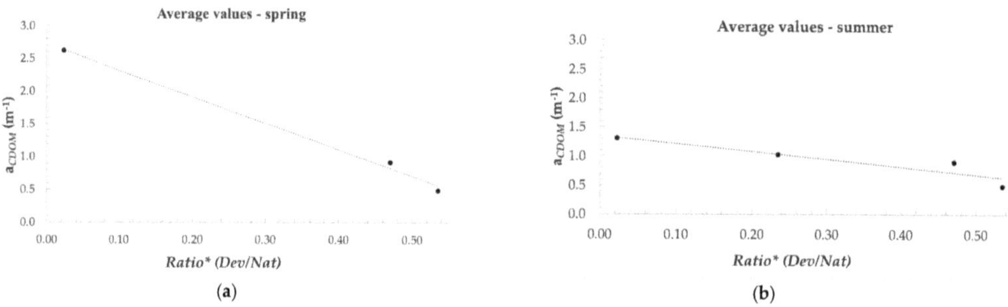

Figure 10. (**a**) CDOM absorption at 440 nm, a_{CDOM}, against the LULC category *Ratio** (*Dev/Nat*), i.e., *the* ratio of developed to natural classes; average value per transect and bay in summer, and (**b**) the average value per transect and bay in spring; Råneå (Rå), Östhammar (Östh), Bråviken (Br) and Himmerfjärden (Hfj).

3.5. Investigating the Nature of Particulate Material with LULC

Overall, the suspended particulate matter (SPM) and turbidity correlation analyses did not show such clear trends as CDOM absorption and slope across the different areas of interest, apart from suspended particulate organic matter, SPOM (Table 6), which showed a positive significant relationship with the LULC classes *Coniferous Forest*, *Mixed Forest* and *Meadow*, and thus with the *Natural** classes, while it showed a negative relationship with *Agriculture*, and the ratio of developed to natural classes (*Ratio**).

Table 6. Correlation matrix for SPM, SPOM and turbidity. Significant correlations are highlighted in bold characters after the p-value test (with significance level $\alpha = 0.05$). Natural classes: *Natural** = *Coniferous Forest* + *Wetland* + *Meadows*; developed classes: *Developed* = Urban + Agriculture; *Ratio** (*Dev/Nat*): the ratio of developed to natural classes (*Dev/Nat*).

	Water	*Coniferous Forest*	*Mixed Forest*	*Meadow*	*Wetland*	*Agriculture*
SPM	−0.274	0.739	0.588	0.538	**−0.991**	−0.895
SPOM	−0.848	**0.999**	**0.979**	**0.964**	−0.824	**−0.963**
Turbidity	0.164	0.379	0.185	0.125	−0.839	0.618
	Urban	*Pasture*	*Discharge*	*Natural**	*Dev**	*Ratio**
SPM	−0.884	0.714	**−0.978**	0.646	−0.925	0.866
SPOM	−0.344	0.060	−0.586	**0.991**	−0.941	**−0.978**
Turbidity	**−0.999**	0.945	**−0.973**	0.256	−0.672	0.672

3.6. Investigating the Dependency of the Chl-a Concentration on LULC

The correlation matrix (Table 7) shows significant negative correlations for *Chl-a* with the classes *Agriculture*, *Wetland*, and developed classes (*Dev**), as well as the ratio of developed to natural classes (*Dev/Nat*).

Table 7. Correlation matrix for *Chl-a* concentration with LULC categories. **Significant correlations are shown in bold numbers** after the *p*-value test (with significance level $\alpha = 0.05$). Natural classes: *Natural* = Coniferous Forest + Wetland + Meadows*; developed classes: *Developed* = Urban + Agriculture; *Ratio* (Dev/Nat)*: the ratio of developed to natural classes (*Dev/Nat*).

	Water	*Coniferous Forest*	*Mixed Forest*	*Heaths*	*Wetland*	*Agriculture*
Chl-*a*	−0.616	0.938	0.849	0.816	**−0.968**	**−0.997**
	Urban	*Pasture*	*Discharge*	*Natural**	*Dev**	*Ratio**
Chl-*a*	−0.644	0.398	−0.828	0.886	**−1.000**	**−0.991**

Figure 11 displays the average Chl-*a* values for a given field campaign plotted against the ratio of developed to natural classes (*Dev/Nat*) LULC. The Råneå bay clearly stands out with its relatively low chlorophyll-a concentrations, both in spring and summer. This is likely to be related to the high CDOM absorption with very large values found in coastal areas of the Bothnian bay (Figure 8a). Indeed, the strong light attenuation by CDOM in this area has been shown to limit the growth of phytoplankton and algae, and instead to favor bacterial production [23]. Since the Råneå measurements clearly stand out when compared to the other bays, it has been rejected from the correlation matrix for the remainder of the analysis. The motivation for this is that the relatively low Chl-a levels here are most likely due to light limitation by CDOM rather than caused by different LULC categories.

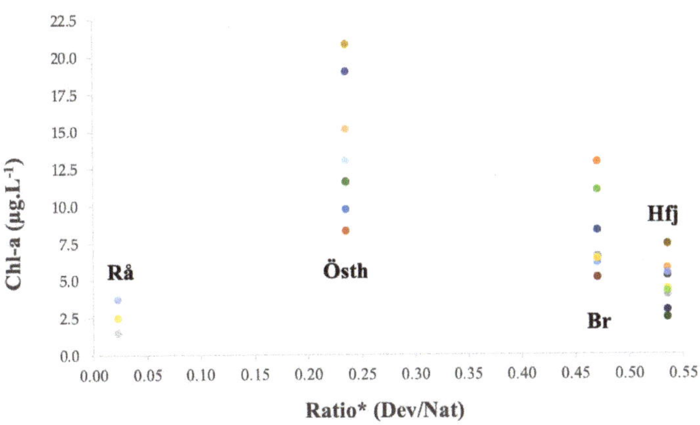

Figure 11. Chl-*a* concentration in four Swedish bays—Råneå (Rå), Östhammar (Östh), Bråviken (BR) and Himmerfjärden (Hfj)—against the ratio of *Developed* to *Natural* (*Dev/Nat*) LULC.

4. Discussion

4.1. LULC Classification

The land use and land cover (LULC) analysis is central in this study, focusing initially on analyzing the variation in LULC from one Swedish bay to another. Our LULC classification (Table 3) gave a similar result as the one performed by Franzén et al. [32] who classified the sub-catchment area around Himmerfjärden with 57% forest, 33% land, 4% lake and 5% urban areas. We decided to use the great catchment in our study, assuming that the optical properties in the whole catchment are carried downstream, and therefore influence the optical properties of the bay. Gullstrand et al. [33] classified Bråviken with

49% forest, 18% agriculture and 20% water which, again is quite similar to our classification (Table 3), even though we decided to include the greater catchment rather than just the surrounding catchment of the bay. As the bays also differ in their surrounding and catchment areas, it is likely that the bio-optical properties are influenced by the variation in the land cover and land use in each area. Table 3 shows that the leading LULC category in each catchment area is the category *Coniferous Forest*, followed by *Mixed Forest*. The dominance of *Coniferous Forest* is very typical for Swedish forests [34]. Our LULC analysis also confirms previous work showing that catchments in boreal regions are mostly covered by *Coniferous* and *Mixed Forest* [35], carrying high concentrations of terrestrial dissolved organic matter (DOM) downstream during the productive seasons. Additionally, there are usually wetlands and bogs in boreal areas, which have the tendency to hold large quantities of organic matter [36]. CDOM usually increases in spring after the ice melt, followed by a decrease in summer and an increase, again, during autumn [5,37]. Said Al-Kharusi [38] proposed that recent advances in remote sensing technology, GIS and modelling could improve our understanding of CDOM in inland waters on large geographic scales. The same has been demonstrated here for the distribution of CDOM in coastal waters.

4.1.1. Influence of LULC Classification on a_{CDOM} and S_{CDOM}

As shown in the results section, a_{CDOM} and S_{CDOM} are both positively related to *Wetland*, and the *Natural** LULC classes while they are negatively correlated to *Agriculture* and the *Developed* classes. Zheng et al. [39] found positive correlations for a_{CDOM} and S_{CDOM} with the natural categories *Forest* and *Grassland* in the yellow river basin. An explanation why we did not find a significant positive correlation with *Forest* might be that our forests are mostly dominated by coniferous trees. Dead plant material is slowly degraded to humic substances by bacterial breakdown [1]. During droughts, it accumulates on land, and during floods and the spring melt it is washed into lakes and coastal waters [38,40,41]. The minimum range of CDOM values (approximately 1 m^{-1}) is found in Himmerfjärden bay (corresponding to the average value in the middle of the transect). Note that Zone I does not exhibit any strong peaks in discharge for the thawing in spring, based on the hydrological data from SMHI [26]; see Figure 5.

The results shown in Figure 10a,b clearly indicate a seasonal variability in the CDOM absorption. The results show a steeper (decreasing) average slope in spring than in summer. According to the correlation matrix (Table 3), *Wetland* and *Meadows* showed significant and positively correlated coefficients with the CDOM slope factor. With an increasing proportion of wetlands and meadows, the humic types of CDOM thus seem to increase as indicated by a CDOM slope closer to −0.0011 [4].

We also find that the categories *Agriculture*, *Developed* and also *Ratio** each show a significant, negative correlation with S_{CDOM}. The influence of developed categories is more difficult to interpret. One explanation could simply be the degradation of CDOM in natural environments, either by UV light (a phenomenon called photobleaching, or fading) and degradation by bacteria [1]. Indeed, CDOM is naturally broken down progressively—chemically or physically—in these ways. For instance, Hulatt et al. [42] found that photo bleaching can cause a significant increase in the spectral slope factor (by 0.004 nm^{-1}), that is to say that CDOM tends to be progressively more broken down into smaller fractions, and thus tends to have a more fulvic profile (i.e., a slope closer to −0.0022). Therefore, Agriculture and other developed categories might influence the CDOM slope factor by enhancing the decomposition of CDOM. Williams et al. [43] found that dissolved organic matter (DOM) in streams influenced by agriculture was more labile and more accessible to microbial degradation than DOM found in wetland streams, supporting lower rates of microbial activity. Such result suggests that the *Developed** classes of LULC will be likely to stimulate bacterial breakdown of CDOM, and thus be more likely to generate a higher content of smaller fulvic acids.

4.1.2. Influence of LULC Classification on SPM

As mentioned before, there were no obvious relationships between total SPM and LULC cover. However, for suspended particulate organic matter (SPOM) significant and robust relationships were found for the classes *Coniferous* and *Mixed Forests*, as well as *Meadows* and the categories *Natural** and *Ratio** (see Table 6) while the *Agricultural* class showed a negative but significant correlation. One could explain the positive correlations with predominantly natural categories by the fact that they generate a lot of pollen in spring and summer. Pollen constitutes indeed an important fraction of SPM in natural waters [44] and in the Baltic Sea [45]. Viennet [44] focused his study on morpho-granulometry analysis as to investigate the nature, origin and transfer of SPM. He found that SPM consists of three main categories: mineral particles (silica sands and clays), organic matter (coal, carbonized OM), peat and loam (humus-containing soil) and other biologic matter (phytoplankton and pollen). This may therefore explain the occurrence of strong relationships between SPOM on the one hand, and forests on the other hand. Pawlik and Ficek [45] found that in many areas of the central Baltic Sea, pine pollen can constitute up to 50% of the SPM in the 1.25–250 μm size range during spring.

Agriculture was found to be negatively correlated with SPOM (but not with SPIM) and agricultural land thus seems to act as sink of organic matter. Wetland showed a significant negative correlation with total SPM. Such a result was also found by Le et al. [19] and is coherent with the hypothesis that bogs may act as sinks for suspended material which may fall out and settle. Furthermore, plants and their roots may act as physical barriers preventing erosion, and thus also decreasing the SPM concentration.

Some of the findings regarding SPM are contradictory to the literature. For example, one may hypothesize that the inorganic fraction (suspended particulate matter, SPIM) would be positively correlated with the *Urban* and *Agriculture* classes as shown in Le et al. [19] in the Gulf of Mexico. This relies on the assumption that developed areas give rise to more exposed soils or barren land, allowing for more runoff of sediments and other particulates into rivers and, eventually the sea. However, considering SPM and SPIM, we did not find any significant relationships to any of the investigated LULC types. This may partially be due to the fact that the *Urban* class only made up a very small percentage in each area of interest (ranging from 0.3% to 4.4%), which is rather low compared to the main categories and therefore seems to have an insignificant effect. Also, given that our study is in a completely different vegetation zone, the results from the study by Le et al. [19] in the Gulf of Mexico, with very large and continued high runoff and larger *Urban* proportions (as well as a lack of seasonal influence), cannot be simply assumed to also apply to high latitudes.

By using SMHI's subbasin division and water course maps it was possible to map the water flow for the four catchments for comparison, and also to use the hydrological information to define the catchment areas. Hannerez and Destouni [46] had shown that approximately 20% of Sweden's coastal catchment areas were unmonitored. However, the catchment areas of the four bays chosen in this study belong to the well-monitored ones. It must also be noted, that the bays in Zone I all had a relatively continues discharge over the year (with some variations), while Råneå showed an extremely strong discharge in spring/early summer (Figure 5a,b), while its urban category is negligible (0.3%). Thus, the nature of our dataset does not allow to properly investigate the effect of the *Urban* class.

4.1.3. Influence of LULC Classification on Chl-*a*

We found a non-significant negative relationship with Chl-*a* concentration for the *Urban* class (Table 5) which, again, may be related to the generally small surface areas of the urban classes in our catchment areas. The negative correlation between *Wetland* and *Chl-a* may indicate that *Wetland* acts as a sink. Indeed, wetlands are often used as a measure to mitigate the effects of eutrophication. For example, wetlands and reed belts are used as buffer zones around the Baltic Sea coasts in order to fix carbon (and nutrients), and to reduce nutrient leakage into the Baltic Sea. Reed has the potential to sequester nutrients and to stabilize soils, as well as to reduce heavy metals [47]. By growing reed one can therefore reduce erosion and drainage of SPM, and at the same time reduce the drainage of nutrients if the reeds are harvested, and for example used as cow feed. Furthermore, underwater seagrass meadows can also act as nutrient and sediment traps [48], and are therefore also important for the reduction in Chl-*a* and SPM contents in coastal bays.

5. Conclusions

It was the aim of this paper to investigate the influence of land use and land cover on the optical properties of four Swedish coastal bays, spanning from the northwestern Baltic proper to the northwestern Bothnian Bay. As there is little information on IOPs available from most of the bays, we focused on those optical properties that have been quite well sampled over the last decade: chlorophyll-a, CDOM, SPM and turbidity. The catchment areas of each bay were defined using GIS in combination with a hydrological model. Correlation analysis was then used to relate land cover type to the optical properties in the coastal bays, and the correlations were tested for significance.

This macroscopic study of the four Swedish bays with their respective catchments lead to some compelling findings. Land use and cover analysis allows to investigate the nature of CDOM in the Baltic Sea. For instance, with a GIS analysis assessing the percentage of land cover, one can then approximate the slope value of CDOM, S_{CDOM}: approximately -0.019 for *Natural/Wetland* covered watershed and approximately -0.015 in more developed catchments. Given that only limited information exists regarding the specific IOPs such as S_{CDOM} in Swedish coastal areas, CDOM estimates based on LULC classification may show to be useful in the local or regional adaptation of bio-optical and/or remote sensing algorithms. This study shows that the CDOM absorption mean value for a given bay can also be approximated by mapping the extent of *Natural* land cover (*Wetland* and *Meadow*). Moreover, the strong correlation found between SPOM and *Meadow*, *Mixed Forest* and *Coniferous Forest* seems to indicate that SPOM mostly originates from pollen. Last but not least, this study also confirms the role of *Wetlands*, which seem to be a primary terrigenous source of CDOM with high humic character and also seems to provide a physical barrier for SPM, acting as buffer zones and trapping sediments. The correlation analysis for Chl-a showed different patterns and were usually opposite to the findings in Le et al. [19], indicating that complex biological factors may rule Chl-a production, and that these results do not generally apply to ecosystems across different vegetation zones (warm-temperate to subtropical vs. boreal) and may also strongly depend on the nature of the dominant LULC categories. Suspended particulate matter and turbidity were the least conclusive parts of this study and require a consideration of other parameters related to physical forcing (e.g., bottom depth of lakes and bays, wind speed and weather condition for a given transect) which was beyond the scope of the study. In further studies, one could also weigh each watershed by its average discharge value, and thereby account for the relative contribution of each area to the total discharge, and also relate it to the volume of the respective bay. Furthermore, one could investigate the influence of different types and locations of crops on the optical properties of coastal waters.

Supplementary Materials: The following supporting information can be downloaded at: https://www.mdpi.com/article/10.3390/rs16010176/s1, Supplementary Figure S1a,b. Enlarged figures of water flow in the catchments of Zone I and Zone II. Supplementary Figure S2a–d: Enlarged figures of land cover maps of all four bays. Table S1: Corine land cover classes. Table S2: Bio-optical dataset used in this study.

Author Contributions: Conceptualization, S.K. and M.A.; methodology, M.A. and S.K.; software, M.A.; validation, M.A. and S.K.; formal analysis, M.A.; investigation, M.A. and S.K.; resources, S.K. and M.A.; data curation, M.A.; writing—original draft preparation, M.A. and S.K.; writing—review and editing, S.K.; visualization, M.A. and S.K.; supervision, S.K.; project administration, S.K.; funding acquisition, S.K and M.A. All authors have read and agreed to the published version of the manuscript.

Funding: S.K. was funded by the Swedish National Space Agency (SNSA), grant number 2021-00064, and by the Swedish Agency for Marine and Water Management (SwAM), Project No. 538-21. M.A. was funded by the EU ERASMUS^{+} program and by the BRMIE grant of the Auvergne-Rhone-Alpes Region, France.

Data Availability Statement: The bio-optical dataset used in this study is available under supplementary material (Supplementary Table S2).

Acknowledgments: The authors would like to thank the pelagic monitoring groups from Stockholm University and Umeå University (UMF) for provision of data and support during field work (UMF). Lots of thanks to Sean O'Kane (Maynooth University, Ireland) and Tom Verheyde (INSA, Lyon, France) for advise on the GIS tools and analysis. Many thanks to SMHI for making the hydrological data available on-line. Thanks to Steve Lyon (Ohio State University, USA) for advise on hydrological analysis, and thanks to Jakob Walve (DEEP, Stockholm University) for providing useful feed-back to the manuscript. Additionally, we would like to thank our three anonymous reviewers who helped to substantially improve the manuscript.

Conflicts of Interest: The authors declare no conflicts of interest.

References

1. Kirk, J.T.O. *Light and Photosynthesis in Aquatic Ecosystems*, 3rd ed.; Cambridge University Press: Cambridge, UK, 2010; 639p, ISBN 978-0-521-15175-7.
2. Schnitzer, M. Humic substances: Chemistry and reactions. In *Developments in Soil Science*; Volume 8: Soil Organic Matter; Schnitzer, M., Khan, S.U., Eds.; Elsevier: Amsterdam, The Netherlands, 1978; pp. 1–64.
3. Preisendorfer, R.W. Application of radiative transfer theory to light measurement in the sea. *Union Geod. Geophys. Inst. Monogr.* **1961**, *10*, 11–30.
4. Carder, K.L.; Steward, R.G.; Harvey, G.R.; Ortner, P.B. Marine humic and fulvic acids: Their effects on remote sensing of ocean chlorophyll. *Limnol. Ocean.* **1989**, *34*, 68–81. [CrossRef]
5. Harvey, T.; Allart, S.; Andersson, A. Relationships between Coloured Dissolved Organic Matter (CDOM) and Dissolved Organic Carbon (DOC) in different coastal gradients of the Baltic Sea. *AMBIO* **2015**, *44*, 392–401. [CrossRef] [PubMed]
6. Kowalczuk, P.; Stedmon, C.A.; Markager, S. Modeling absorption by CDOM in the Baltic Sea from season, salinity and chlorophyll. *MAR CHEM* **2006**, *101*, 1–11. [CrossRef]
7. Kratzer, S.; Tett, P. Using bio-optics to investigate the extent of coastal waters: A Swedish case study. *Hydrobiologia* **2009**, *629*, 169–186. [CrossRef]
8. Robinson, I.S. Satellite observations of ocean colour. *Philos. Trans. R. Soc. London. Ser. A Math. Phys. Sci.* **1983**, *309*, 415–432. [CrossRef]
9. Wild-Allen, K.; Lane, A.; Tett, P. Phytoplankton, sediment and optical observations in Netherlands coastal water in spring. *J. Sea Res.* **2002**, *47*, 303–315. [CrossRef]
10. van de Hulst, H.C.; Twersky, V. *Light Scattering by Small Particles*; Dover Publications, Inc.: New York, NY, USA, 1981; 470p, ISBN 0-486-64228-3.
11. Carder, K.L.; Betzer, P.R.; Eggimann, D.W. Physical, chemical, and optical measures of suspended-particle concentrations: Their intercomparison and application to the West African Shelf. In *Suspended Solids in Water*; Gibbs, R.J., Ed.; Springer: Boston, MA, USA, 1974; pp. 173–181. [CrossRef]
12. Aas, E. Refractive index of phytoplankton derived from its metabolite composition. *J. Plankton Res.* **1996**, *18*, 2223–2249. [CrossRef]
13. Lide, D.R. *Physical and optical properties of minerals. CRC Handbook of Chemistry and Physics*, 77th ed.; CRC Press: Boca Raton, FL, USA; Taylor & Francis: Boca Raton, FL, USA, 1996; p. 2608, ISBN 10-0849305969.
14. Boss, E.; Pegau, W.S.; Lee, M.; Twardowski, M.; Shybanov, E.; Korotaev, G.; Baratange, F. Particulate backscattering ratio at LEO 15 and its use to study particle composition and distribution. *J. Geophys. Res. Oceans* **2004**, *109*. [CrossRef]
15. Kratzer, S.; Kyryliuk, D.; Brockmann, C. Inorganic suspended matter as an indicator of terrestrial influence in Baltic Sea coastal areas—Algorithm development and validation, and ecological relevance. *Remote Sens. Environ.* **2020**, *237*, 111609. [CrossRef]
16. Kari, E.; Kratzer, S.; Beltrán-Abaunza, J.M.; Harvey, E.T.; Vaičiūtė, D. Retrieval of suspended particulate matter from turbidity–model development, validation, and application to MERIS data over the Baltic Sea. *Int. J. Remote Sens.* **2017**, *38*, 1983–2003. [CrossRef]
17. Miller, R.L.; McKee, B.A. Using MODIS Terra 250 M Imagery to Map Concentrations of Total Suspended Matter in Coastal Waters. *Remote Sens. Environ.* **2004**, *93*, 259–266. [CrossRef]
18. Kratzer, S.; Moore, G. Inherent Optical Properties of the Baltic Sea in Comparison to Other Seas and Oceans. *Remote Sens.* **2018**, *10*, 418. [CrossRef]
19. Le, C.; Lehrter, J.C.; Hu, C.; Schaeffer, B.; MacIntyre, H.; Hagy, J.D.; Beddick, D.L. Relation between inherent optical properties and land use and land cover across Gulf Coast estuaries. *Limnol. Ocean.* **2015**, *60*, 920–933. [CrossRef]
20. HELCOM (Helsinki Commission) Sub-Basins. 2018. Available online: http://metadata.helcom.fi (accessed on 15 August 2022).
21. Natural Earth, Free Vector and Raster Map Data. Available online: https://www.naturalearthdata.com (accessed on 11 September 2022).
22. European Environment Agency (EEA) European Shapefile. Available online: https://www.eea.europa.eu (accessed on 11 September 2023).

23. SMHI (Swedish Meteorological and Hydrological Institute), SVAR (Svenskt VattenARkiv). 2021. Available online: https://www.smhi.se/data/hydrologi/sjoar-och-vattendrag/ladda-ner-data-fran-svenskt-vattenarkiv-1.20127 (accessed on 11 September 2023).
24. Wikner, J.; Andersson, A. Increased freshwater discharge shifts the trophic balance in the coastal zone of the northern Baltic Sea. *Glob. Change Biol.* **2012**, *18*, 2509–2519. [CrossRef]
25. Pekel, J.F.; Cottam, A.; Gorelick, N.; Belward, A.S. High-resolution mapping of global surface water and its long-term changes. *Nature* **2016**, *540*, 418–422. [CrossRef]
26. SMHI, Hydrologiskt Nuläge, Vattenwebb (In English: Current Hydrological Status, Water Web). Available online: https://vattenwebb.smhi.se/hydronu/ (accessed on 11 September 2023).
27. Kratzer, S.; Harvey, E.T.; Canuti, E. International Intercomparison of In Situ Chlorophyll-a Measurements for Data Quality Assurance of the Swedish Monitoring Program. *Front. Remote Sens.* **2022**, *3*, 866712. [CrossRef]
28. Strickland, J.H.D.; Parsons, T.R. *A Practical Hand-Book of Sea-Water Analysis*; Bulletin Journal of the Fisheries Research Board of Canada; Pergamon Press: Oxford, UK, 1972; Volume 167, pp. 185–203, ISBN 0-08-030-288-2.
29. Parsons, T.R.; Maita, Y.; Lalli, C.M. *A Manual of Chemical and Biological Methods for Seawater Analysis*; Pergamon Press: Oxford, UK, 1984; Volume 173, ISBN 0-08-030288-2.
30. Sørensen, K.; Grung, M.; Röttgers, R. An intercomparison of in vitro chlorophyll a determinations for MERIS level 2 data validation. *Int. J. Remote Sens.* **2007**, *28*, 537–554. [CrossRef]
31. Cema, G.; Płaza, E.; Trela, J.; Surmacz-Górska, J. Dissolved oxygen as a factor influencing nitrogen removal rates in a one-stage system with partial nitritation and Anammox process. *Water Sci. Technol.* **2011**, *64*, 1009–1015. [CrossRef]
32. Franzén, F.; Kinell, G.; Walve, J.; Elmgren, R.; Söderqvist, T. Participatory social-ecological modeling in eutrophication management: The case of Himmerfjärden, Sweden. *Ecol. Soc.* **2011**, *16*, 27. [CrossRef]
33. Gullstrand, M.; Löwgren, M.; Castensson, R. Water issues in comprehensive municipal planning: A review of the Motala River Basin. *J. Environ. Manag.* **2003**, *69*, 239–247. [CrossRef] [PubMed]
34. SLU. *Official Statistics of Sweden*; Publikationsservice, Uppsala; Swedish University of Agricultural Sciences: Umeå, Sweden, 2017; ISSN 0280-0543.
35. Laudon, H.; Berggren, M.; Ågren, A.; Buffam, I.; Bishop, K.; Grabs, T.; Jansson, M.; Köhler, S. Patterns and dynamics of Dissolved Organic Carbon (DOC) in boreal streams: The role of processes, connectivity, and scaling. *Ecosystems* **2011**, *14*, 880–893. [CrossRef]
36. Mzobe, P.; Berggren, M.; Pilesjö, P.; Lundin, E.; Olefeldt, D.; Roulet, N.T.; Persson, A. Dissolved organic carbon in streams within a subarctic catchment analysed using a GIS/remote sensing approach. *PLoS ONE* **2018**, *13*, e0199608. [CrossRef] [PubMed]
37. Ågren, A.; Buffam, I.; Berggren, M.; Bishop, K.; Jansson, M.; Laudon, H. Dissolved organic carbon characteristics in boreal streams in a forest-wetland gradient during the transition between winter and summer. *J. Geophys. Res. Biogeosci.* **2008**, *113*, G03031. [CrossRef]
38. Said Al-Kharusi, E. Broad-Scale Patterns in CDOM and Total Organic Matter Concentrations of Inland Waters—Insights from Remote Sensing and GIS. Ph.D. Thesis, Department of Physical Geography and Ecosystem Science, Faculty of Science, Lund University, Lund, Sweden, 2021.
39. Zheng, K.; Shao, T.; Ning, J.; Zhuang, D.; Liang, X. Water quality, basin characteristics, and discharge greatly affect CDOM in highly turbid rivers in the Yellow River Basin, China. *J. Clean Prod.* **2023**, *4*, 136995. [CrossRef]
40. Toming, K.; Arst, H.; Paavel, B.; Laas, A.; Nõges, T. Spatial and temporal variations in coloured dissolved organic matter in large and shallow Estonian waterbodies. *Bor. Environ. Res.* **2009**, *14*, 959–970.
41. Kari, E.; Merkouriadi, I.; Walve, J.; Leppäranta, M.; Kratzer, S. Development of under-ice stratification in Himmerfjärden bay, North-Western Baltic proper, and their effect on the phytoplankton spring bloom. *J. Mar. Syst.* **2018**, *186*, 85–95. [CrossRef]
42. Hulatt, C.J.; Thomas, D.N.; Bowers, D.G.; Norman, L.; Zhang, C. Exudation and decomposition of chromophoric dissolved organic matter (CDOM) from some temperate macroalgae. *Estuarine Coast. Shelf Sci.* **2009**, *84*, 147–153. [CrossRef]
43. Williams, C.J.; Yamashita, Y.; Wilson, H.F.; Jaffé, R.; Xenopoulos, M.A. Unraveling the role of land use and microbial activity in shaping dissolved organic matter characteristics in stream ecosystems. *Limnol. Oceanogr.* **2010**, *55*, 1159–1171. [CrossRef]
44. Viennet, D. The Use of Morpho Granulometry in Source to Sink Monitoring of Particle Transfer in Watersheds. Ph.D. Thesis, École Doctorale Normande de Biologie Intégrative, Santé, Environnement, Mont-Saint-Aignan CEDEX, France, 2020. Available online: http://viaf.org/viaf/408160483843904992099 (accessed on 1 December 2023).
45. Pawlik, M.M.; Ficek, D. Spatial Distribution of Pine Pollen Grains Concentrations as a Source of Biologically Active Substances in Surface Waters of the Southern Baltic Sea. *Water Sui* **2023**, *15*, 978. [CrossRef]
46. Hannerz, F.; Destouni, G. Spatial characterization of the Baltic Sea drainage basin and its unmonitored catchments. *AMBIO* **2006**, *35*, 214–219. [CrossRef] [PubMed]
47. Karstens, S.; Buczko, U.; Jurasinski, G.; Peticzka, R.; Glatzel, S. Impact of adjacent land use on coastal wetland sediments. *Sci. Total Environ.* **2016**, *550*, 337–348. [CrossRef] [PubMed]
48. Dahl, M.; Asplund, M.E.; Björk, M.; Deyanova, D.; Infantes, E.; Isaeus, M.; Nyström Sandman, A.; Gullström, M. The influence of hydrodynamic exposure on carbon storage and nutrient retention in eelgrass (*Zostera marina* L.) meadows on the Swedish Skagerrak coast. *Sci. Rep.* **2020**, *10*, 13666. [CrossRef] [PubMed]

remote sensing

MDPI

Review

Satellite Altimetry for Ocean and Coastal Applications: A Review

Margaret Srinivasan * and Vardis Tsontos

Jet Propulsion Laboratory, California Institute of Technology, Pasadena, CA 91109, USA; vardis.m.tsontos@jpl.nasa.gov
* Correspondence: margaret.srinivasan@jpl.nasa.gov

Abstract: More than 30 years of observations from an international suite of satellite altimeter missions continue to provide key data enabling research discoveries and a broad spectrum of operational and user-driven applications. These missions were designed to advance technologies and to answer scientific questions about ocean circulation, ocean heat content, and the impact of climate change on these Earth systems. They are also a valuable resource for the operational needs of oceanographic and weather forecasting agencies that provide information to shipping and fishing vessels and offshore operations for route optimization and safety, as well as for other decision makers in coastal, water resources, and disaster management fields. This time series of precise measurements of ocean surface topography (OST)—the "hills and valleys" of the ocean surface—reveals changes in ocean dynamic topography, tracks sea level variations at global to regional scales, and provides key information about ocean trends reflecting climate change in our warming world. Advancing technologies in new satellite systems allows measurements at higher spatial resolution ever closer to coastlines, where the impacts of storms, waves, and sea level rise on coastal communities and infrastructure are manifest. We review some collaborative efforts of international space agencies, including NASA, CNES, NOAA, ESA, and EUMETSAT, which have contributed to a collection of use cases of satellite altimetry in operational and decision-support contexts. The extended time series of ocean surface topography measurements obtained from these satellite altimeter missions, along with advances in satellite technology that have allowed for higher resolution measurements nearer to coasts, has enabled a range of such applications. The resulting body of knowledge and data enables better assessments of storms, waves, and sea level rise impacts on coastal communities and infrastructure amongst other key contributions for societal benefit. Although not exhaustive, this review provides a broad overview with specific examples of the important role of satellite altimetry in ocean and coastal applications, thus justifying the significant resource contributions made by international space agencies in the development of these missions.

Keywords: satellite altimetry; radar altimetry; applications; operational oceanography; user communities; ocean surface topography; SWOT; Sentinel-6; Jason-3

1. Introduction

For nearly 30 years, a series of satellite altimeter missions, led by the National Aeronautics and Space Administration (NASA) and its partners at the French space agency, Centre National d'Etudes Spatial (CNES), have measured ocean surface topography—the "hills and valleys" of the ocean surface influenced by gravity, ocean currents, heat content, and other dynamic geostrophic forces—to produce a continuous time series of data. Along the way, other national and international partners, including the National Oceanographic and Atmospheric Administration (NOAA), the European Space Agency (ESA), and the European Organisation for the Exploitation of Meteorological Satellites (EUMETSAT), have joined the missions. These and other partnering organizations continue to make important contributions to the missions that strengthen and enhance the science and operational returns of the significant investments made in these satellite systems. With scheduled launches through 2025, and future missions planned in the next decade, the benefit to research fields and to applications for the societal benefit of these continued missions and partnerships can be expected to grow.

Future missions in development, with advancing technologic capabilities, will extend the continuous record of consistent and calibrated data into the future. Applications for societal benefit that leverage these valuable, long-term datasets continue to expand beyond the science objectives that drove the initial development of the missions. An effort is underway by NASA and its partners to advance awareness of the utility of satellite data assets and to grow the user community. The benefits to applications from these altimetry mission data and information products include weather prediction, coastal impact (storm surges and coastal currents) assessments, fisheries management, marine transport, and disaster risk management related to sea level change and flooding (both coastal and inland), among many others. Identifying existing and potential uses of the data in operational, scientific, and other realms validates the significant resources dedicated by international space agencies to these missions. As the impacts of climate change continue to affect coastal regions around the world, the importance of satellite altimetry and its user communities can only be expected to increase.

This paper will communicate and illustrate the value of these missions to decision makers and scientific and operational organizations. We provide historical background on a joint U.S. and European satellite mission series primarily utilizing altimetry measurement systems for ocean and land surface water body targets, describing the impetus of these missions in the larger space agency exploration and technology development context. We then focus on the user communities and specific operational and non-research "applications"—that is, the utilization of the data in various operational and/or practical modalities—to illustrate the broad use categories and specific examples of these applications. We conclude with a view to potential contributions of the extended altimeter and new technologies missions planned.

2. A Legacy of Ocean Altimetry Missions

The heritage of missions building this long-term record of data began with the launch of the flagship TOPEX/Poseidon in 1992. Jason-1 followed in 2001, with Jason-2 launching in 2008. Between three and five years of overlap in the satellite operational lifetimes of these missions provided for continuity and validation of the data record. Jason-3 launched in 2016 and remains operational. Sentinel-6 Michael Freilich (S6MF) launched in 2020 and is now the reference mission for the Jason series of satellite altimetry missions. Launched in December 2022, the Surface Water and Ocean Topography (SWOT) satellite contains advanced technologies designed to provide significantly higher resolution observations. The next satellite in this series, Jason-CS/Sentinel-6B, is planned for launch in 2025 with NASA, NOAA, ESA, EUMETSAT, and CNES continuing their partnership (Figure 1).

The measurement method of radar altimetry involves two geometric measurements: (1) the distance between the satellite and the sea surface as determined by the round-trip travel time of a microwave pulse emitted downward from the radar and reflected back to the spacecraft from the ocean and (2) multiple independent tracking systems precisely computing the satellite's location relative to a fixed coordinate system on Earth. The resulting data yield highly accurate measurements of sea surface topography relative to a reference ellipsoid. Sophisticated instrumentation, processing, and modeling systems, along with complex data assimilation techniques, are required to produce validated data products [1].

In addition to sea surface height (SSH), these missions also provide surface height data on large lakes and rivers that fall under the satellite ground tracks, as well as significant wave height, surface wind speed, and some sea ice height and thickness information. More information on these missions, including links to mission partner websites, can be found at sealevel.jpl.nasa.gov (accessed on 3 August 2023).

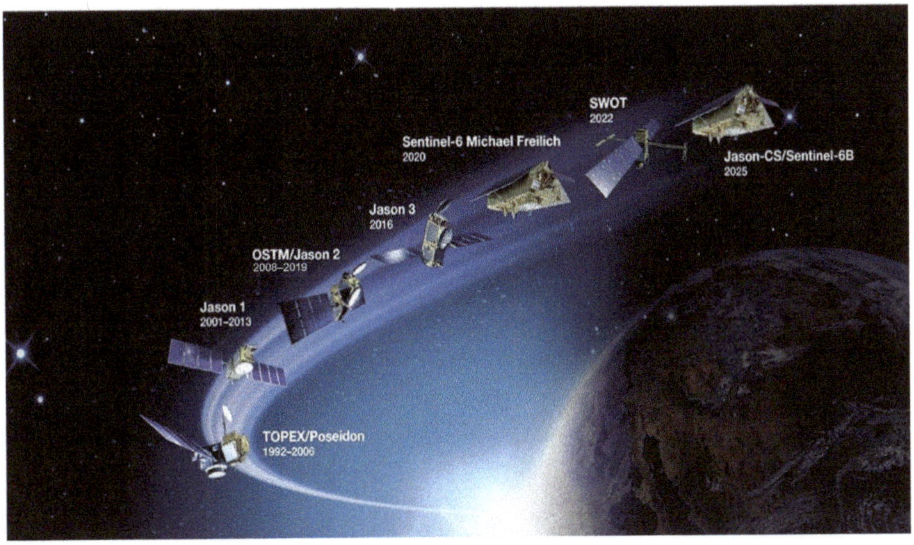

Figure 1. NASA, along with CNES, NOAA, EUMETSAT, ESA, and other international partners, developed and launched a series of satellite altimetry missions beginning in 1992 with the launch of TOPEX/Poseidon. This was followed by the Jason series of missions, including Jason-1 (launched 2001), the Ocean Surface Topography Mission/Jason-2 (OSTM/Jason-2, launched 2008), and Jason-3 (2016), S6MF (2020), and most recently, the SWOT mission (2022). International partnerships will extend the time series of ocean surface topography as well as land water measurements into the next decade with the launch of Sentinel-6B (planned launch 2025) and beyond with new planned missions. (Image: NASA.)

S6MF ensures the continuity of sea level observations by providing ongoing measurements of global sea level rise—a key indicator of climate change. It also supports operational oceanography by improving forecasts of ocean currents, as well as wind and wave conditions. Due to the nature of the nadir altimeters on the heritage satellite missions, data in coastal regions have been limited to approximately 40 km from the coast [2]. With new digital altimeter technology and dedicated onboard processing, more precise measurements of SSH are possible. The dual frequency (C- and Ku-band) Poseidon-4 nadir radar altimeter on this satellite uses an interleaved mode to improve performance compared to the heritage instruments. The special processing of data from this instrument (unfocused synthetic aperture radar processing) will provide information on coastal ocean dynamics (coastal currents, for example) within a few kilometers of the coast.

The recently launched SWOT mission contains an exciting new technology—a Ka-band radar interferometer (KaRIn)—that will collect data across a 120 km wide swath, with a gap in the center for a nadir altimeter track. The nadir altimeter will continue the data record from the Jason series satellites. Globally, measurements will be taken both over the ocean and freshwater areas. SWOT will resolve lakes that are at least 250 m $\times$ 250 m in size, rivers 100 m wide and larger, and ocean features in the sub-mesoscale (10–100 km in extent). Improvements in resolution with SWOT will allow researchers and application user communities to monitor changes in land surface hydrology between 78 degrees north and south latitudes at unprecedented levels of detail—up to 10 times higher than the current 25 km multi-mission nadir-altimeter-based ocean data products (Figure 2). SWOT will provide the first measurements of both of these parameters with a single satellite mission.

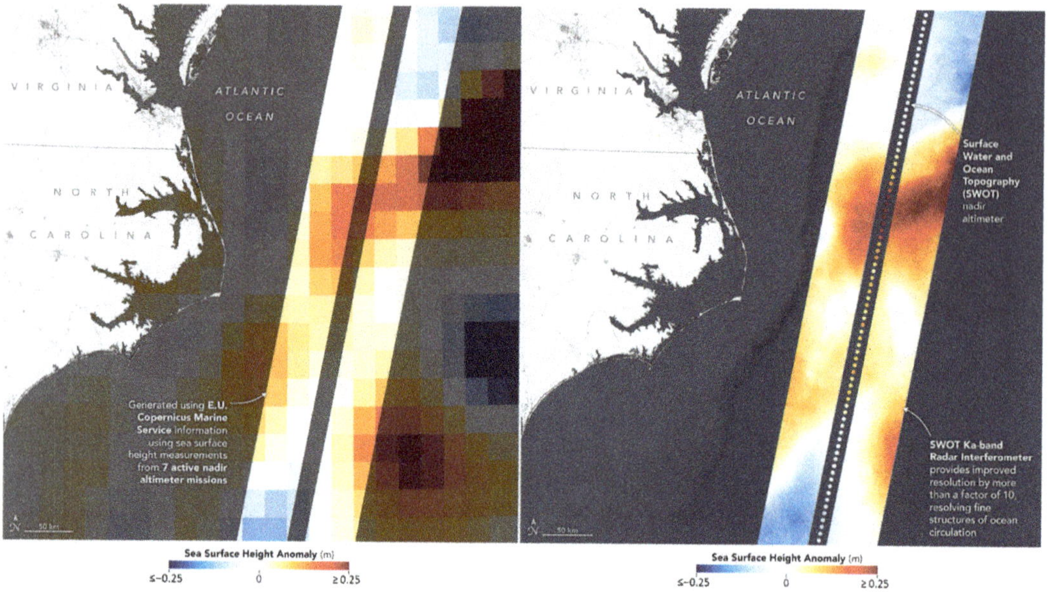

Figure 2. Comparison of the 25 km merged, multi-mission sea surface height anomaly (SSH-A) Level 4 data product from Copernicus Marine Environment Monitoring Service (CMEMS) with SWOT first light SSH-A imagery on 21 January 2023 shows the detection of structural features at about a 10-times-higher resolution for an area of the Gulf Stream off North Carolina. Image credit: NASA/JPL-Caltech, CNES, and CMEMS.

3. From Research to Applications

Satellite missions are generally conceived of and funded with research and/or technology development objectives as a focus. They are designed to advance technologies and to address scientific questions related to ocean circulation, ocean heat potential, tides, and climate change and to advance and improve models in these and other areas. They can also be used to address operational needs of oceanographic and meteorological institutions that provide information to a wide variety of user communities, such as fishing and shipping fleets who can use it to optimize routes and operations, and for disaster management.

Global sea level rise exemplifies and is one of the more obvious and impactful consequences of climate change in the ocean. Coastal communities worldwide are already experiencing significant effects from both event-driven ocean inundation (i.e., storms at sea), as well as from high tide flooding or "sunny day" flooding where average sea levels in low-lying regions are simply more prone to flooding as sea levels continue to rise. Impacts to coastal regions from climate-driven increases in the frequency of storms and associated storm surges are exacerbated by rising sea levels and have a real cost to coastal communities [3]. Neumann et al. [4] estimated that coastal infrastructure adaptation costs for shoreline protection and nourishment along the continental U.S. coastline could reach as high as $254 billion through 2100. Community resilience planning to address these increasing threats may be improved by access to better data and models of coastal ocean dynamics, the development of which is heavily reliant on satellite remote sensing [5,6].

Before the launch of TOPEX/Poseidon, tide gauges were the only reliable mechanism for measuring trends in global sea levels. Although the tide gauge record over the last century is not definitive (due to poor spatial sampling, sensitivity to land motion and coastal effects, and other factors), some studies suggest a rate of 1.5 mm per year over this period, indicating a global average sea level rise of 15 cm (6 in) in the 20th century [7]. The satellite record (Figure 3) is now observing up to 4 mm (0.16 in) per year, resulting in nearly a 10 cm (4 in) rise since 1992 alone.

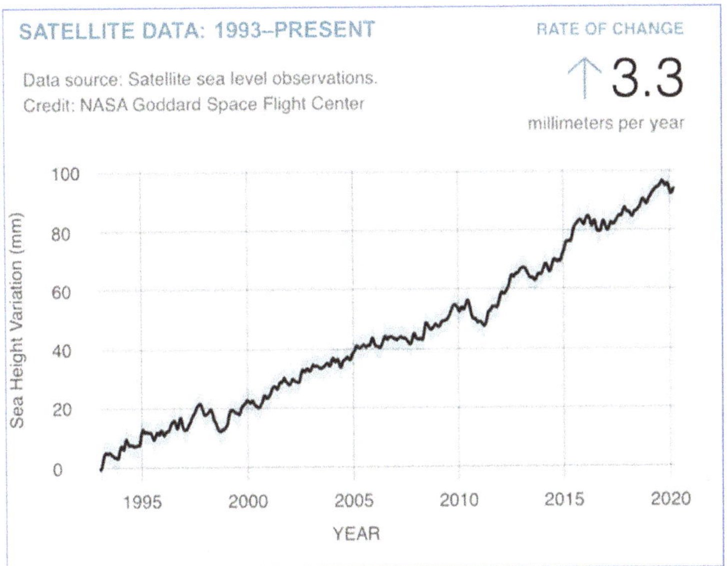

Figure 3. Global average sea level change from 1992 to 2022. The rate of rise in sea level accelerated from 2.8 mm per year in the 1990s to 4.0 mm/year in 2022 as measured by satellite altimeters. Credit: NASA.

The implications of this dramatic increase in both the absolute rise and the rate of rise in sea level on coastal communities can be significant, particularly in developing countries where the adaptive capacity is more limited [8,9]. Some factors are outlined below:

- Increased flooding: Rising sea levels exacerbate the risk of coastal flooding, making low-lying areas more vulnerable to storm surges and high tides. This puts coastal properties, infrastructure, and human lives at greater risk [9].
- Coastal erosion: Rising sea levels contribute to the loss of beaches, dunes, and coastal ecosystems from erosion. This can impact ecosystem health, tourism, and recreational activities, as well as the beauty and economic viability of coastal regions [10,11].
- Infrastructure vulnerability: Critical infrastructure such as roads, bridges, ports, and other utilities in coastal regions are at higher risk from rising sea levels. These assets may require costly upgrades, relocation, or protection measures to mitigate impacts and ensure long-term functionality [8,12].
- Displacement and relocation: More coastal and island communities will face the daunting task of relocation due to increased flooding and the loss of habitable land as sea levels rise. Displaced populations face challenges in finding alternative housing, as well as potential social and economic disruptions [12].
- Environmental impacts: Coastal ecosystems (i.e., wetlands and estuaries) are critical habitats for numerous species and provide valuable ecosystem services. Threats from sea level rise include habitat loss, altered biodiversity, and possible cascading effects on marine and terrestrial ecosystems [13].
- Socioeconomic consequences: Coastal communities are often centers of economic activity (tourism, fisheries, and commerce) that can be disrupted, leading to financial loss, job reduction, and decreased property value. Strain on local civic budgets can result from the need for investment in adaptation measures and disaster recovery [11,14].

Addressing these issues requires comprehensive strategies that include coastal planning, adaptation and mitigation measures, and sustainable development practices [15–17]. Collaboration between governments, communities, and stakeholders is crucial to effectively manage the challenges posed by accelerating sea level rise [9,18]. And, importantly, informing and educating people in these communities on the connection of these impacts to long-term climate change drivers may promote better governance and more thoughtful personal and civic choices and foster a sense of collective responsibility in addressing the underlying causes of sea level rise and climate change [19].

4. User Communities

Traditional satellite altimeter missions offer valuable information on water heights over the ocean and over large lakes, reservoirs, and rivers. The new technology and higher resolution measurements from SWOT will greatly expand the number of surface water bodies that can be viewed and monitored [20], as well as vastly improve the resolution of ocean circulation features [21]. The data can be used to assess and monitor sea level change and the coastal impacts that result but can only be truly impactful if they reach and are successfully utilized by relevant user communities. NASA and its partners recognize the importance of engaging and supporting both (1) communities of practice—those who already possess capacities for using remote sensing data and, in particular, altimetry data— and (2) communities of potential—those who could benefit from the use of satellite altimetry data but are unaware of, or lack the knowledge, technical skill, or other resources to best leverage, these observations. Investment in outreach, training, educational resources, and tools to support analysis and interpretation of these data is of utmost importance in a successful data user engagement strategy.

NASA and its partners can help to inform communities about challenges such as sea level rise, coastal flooding, and other impacts related to climate change and natural processes affecting regional coastal communities globally. One key approach is the continual measurement of water heights over the ocean and large lakes and rivers. With the launch of SWOT, monitoring of dynamic ocean topography features and an even greater number of lakes and reservoirs at significantly finer scale will be possible.

The user communities for these satellite missions include oceanographers, climatologists, coastal managers, private sector organizations, and decision makers at various levels of government. These groups can make informed decisions related to climate change adaptation and mitigation, such as identifying coastal areas at risk from sea level rise (both directly from wave action and indirectly from saltwater intrusion into coastal groundwater aquifers) and planning coastal infrastructure projects, including those designed to protect coastal communities. A key component of outreach to data users is encouraging the use of data and information products from these missions via active engagement and by supporting and training individuals and organizations in the existing and potential altimetry data user communities.

S6MF is continuing to chart the rise of sea level more precisely than previously possible, allowing researchers to understand how climate change is reshaping coastlines and the accelerating rate at which this is occurring. As more than 600 million people live in coastal areas that are lower than 10 m above sea level—a number projected to rise to one billion people by 2060 [22]—and given that many coastal regions include facilities and infrastructure critical to commerce, recreation, military installations, and transportation, understanding the impacts and trends of sea level rise will allow improved assessments of threats to vulnerable coastal regions.

Impacts from Hurricane Sandy in 2012, described as one of the most damaging hurricanes ever to make landfall in the U.S., and Hurricane Harvey in 2017, which caused $125 billion in damage (second only to Hurricane Katrina, 2005, in cost) were exacerbated by intensified storms coupled with sea level rise. Recent examples of high tide flooding along the Atlantic coast of the United States in November 2021 [23] illustrate a troubling trend that is increasingly common due to a higher relative sea level in many coastal communities.

This "nuisance flooding" is expected to increase in frequency over the coming decades, according to the NOAA report "State of High Tide Flooding and Annual Outlook" [24]. For U.S. coastal communities, high tide flooding events are likely to reach 7–15 days by 2030 and 25–27 days by 2050 [25].

The data collected from satellite altimeter missions have tremendous potential to inform decision making and improve the quality of life for people around the world. Measuring water surface topography over the globe enables a wide range of practical applications with tangible benefits to society. These include planning for the impacts of sea level rise on coastal communities as discussed above, supporting operational oceanography and safety at sea, improved flood modeling, transboundary water information sharing, weather and climate forecasting, water resources management, and decision support, among others.

In 2014, a U.S. presidential executive order established the National Plan for Civil Earth Observations (the National Plan) to promote the use of observing system data and information products in 12 identified societal benefit areas [26]. Altimetry data may be useful in informing at least half of these. The use of altimetry data contributes to the following societal benefit areas of the National Plan:

- Biodiversity—understanding and conservation of biodiversity, fisheries management, and marine protected areas.
- Climate—understanding and assessment of sea level rise and global ocean heat content using climate records from altimetry.
- Disasters (hazards)—storm surge from coastal storms, hurricane intensity forecasts, and improved tsunami wave models.
- Ocean and coastal resources—storm surge modeling, sediment transport, and water quality.
- Water resources—climate-related impacts to the Earth's water cycle and resources.
- Weather—seasonal forecasts of the numbers and strengths of hurricanes expected in a given hurricane season, as well as intensity forecasts of individual hurricanes.

5. Applications Areas

Select applications relevant to coastal and ocean management issues are highlighted below, with some examples of user communities already engaged in the use of the data.

5.1. Operational Oceanography—Simultaneous Operation of Multiple Missions for Operations

Unique capabilities for ocean operations are enabled by the long data record and the development of data merging techniques. Operational oceanography is a critical application of satellite altimetry missions and provides essential support to a range of maritime operations. Near-real-time, high-resolution global sea level data products and maps are routinely produced by CNES via a multi-mission production system [27]. Figure 4 illustrates an example of this operational data system, called SSALTO/DUACS (the CNES/CLS multi-mission ground system/operational data system). These near-real-time, high-resolution global sea level anomaly maps from the simultaneous operation of multiple missions are routinely produced and distributed publicly via the CMEMS data portal as part of the European Copernicus Program (marine.copernicus.eu, accessed on 3 August 2023). The system provides real-time information about ocean currents, sea surface temperature, and other important variables and can, thus, help to improve maritime safety and support search and rescue operations. In addition, the data can be used to improve weather and climate forecasting, which is critical for agriculture, transportation, and other industries that are heavily dependent on accurate weather predictions both in coastal and inland zones.

The SSALTO/DUACS system also allows for the development of unique capabilities that can help address operational challenges in marine sectors. The generated maps are routinely assimilated in ocean models and used in setting initial and boundary conditions. The resulting forecasts of ocean state are then utilized to forecast conditions for maritime operations, including the optimization of shipping routes, the assessment of sea state for risk to offshore infrastructure, and oil spill tracking.

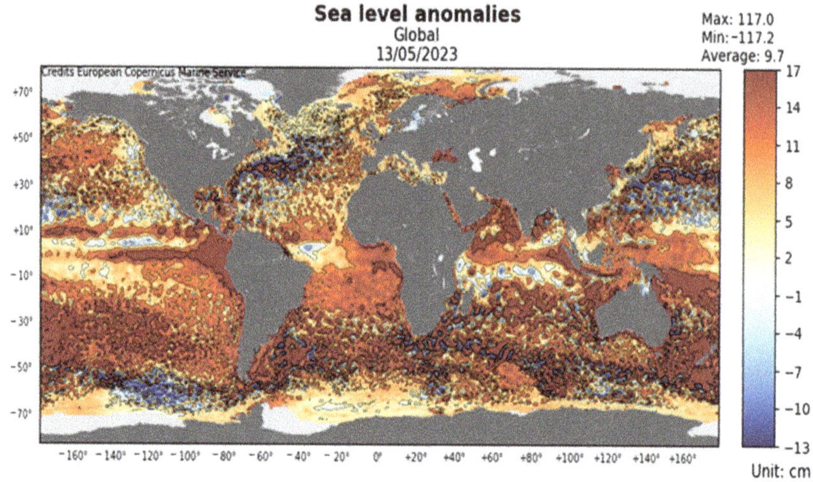

Figure 4. Global sea level anomaly map produced as part of an operational system developed by the French space agency (CNES) and the Collecte Localisation Satellite (CLS) and distributed by the European Copernicus Marine Service. The SSALTO/DUACS system processes data from the altimeter missions Jason-3, Sentinel-3A, Satellite with ARgos and ALtiKa (SARAL), CryoSat-2, Jason-1 and -2, TOPEX/Poseidon, Envisat, Geosat Follow-On (GFO), European Remote-Sensing Satellite-1 and -2 (ERS-1 and -2), and Geodetic Satellite (Geosat) to provide a consistent and homogeneous catalog of products for applications, both for near-real-time applications and research studies. Credit: European Copernicus Marine Service.

5.2. Fisheries Management and Biodiversity—Tracking Marine Life

Over the past two decades, there has been a proliferation in the use of remote sensing data in a range of marine ecological management applications focusing especially on fisheries [28–34], aquaculture [35], biodiversity conservation and marine protected areas [36–38], coastal ecosystem monitoring, and marine spatial planning [6,39]. Fisheries and biodiversity applications in particular have involved the integration of key ocean variables from a series of multi-sensor satellite measurements with biological data to (1) characterize species habitat suitability and preferences based on observed distributions and environmental variable value ranges [40–45]; (2) quantify relationships between environmental factors and spatiotemporal variability in species abundance distributions [46–53]; and (3) identify species associations with dynamic mesoscale oceanographic features that serve as hotspots of enrichment and biological productivity that are the target of commercial fishing activity [54–56]. The latter includes eddies [57–59], Lagrangian fronts [32,60], geostrophic currents [61–63], and convergence and upwelling zones [64–66].

Such applications have been enabled by the increased availability of an extensive series of global ocean surface topography measurements that are the consequence of continuity in satellite altimetry missions conducted by multiple agencies over 30 years. These include data on sea surface height anomaly (SSH-A), geostrophic current velocity, eddy kinetic energy, and derived products on Lagrangian fronts, filaments, and gradients that are applied synergistically with complementary satellite observations on sea surface temperature, chlorophyll-A concentration, surface wind velocity and curl, and satellite surface salinity. The extent of the coverage of this suite of remotely sensed essential ocean variables has facilitated fishery applications spanning the world's oceans basins, including multiple regions of the Pacific [45,47,67–69], Atlantic [59,64,70], and Indian Oceans [41,71]; the Southern Ocean [61,72]; inland seas such as the Mediterranean sea [57,62,73] and the

Arabian Sea [48]; and the Gulfs of Mexico [50] and California [74]. They involve the integration of remote sensing environmental observations with in situ biological data of different kinds, including predominantly fishery-dependent species catch; fishing effort; derived indices of relative population abundance (catch per unit effort—CPUE); systematic scientific surveys undertaken predominantly for early life history and recruitment studies of larval ichthyoplankton stages; electronic satellite tagging data of individual animal movements [51,67,68,75]; automatic identification system (AIS) and vessel monitoring system (VMS) data on fishing fleet dynamics [52,76]; and even fish population genetic information [74]. Analyses of collocated satellite environmental and in situ biological data for fisheries and other ecological applications predominantly involve the use of statistical modeling techniques, such as general additive models (GAM); Bayesian approaches, non-linear time series, non-parametric approaches, multivariate methods, the computation of synoptic habitat suitability indices (HSI), and the application of GIS tools [49,70].

Fishery applications involve the use of satellite altimetry data that span a wide range of biological taxa and include both target and incidental bycatch species. They range principally from highly migratory large pelagic fish species such as tunas [41,42,46,53,77] and swordfish [45,70] to small pelagics like sardine [47,57], mackerel, [44,45], and saury [60], in addition to invertebrate species of commercial importance such as squid [40,43,78] and lobster [79,80]. Habitat analyses of by-catch species at risk and the degree of their population overlap with capture fisheries have included elasmobranch species [38,51,75], marine turtles [67,68], marine mammals [59,81], albatross, and petrel bird species [72]. Multivariate remote sensing data, often in conjunction with regional ocean and coastal circulation models that assimilate altimetry, have also been applied for some demersal species such as rockfish, cod, and walleye pollock to examine larval transport, reproductive subsidy, and recruitment from spawning areas and thus the extent of coupling between spatially extended metapopulations in relation to the location of marine protected areas [61,63,74,82,83].

Successful operational uses of altimetry in conjunction with data from other satellite sensors in fisheries and ecological conservation applications include the routine provisioning of potential fishing zone (PFZ) advisories and map products indicating by-catch risk likelihood for both target species and those protected or vulnerable to incidental capture in a given fishery. Such forecasts are based on the availability of near-real-time satellite data and empirical quantitative relationships between species abundance distributions and remotely sensed environmental variable ranges and mesoscale oceanographic features, such as eddies and fronts. PFZs have been developed for both offshore and coastal fisheries in several regions and have shown to be effective in directing fishing effort to areas with higher resource concentrations, thus maximizing the efficiency and revenue of fishing operations by reducing fuel costs [31,40,69,84–87]. PFZ advisories for the Bay of Bengal and Andaman Sea indicating the locations of eddies from sea level anomaly data were shown by experimental fishing to significantly increase catch rates within eddies relative to traditional control fishing areas [58]. Similar approaches have been used to mitigate by-catch and have involved the use of remote sensing data to help minimize the overlap of fishery operations with likely times and areas of aggregation of protected species. The NOAA TurtleWatch product [67], focused on the North Pacific subtropical frontal zone, is an example that has been successfully adopted and applied to reduce Hawaii's longline fishery interactions with protected loggerhead turtle populations in that region. NOAA's WhaleWatch [88] and EcoCast [89] tools are successful examples of a more generalized framework developed for the California Current ecosystem that routinely make available remote sensing data and habitat distribution model results to reduce the by-catch of several pelagic species, including sharks, turtles, seals, and whales.

The breadth and growing extent of applications discussed indicate that remote sensing more generally has an important role to play in supporting emerging ecosystem-based fisheries management and dynamic ocean management frameworks that explicitly incorporate information on environmental variability. The potential utility of satellite ocean surface topography data in this application space is expected to increase further with new missions providing observations at unprecedented spatial resolution. Future data from the SWOT mission, launched in December 2022, will enable the detection and tracking of sub-mesoscale features, providing unique insights on the spatial dynamics of marine populations in relation to their environment that will further enable ecological assessment and management applications.

5.3. *Weather and Climate Forecasting—Improved Accuracy*

Data from altimetry missions have proved to be a critical component of global climate studies and key to understanding the Earth's delicate climate balance. They provide information and insight on short-term climate events such as the El Niño and La Niña phases of the El Niño Southern Oscillation (ENSO), as well as longer-term climate events such as the Pacific Decadal Oscillation (PDO). Figure 5 illustrates the ENSO cycle pattern from the altimeter era. Altimeter data products are used by hundreds of researchers and operational users around the world to monitor ocean circulation and improve our understanding of the role of the ocean in climate and weather [90]. Initial and boundary conditions derived from the state of the ocean surface and the hydrologic conditions of catchment areas are used in numerical models to improve the quality of weather and climate forecasting. SWOT will enable more accurate weather and climate forecasting, especially seasonally.

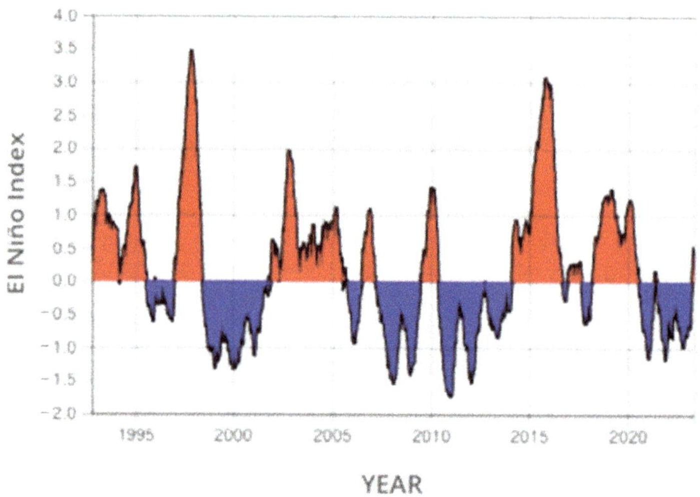

Figure 5. El Niño and La Niña are two phases of the ENSO, a climate pattern in the Pacific Ocean that results from interactions between wind patterns and ocean circulation. This graph shows the ENSO phase in the Pacific Ocean. Red shading indicates El Niño events when water in the index area is generally warmer, and the value of the time series is positive, and blue shading indicates La Niña conditions when the water is cooler, and the value is negative. Credit: NASA MEaSUREs/PO.DAAC.

Satellite altimetry missions are vital for the monitoring and forecasting of extreme weather events such as tropical cyclones and hurricanes. Near-real-time data from these missions are used to produce sea level anomaly maps, which can provide critical information for forecasting storm surges and predicting the impact of these events on coastal communities [91]. These data are used to assess and monitor the impacts of severe weather events such as hurricanes and floods and to provide critical information that can support emergency response efforts. Altimetry data are routinely assimilated into weather models to forecast ENSO events, as well as at the National Hurricane Center to forecast hurricane intensity [92]. Global water resource impacts from climate change can be measured and modeled with decades-long climate data records from nadir altimeter missions.

5.4. Improved Flood Modeling—Coastal Flooding from Upstream and Downstream

The satellite radar altimeter record demonstrates the ability to observe water level variations of lakes, rivers, and floodplains on land, including in coastal regions. River discharge from these regions to and across the coastline can be estimated from a combination of satellite estimates of rainfall and hydrologic modeling [93]. Subsequent interactions with river delta and estuary systems, which contain complex hydrologic regimes resulting from upstream flow and downstream tidal influences, can also be assessed. Combining satellite altimetry in data assimilation frameworks can improve predictive abilities of models to help assess flood hydraulics and the risk and impacts to coastal regions from upstream and downstream conditions [94]. Reprocessing Jason-1 and Jason-2 data has provided a gridded sea level anomaly data set close to the coast (within 20 km) to better assess coastal impacts of sea level change [95]. SWOT will bring the data to within a few kilometers of the coast. Nuisance flooding, flooding from land, and coastal erosion during extreme events can be assessed using these tools and can provide valuable information for coastal planners. These methods can be made more effective by combining satellite with in situ data and models to improve forecasts.

Satellite remote sensing also provides indirect measurements to predict or forecast river discharge. When information about upstream water levels is available, the accuracy of flood potential on downstream river reaches can be better assessed. Hossain et al. [93] describe a system developed for the Flood Forecasting and Warning Center (FFWC; http://www.ffwc.gov.bd, accessed on 3 August 2023) of the Bangladesh government to use satellite altimetry as "virtual" gauging stations that can supplement in situ gauging capacity. Data from the Jason series satellites was used to demonstrate a higher sampling of data from upstream locations to improve the accuracy of river levels downstream.

As sea levels continue to rise and the frequency of coastal flooding increases, the use of satellite altimetry data is becoming increasingly important to understand the impacts and trends that will affect coastal regions. By engaging with user communities, NASA and its partners can ensure that the benefits of satellite altimetry are realized by a wide range of stakeholders, including those in coastal communities, who are particularly vulnerable to the impacts of climate change.

5.5. Hazards—Floods and Insurance

In addition to improvements in flood modeling, assessments of real or potential property damage and economic losses from floods and coastal impacts of sea level rise can be mitigated when hazards are better anticipated. Risk assessment methods used by property insurers can be informed by satellite remote sensing. Satellite altimetry data can be used to assess flood risk and determine insurance premiums for properties located in flood-prone areas, for example. Insurance companies can gather information about the elevation of the land and coastal and inland water levels to identify potential flood zones.

Several scientific studies have explored the role of satellite altimetry in decision support for coastal hazards and change. For example, a study by Woodworth et al. [96] used satellite altimeter data to investigate global and regional sea level changes and their implications for coastal flooding. Cazenave and Llovel [97] explored the role of satellite altimetry in the measurement of global sea level rise and its implications for coastal hazards and change. These changes include the impact of human activity on deltaic systems, including dam and reservoir construction and river diversion for agriculture. These activities upset the natural equilibrium of many deltas, exacerbating impacts from sea level rise, among other complications. Similarly, the International Altimetry Team [98] highlighted the importance of satellite altimetry data in supporting decision making for coastal hazards and change, including the identification of areas at risk from sea level rise and the development of adaptive management strategies.

Hurricanes and typhoons represent many challenges to coastal communities including extreme winds, rain, storm surges, and flooding. These, typically tropical, storms form over warm water and are the most powerful storms on Earth. Forecasting the impact of these events can save lives and mitigate losses to property in coastal regions by providing accurate predictions of storm structure and storm-induced surface waves. The societal impact of hurricanes is strongly related to the strength of the storm measured in "categories" from 1 (with sustained winds between 74–95 miles per hour) to 5 (sustained winds of 157 mph or greater). Higher-category storms have a greater potential to cause significant damage to property and infrastructure and to threaten lives and public safety.

OST data are used for both seasonal forecasts of the numbers and strengths of hurricanes expected during a hurricane season and for forecasts of the strength of individual hurricanes. When hurricanes pass over warm ocean features, they can strengthen. Altimetry data are routinely used by the U.S. NOAA National Hurricane Center to improve hurricane intensity forecasts. A NOAA research area that assesses tropical cyclone heat potential (TCHP) shows that storms and hurricanes can intensify when they travel over warm ocean features (Figure 6) [99].

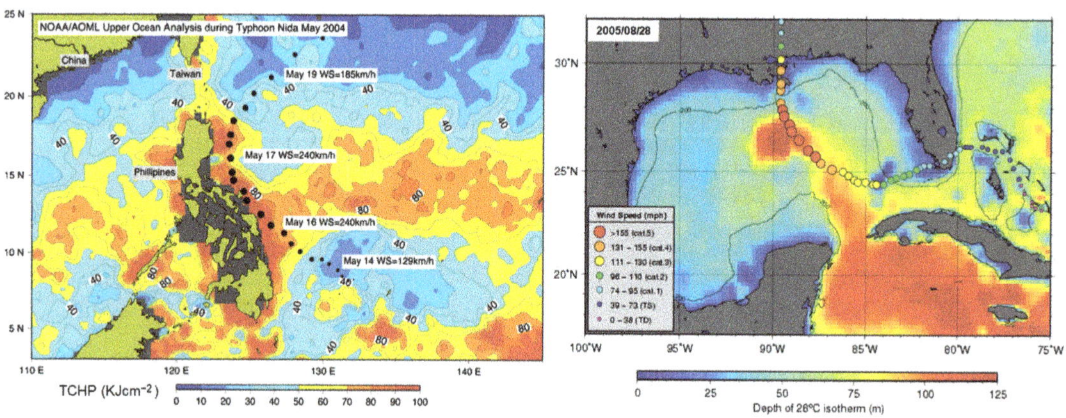

Figure 6. These images illustrate altimetry combined with sea surface temperature data and a two-layer model to show ocean heat potential. On the left is Typhoon Nida (2004), which intensified from a category 2 to a category 4 in a 10 h period when it crossed a region of warm water off the east coast of the Philippines. This warm water essentially fueling the storm can be characterized by its tropical cyclone heat potential (TCHP) derived from sea height anomaly and sea surface temperature fields. On the right, Hurricane Katrina (2005) ramped from category 2 to category 5 as its path passed over the warm water of the Gulf of Mexico Loop Current. TCHP maps are produced in near real time in all seven basins where tropical cyclones occur and are distributed daily on the web. Credit: NOAA AOML.

Researchers studying Hurricane Ike (2008) and Superstorm Sandy (2012) compared an atmosphere–wave–ocean model to Jason-1 and OSTM/Jason-2 satellite data and were able to determine the processes that influence hurricane-generated surface waves near the coast. This provided information about changes in ocean waves with water depth, directional changes in waves as they were deflected off the bottom, and changes in wind fetch (the distance over which wind has blown unobstructed) over land. The information could then be used to predict ocean surface waves at landfall three to five days in advance of a storm hitting the coast, thus providing a valuable storm surge forecast tool [100].

5.6. Additional Applications

In addition to the coastal and ocean practical and operational applications described, there are several important research applications of satellite altimetry data that have operationalization potential.

5.6.1. Tsunami Detection

Satellite altimeter data can support tsunami detection and tracking as they provide valuable information about sea surface height [101]. By continuously monitoring the ocean surface, satellite altimeters can detect and measure anomalies in sea level caused by underwater earthquakes and landslides, volcanic eruptions, or other seismic events that may trigger a tsunami [102]. Altimetry data can be used by scientists and meteorologists to assess the size and energy of tsunami-triggered ocean surface disturbances when the spacecraft ground track coincides with an event. The tsunami propagation can be modeled and combined with other data (i.e., seismographic data and ocean buoys) to provide coastal authorities an opportunity to issue timely and accurate warnings [103]. Despite limitations due to spatial and temporal sampling, it has been found that satellite altimetry could be used to directly detect tsunamis in the open ocean and to improve model predictions [104]. Given higher spatial and temporal resolution from their combined measurements, the S6MF, Jason 6B and C, and SWOT missions may be able to further improve on the current state of tsunami detection.

5.6.2. Geodetic Applications

High-precision sea surface height measurements from satellite altimetry combined with satellite-derived gravity data contribute to ocean geodetic studies and mapping the ocean floor. By combining altimetry data with other satellite-based positioning systems (i.e., GPS), marine navigation can be optimized to avoid hazards such as shallow waters or underwater obstacles [105]. Additionally, ocean circulation is influenced by ocean bottom topography [106]. Coastal bathymetry influences currents and other near-shore processes, including erosion, sediment, and even larval transport. Such information is thus integral to marine spatial planning and the management of coastal zones (see Section 5.2).

5.7. Decision Support—Reducing Environmental Risk and Contributing to Public Policymaking

Decision support is a vital application of satellite altimeter data, with the potential to reduce environmental risk and contribute to public policymaking. With the inevitable effects of climate change such as altered weather patterns and rising sea level, coastal and inland environments, infrastructure, and ecosystems are expected to be significantly impacted. Predictions of a steady increase in sea level rise and storm frequency and intensity [107], coupled with changes in land use and population increases, pose significant planning challenges for coastal communities. While significant resources have been directed to predicting the potential consequences of climate change, emphasis is required to develop rational approaches to guide decision making under uncertainty. There is also a need to develop and assess alternative adaptive strategies to manage coastal hazards and change. The use of measurements from land hydrological systems and from ocean data closer to the coasts, combined with in situ and other remote sensing assets, can support the development of data and information products that meet evolving user requirements [98].

Satellite altimetry data can provide an essential foundation for effective coastal management. The data can be particularly useful in supporting decision making related to coastal issues. By providing accurate information about sea level rise and coastal flooding, these data can help to support coastal planning and infrastructure development [108]. For example, the data can be used to identify areas that are most at risk of flooding and to develop strategies for protecting critical infrastructure such as ports, airports, and power plants. They can be used to facilitate the development of early warning systems for coastal hazards such as storm surges and flooding, which can help to save lives and minimize economic losses. Using the data to monitor changes in sea level and combining that information with coastal erosion over time can provide critical information for long-term planning and adaptation strategies. User communities, including coastal planners, emergency management officials, and infrastructure developers, among others, can benefit greatly from these data for use in their management activities.

6. Conclusions

Continuity in satellite altimetry missions via partnerships among international space agencies for over three decades has been instrumental to the development of an essential climate data record on aspects of land and ocean water surface topography. Such observations have been fundamental to our improved scientific understanding of global ocean circulation, sea level rise, and water cycle processes more generally. However, they are also increasingly contributing to a growing number of practical decision-support and operational uses of altimetry data that extend the primary technology development and research objectives of these satellite missions.

As we have discussed here, such applications for societal benefit span a number of topical areas including ocean state for marine transport, coastal impact assessment and planning, fisheries management, weather prediction and climate studies, flood hazards, tsunami early warning, and disaster risk management, among others. Several of these applications are quite mature, leveraging over 30 years of altimetry data. Particularly when used synergistically with time series of observations on essential ocean variables from other Earth-observing satellite constellations (e.g., sea surface temperature, ocean color radiometry, ocean vector winds, ocean surface salinity, and gravimetry) for direct use or as model inputs, the potential to support a larger number of practical operational applications involving improved predictive capabilities increases further. Study regions can extend beyond coastal and oceanic realms to valuable hydrology and land-based applications of satellite altimetry, which can be dedicated review topics in their own right. Both existing and emerging applications relating to surface water topography are likely to be further enabled by the suite of significantly higher spatial resolution data over the oceans, coasts, and land (rivers, lakes, and wetlands) from the recently launched SWOT mission with its novel, wide-swath KaRIn interferometric radar that complements and extends the important contributions of nadir altimeter measurements.

It is important to acknowledge that there are certain limitations to the use of satellite altimetry for coastal applications and studies of sub-mesoscale processes even for the open ocean. While some of the constraints to legacy nadir altimeter spatial coverage are addressed via the implementation of merged, multi-mission satellite altimetry data products from systems such as SSALTO/DUACS, coverage is still limited off of coasts, and the native resolution of altimetry datasets (typically one quarter degree) restricts the size of features that may be resolved or modeled. The coastal proximity of data coverage of the pre-Sentinel 6 missions is limited to about 20 km from the coast [108]. However, data from the newer satellite altimetry missions seek to address these limitations, thus further catalyzing both innovative science and applications. The U.S.–European Sentinel 6 Michael Freilich satellite uses advanced radiometer techniques and instrumentation to make observations closer to the coast, and SWOT will reach as close as a few kilometers from many coastlines and resolve sub-mesoscale features in unprecedented detail. Future missions and technological advances combined with data processing and model development may narrow the gap between data coverage and the coast even further.

Rising sea levels are widely acknowledged to be linked to climate change and are acutely impactful in coastal regions. The improved monitoring of coastal zones will help decision makers better understand changes affecting these highly dynamic areas stemming from sea level rise, local oceanic and atmospheric processes, ground subsidence, and other anthropogenic-forcing factors [109]. The continued monitoring of coastal and ocean areas from satellites, including altimeter observations, will help improve our knowledge, more generally, of Earth's water cycle and ocean circulation. As observational and modeling capacities and collections of unique data on water storage and fluxes are enhanced and made more available, an improved understanding of the physics that drives surface water dynamics and related other ecosystem processes will result. Water resources, natural hazards (hurricane forecasting, floods, climate change, etc.), health (threats of water-borne diseases), biodiversity, the agricultural sector, energy (including electricity production, offshore production facilities, and renewables)—these areas and many more can benefit from both continued satellite monitoring and new altimetry mission sensor technologies that advance our understanding and capabilities even further.

As we continue to gather more data and refine our understanding of the world's oceans and coastal areas, remote sensing data will play an increasingly important role in improving our ability to manage and protect our critical natural resources, manage responses to climate change, and support operational and private sector activities. Identifying existing and potential uses of the data and information products from these missions in both scientific and operational capacities validates the significant resources dedicated by international space agencies to these projects. Here, we have sought to communicate and illustrate the value of these missions to decision makers and scientific and operational organizations.

Author Contributions: M.S. and V.T. collaboratively wrote this manuscript from the original draft through the final version. All authors have read and agreed to the published version of the manuscript.

Funding: This work was funded with support from the NASA Applied Sciences Program and by the Jason-3, S6MF, and SWOT projects at the Jet Propulsion Laboratory, California Institute of Technology.

Data Availability Statement: Not applicable.

Acknowledgments: The authors would like to acknowledge funding and support from NASA's Applied Sciences Program; the Jason-3, S6MF, and SWOT projects; and the French Investment Program. Some of the work reported here was performed at the Jet Propulsion Laboratory, California Institute of Technology, under contract with the National Aeronautics and Space Administration.

Conflicts of Interest: The authors declare no conflict of interest. The funders had no role in the design of the study; in the collection, analyses, or interpretation of data; in the writing of the manuscript; or in the decision to publish the results.

References

1. Cheney, R.E. Satellite Altimetry. In *Encyclopedia of Ocean Sciences*; Steele, J.H., Thorpe, S.A., Turekian, K.K., Eds.; Elsevier Press: Amsterdam, The Netherlands, , 2001; pp. 2504–2510.
2. Vignudelli, S.; Kostianoy, A.G.; Cipollini, P.; Benveniste, J. (Eds.) *Coastal Altimetry*; Springer: Berlin/Heidelberg, Germany, 2011; 566p. [CrossRef]
3. Padgett, J.E.; Panakkal, P.; González-Dueñas, C. Infrastructure impacts and vulnerability to coastal flood events. In *Coastal Flood Risk Reduction*; Brody, S., Lee, Y., Kothuis, B.B., Eds.; Elsevier Press: Amsterdam, The Netherlands, 2022; pp. 151–165. [CrossRef]
4. Neumann, J.; Hudgens, D.; Herter, J.; Martinich, J. The Economics of Adaptation along Developed Coastlines. *Wiley Interdiscip. Rev. Clim. Chang.* **2011**, *2*, 89–98. [CrossRef]
5. Vitousek, S.; Buscombe, D.; Vos, K.; Barnard, P.; Ritchie, A.; Warrick, J. The future of coastal monitoring through satellite remote sensing. *Camb. Prism. Coast. Futures* **2023**, *1*, E10. [CrossRef]
6. Ouellette, W.; Getinet, W. Remote sensing for Marine Spatial Planning and Integrated Coastal Areas Management: Achievements, challenges, opportunities and future prospects. *Remote Sens. Appl. Soc. Environ.* **2016**, *4*, 138–157.
7. Nerem, R.S.; Fasullo, J. Observations of the Rate and Acceleration of Global Mean Sea Level Change. *Bull. Amer. Meteor. Soc.* **2019**, *100*, S15–S18. [CrossRef]
8. Appeaning Addo, K.; Larbi, L.; Amisigo, B.; Kwabena Ofori-Danson, P. Impacts of Coastal Inundation Due to Climate Change in a Cluster of Urban Coastal Communities in Ghana, West Africa. *Remote Sens.* **2011**, *3*, 2029–2050. [CrossRef]
9. Johnson, E.; Bell, J.; Coker, D.; Hertz, E.; Labarge, N.; Blake, G. A lifeline and social vulnerability analysis of sea level rise impacts on rural coastal communities. *Shore Beach* **2018**, *86*, 36–44.
10. Periasamy, A. Effects of coastal erosion due to climate change on fishermen communities in Tamil Nadu. *Disaster Adv.* **2023**, *16*, 68–74. [CrossRef]
11. Scyphers, S.; Beck, M.; Furman, K.; Haner, J.; Josephs, L.; Lynskey, R.; Keeler, A.; Landry, C.; Powers, S.; Webb, B.; et al. A Waterfront View of Coastal Hazards: Contextualizing Relationships among Geographic Exposure, Shoreline Type, and Hazard Concerns among Coastal Residents. *Sustainability* **2019**, *11*, 6687. [CrossRef]

12. Gomaa, M.M. Assessing the Impacts of Population Relocation Induced by Future Sea-Level Rise Scenarios on Transportation Systems in Coastal Communities. *Sustain. Urban Plan. Divid. Cities* **2022**, *7*, 68–83. [CrossRef]
13. Ward, R.D.; Burnside, N.G.; Joyce, C.B.; Sepp, K.; Teasdale, P.A. Improved modelling of the impacts of sea level rise on coastal wetland plant communities. *Hydrobiologia* **2016**, *774*, 203–216. [CrossRef]
14. Makame, M.O.; Mwevura, H. Vulnerability and Adaptation Strategies of Coastal Communities to the Associated Impacts of Sea Level Rise and Coastal Flooding. In *Climate Change and Coastal Resources in Tanzania*; Yanda, P., Bryceson, I., Mwevura, H. Mung'ong'o, C., Eds.; Springer Climate: Cham, Switzerland, 2019. [CrossRef]
15. D'Alessandro, F.; Tomasicchio, G.; Francone, A.; Leone, E.; Frega, F.; Chiaia, G.; Saponieri, A.; Damiani, L. Coastal sand dune restoration with an eco-friendly technique. *Aquat. Ecosyst. Health Manag.* **2020**, *23*, 417. [CrossRef]
16. Guthrie, A.; Stafford, S.; Scheld, A.M.; Nunez, K.; Bilkovic, D.M. Property owner shoreline modification decisions vary based on their perceptions of shoreline change and interests in ecological benefits. *Front. Mar. Sci.* **2023**, *10*, 1031012. [CrossRef]
17. Sauvé, P.; Bernatchez, P.; Moisset, S.; Glaus, M.; Goudreault, M.-O. A need to better monitor the effects of coastal defense measures on coastal socio-ecological systems to improve future adaptation solutions. *Ocean Coast. Manag.* **2023**, *239*, 106599. [CrossRef]
18. Smith, E.A.; Sweet, W.; Mitchell, M.; Domingues, R.; Weaver, C.P.; Baringer, M.; Goni, G.; Haines, J.; Loftis, J.D.; Boon, J.; et al Treading Water: Tools to Help US Coastal Communities Plan for Sea Level Rise Impacts. *Front. Mar. Sci.* **2019**, *6*, 300. [CrossRef]
19. Vázquez, L.; Vandergeest, P. Coastal erosion narratives in the Gulf of Mexico: Implications for climate change governance *J. Political Ecol.* **2022**, *29*, 705–724. [CrossRef]
20. Altenau, E.H.; Pavelsky, T.M.; Durand, M.T.; Yang, X.; Frasson, R.P.d.M.; Bendezu, L. The surface water and ocean topography (SWOT) mission river database (SWORD): A global river network for satellite data products. *Water Resour. Res.* **2021**, *57*, e2021WR030054. [CrossRef]
21. Cazenave, A. Satellite Altimetry. In *Encyclopedia of Ocean Sciences*, 3rd ed.; Cochran, J.K., Bokuniewicz, H.J., Yager, P.L., Eds.; Academic Press: San Diego, CA, USA, 2019; pp. 397–401. [CrossRef]
22. Neumann, B.; Vafeidis, A.T.; Zimmermann, J.; Nicholls, R.J. Future Coastal Population Growth and Exposure to Sea-Level Rise and Coastal Flooding—A Global Assessment. *PLoS ONE* **2015**, *10*, e0118571. [CrossRef]
23. US Department of Commerce (USDC) 2021. November 6–8, 2021 Coastal Flooding. *National Weather Service*, 22 December 2021. Available online: https://www.weather.gov/ilm/Nov2021CoastalFlood(accessed on 3 August 2023).
24. Sweet, W.; Simon, S.; Dusek, G.; Marcy, D.; Brooks, W.; Pendleton, M.; Marra, J. The State of High Tide Flooding and Annual Outlook, NOAA Tides & Currents. Retrieved November 30, 2021. Available online: https://tidesandcurrents.noaa.gov/publications/2021_State_of_High_Tide_Flooding_and_Annual_Outlook_Final.pdf(accessed on 3 August 2023).
25. US Harbors. The State of High Tide Flooding and Annual NOAA Outlook. 2022. USHarbors.com. Available online: https://www.usharbors.com/2022/05/the-state-of-high-tide-flooding-and-annual-noaa-outlook/ (accessed on 3 August 2023).
26. National Science and Technology Council. National Plan for Civil Earth Observations. 2014. Available online: https://obamawhitehouse.archives.gov/sites/default/files/microsites/ostp/NSTC/2014_national_plan_for_civil_earth_observations.pdf (accessed on 3 August 2023).
27. Dibarboure, G.; Dorandeu, J.; LeTraon, P.-Y.; Picot, N. SSALTO/DUACS: 15 Years of Precise and Consistent Multi-Mission Altimetry Data. European Space Agency, (Special Publication) ESA SP. 2006. Available online: https://www.researchgate.net/publication/240622367_SSALTODUACS_15_years_of_precise_and_consistent_multi-mission_altimetry_data (accessed on 3 August 2023).
28. Santos, A.M.P. Fisheries oceanography using satellite and airborne remote sensing methods: A review. *Fish. Res.* **2000**, *49*, 1–20. [CrossRef]
29. Chassot, E.; Bonhommeau, S.; Reygondeau, G.; Nieto, K.; Polovina, J.J.; Huret, M.; Dulvy, N.K.; Demarcq, H. Satellite remote sensing for an ecosystem approach to fisheries management. *ICES J. Mar. Sci.* **2011**, *68*, 651–666. [CrossRef]
30. Stuart, V.; Platt, T.; Sathyendranath, S.; Pravin, P. Remote sensing and fisheries: An introduction. *ICES J. Mar. Sci.* **2011**, *68*, 639–641. [CrossRef]
31. Klemas, V. Fisheries applications of remote sensing: An overview. *Fish. Res.* **2013**, *148*, 124–136. [CrossRef]
32. Belkin, I.M. Remote Sensing of Ocean Fronts in Marine Ecology and Fisheries. *Remote Sens.* **2021**, *13*, 883. [CrossRef]
33. Yen, K.W.; Chen, C.H. Research Gap Analysis of Remote Sensing Application in Fisheries: Prospects for Achieving the Sustainable Development Goals. *Remote Sens.* **2021**, *13*, 1013. [CrossRef]
34. Schwing, F.B. Modern technologies and integrated observing systems are "instrumental" to fisheries oceanography: A brief history of ocean data collection. *Fish. Oceanogr.* **2023**, *32*, 28–69. [CrossRef]
35. Ottinger, M.; Clauss, K.; Kuenzer, C. Aquaculture: Relevance, distribution, impacts and spatial assessments—A review. *Ocean Coast. Manag.* **2016**, *119*, 244–266. [CrossRef]
36. Kachelriess, D.; Wegmann, M.; Gollock, M.; Pettorelli, N. The application of remote sensing for marine protected area management. *Ecol. Indic.* **2014**, *36*, 169–177. [CrossRef]
37. Kuenzer, C.; Ottinger, M.; Wegmann, M.; Guo, H.D.; Wang, C.L.; Zhang, J.Z.; Dech, S.; Wikelski, M. Earth observation satellite sensors for biodiversity monitoring: Potentials and bottlenecks. *Int. J. Remote Sens.* **2014**, *35*, 6599–6647. [CrossRef]
38. Williamson, M.J.; Tebbs, E.J.; Dawson, T.P.; Jacoby, D.M.P. Satellite Remote Sensing in Shark and Ray Ecology, Conservation and Management. *Front. Mar. Sci.* **2019**, *6*, 135. [CrossRef]

39. McCarthy, M.J.; Colna, K.E.; El-Mezayen, M.M.; Laureano-Rosario, A.E.; Mendez-Lazaro, P.; Otis, D.B.; Toro-Farmer, G.; Vega-Rodriguez, M.; Muller-Karger, F.E. Satellite Remote Sensing for Coastal Management: A Review of Successful Applications. *Environ. Manag.* **2017**, *60*, 323–339. [CrossRef]
40. Chen, X.J.; Tian, S.Q.; Chen, Y.; Liu, B.L. A modeling approach to identify optimal habitat and suitable fishing grounds for neon flying squid (*Ommastrephes bartramii*) in the Northwest Pacific Ocean. *Fish. Bull.* **2010**, *108*, 1–14.
41. Song, L.M.; Zhou, Y.Q. Developing an integrated habitat index for bigeye tuna (*Thunnus obesus*) in the Indian Ocean based on longline fisheries data. *Fish. Res.* **2010**, *105*, 63–74. [CrossRef]
42. Yen, K.; Lu, H.; Chang, Y.; Lee, M. Using remote-sensing data to detect habitat suitability for yellowfin tuna in the Western and Central Pacific Ocean. *Int. J. Remote Sens.* **2012**, *33*, 7507–7522. [CrossRef]
43. Yu, W.; Chen, X.J.; Yi, Q.; Chen, Y.; Zhang, Y. Variability of Suitable Habitat of Western Winter-Spring Cohort for Neon Flying Squid in the Northwest Pacific under Anomalous Environments. *PLoS ONE* **2015**, *10*, e0122997. [CrossRef] [PubMed]
44. Li, G.; Cao, J.; Zou, X.R.; Chen, X.J.; Runnebaum, J. Modeling habitat suitability index for Chilean jack mackerel (*Trachurus murphyi*) in the South East Pacific. *Fish. Res.* **2016**, *178*, 47–60. [CrossRef]
45. Su, N.-J.; Chang, C.-H.; Hu, Y.-T.; Chiang, W.-C.; Tseng, C.-T. Modeling the Spatial Distribution of Swordfish (*Xiphias gladius*) Using Fishery and Remote Sensing Data: Approach and Resolution. *Remote Sens.* **2020**, *12*, 947. [CrossRef]
46. Zainuddin, M.; Saitoh, K.; Saitoh, S. Albacore (*Thunnus alalunga*) fishing ground in relation to oceanographic conditions in the western North Pacific Ocean using remotely sensed satellite data. *Fish. Oceanogr.* **2008**, *17*, 61–73. [CrossRef]
47. Zwolinski, J.P.; Emmett, R.L.; Demer, D.A. Predicting habitat to optimize sampling of Pacific sardine (*Sardinops sagax*). *ICES J. Mar. Sci.* **2011**, *68*, 867–879. [CrossRef]
48. Solanski, H.U.; Bhatpuria, D.; Chauhan, P. Signature analysis of satellite derived SSHa, SST and chlorophyll concentration and their linkage with marine fishery resources. *J. Mar. Syst.* **2015**, *150*, 12–21. [CrossRef]
49. Setiawati, M.D.; Sambah, A.; Miura, F.; Tanaka, T.; As-syakur, A. Characterization of bigeye tuna habitat in the Southern Waters off Java-Bali using remote sensing data. *Adv. Space Res.* **2015**, *55*, 732–746. [CrossRef]
50. Cornic, M.; Smith, B.L.; Kitchens, L.L.; Bremer, J.R.A.; Rooker, J.R. Abundance and habitat associations of tuna larvae in the surface water of the Gulf of Mexico. *Hydrobiologia* **2018**, *806*, 29–46. [CrossRef]
51. Guzman, H.M.; Collatos, C.M.; Gomez, C.G. Movement, Behavior, and Habitat Use of Whale Sharks (Rhincodon typus) in the Tropical Eastern Pacific Ocean. *Front. Mar. Sci.* **2022**, *9*, 793248. [CrossRef]
52. Fei, Y.; Yang, S.; Fan, W.; Shi, H.; Zhang, H.; Yuan, S. Relationship between the Spatial and Temporal Distribution of Squid-Jigging Vessels Operations and Marine Environment in the North Pacific Ocean. *J. Mar. Sci. Eng.* **2022**, *10*, 550. [CrossRef]
53. Mugo, R.; Seitoh, S. Ensemble Modelling of Skipjack Tuna (Katsuwonus pelamis) Habitats in the Western North Pacific Using Satellite Remotely Sensed Data; a Comparative Analysis Using Machine-Learning Models. *Remote Sens.* **2020**, *12*, 2591. [CrossRef]
54. Falkowski, P.; Ziemann, D.; Kolber, Z.; Bienfang, P.K. Role of eddy pumping in enhancing primary production in the ocean. *Nature* **1991**, *352*, 55–58. [CrossRef]
55. McGillicuddy, D.J.; Robinson, A.R.; Siegel, D.A.; Jannasch, H.W.; Johnson, R.; Dickeys, T.; McNeil, J.; Michaels, A.F.; Knap, A.H. Influence of mesoscale eddies on new production in the Sargasso Sea. *Nature* **1998**, *394*, 263–266. [CrossRef]
56. Bakun, A. Fronts and eddies as key structures in the habitat of marine fish larvae: Opportunity, adaptive response and competitive advantage. *Sci. Mar.* **2006**, *70*, 105–122. [CrossRef]
57. Sabates, A.; Salat, J.; Raya, V.; Emelianov, M. Role of mesoscale eddies in shaping the spatial distribution of the coexisting Engraulis encrasicolus and Sardinella aurita larvae in the northwestern Mediterranean. *J. Mar. Syst.* **2013**, *111*, 108–119. [CrossRef]
58. Arur, A.; Krishnan, P.; Kiruba-Sankar, R.; Suryavanshi, A.; Kumar, K.L.; Kantharajan, G.; Choudhury, S.B.; Manjulatha, C.; Babu, D.E. Feasibility of targeted fishing in mesoscale oceanic eddies: A study from commercial fishing grounds of Andaman and Nicobar Islands, India. *Int. J. Remote Sens.* **2020**, *41*, 5011–5045. [CrossRef]
59. Gilbert, L.; Rouby, E.; Tew-Kai, E.; Spitz, J.; Peltier, H.; Quilfen, V. Spatiotemporal models highlight influence of oceanographic conditions on common dolphin bycatch risk in the Bay of Biscay. *Mar. Ecol. Prog. Ser.* **2021**, *679*, 195–212. [CrossRef]
60. Prants, S.; Budyansky, M.; Uleysky, M. Identifying Lagrangian fronts with favourable fishery conditions. *Deep Sea Res. Part I Oceanogr. Res. Pap.* **2014**, *90*, 27–35. [CrossRef]
61. Mori, M.; Corney, S.P.; Melbourne-Thomas, J.; Welsford, D.C.; Klocker, A.; Ziegler, P.E. Using satellite altimetry to inform hypotheses of transport of early life stage of Patagonian toothfish on the Kerguelen Plateau. *Ecol. Model.* **2016**, *340*, 45–56. [CrossRef]
62. Cuttitta, A.; Torri, M.; Zarrad, R.; Zgozi, S.; Jarboui, O.; Quinci, E.M.; Hamza, M.; Abdulfatah, E.; Haddoud, D.; El Turki, A.; et al. Linking surface hydrodynamics to planktonic ecosystem: The case study of the ichthyoplanktonic assemblages in the Central Mediterranean Sea. *Hydrobiologia* **2018**, *821*, 191–214. [CrossRef]
63. Budyansky, M.V.; Kulik, V.V.; Kivva, K.K.; Uleysky, M.Y.; Prants, S.V. Lagrangian Analysis of Pacific Waters in the Sea of Okhotsk Based on Satellite Data in Application to the Walleye Pollock Fishery. *Izv. Atmos. Ocean. Phys.* **2022**, *58*, 1427–1437. [CrossRef]
64. Zagaglia, C.R.; Lorenzzetti, J.A.; Stech, J.L.; Zagaglia, C.R.; Lorenzzetti, J.A.; Stech, J.L. Remote sensing data and longline catches of yellowfin tuna (*Thunnus albacares*) in the equatorial Atlantic. *Remote Sens. Environ.* **2004**, *93*, 267–281. [CrossRef]
65. Meinert, C.R.; Clausen-Sparks, K.; Cornic, M.; Sutton, T.T.; Rooker, J.R. Taxonomic Richness and Diversity of Larval Fish Assemblages in the Oceanic Gulf of Mexico: Links to Oceanographic Conditions. *Front. Mar. Sci.* **2020**, *7*, 579. [CrossRef]

66. Prants, S.; Budyansky, M.; Uleysky, M.; Kulik, V. Lagrangian fronts and saury catch locations in the Northwestern Pacific in 2004–2019. *J. Mar. Syst.* **2021**, *222*, 103605. [CrossRef]
67. Howell, E.; Kobayashi; Parker, D.; Balazs, G.; Polovina, A. TurtleWatch: A tool to aid in the bycatch reduction of loggerhead turtles *Caretta caretta* in the Hawaii-based pelagic longline fishery. *Endanger. Species Res.* **2008**, *5*, 267–278. [CrossRef]
68. Benson, S.R.; Eguchi, T.; Foley, D.G.; Forney, K.A.; Bailey, H.; Hitipeuw, C.; Samber, B.P.; Tapilatu, R.F.; Rei, V.; Ramohia, P.; et al. Large-scale movements and high-use areas of western Pacific leatherback turtles, *Dermochelys coriacea. Ecosphere* **2011**, *2*, 1–27. [CrossRef]
69. Jishad, M.; Sarangi, R.K.; Ratheesh, S.; Ali, S.M.; Sharma, R. Tracking fishing ground parameters in cloudy region using ocean colour and satellite-derived surface flow estimates: A study in the Bay of Bengal. *J. Oper. Ocean.* **2021**, *14*, 59–70. [CrossRef]
70. Chang, Y.; Sun, C.; Chen, Y.; Yeh, S.; Dinardo, G. Habitat suitability analysis and identification of potential fishing grounds for swordfish, *Xiphias gladius*, in the South Atlantic Ocean. *Int. J. Remote Sens.* **2012**, *33*, 7523–7541. [CrossRef]
71. Kamaruzzaman, Y.N.; Mustapha, M.A. An overview Assessment of the Effectiveness of Satellite Images and Remote Sensing in Predicting Potential Fishing Grounds and its Applicability for *Rastrelliger kanagurta* in the Malaysian EEZ off the South China Sea. *Rev. Fish. Sci. Aquac.* **2023**, *31*, 320–341. [CrossRef]
72. Kruger, L.; Ramos, J.A.; Xavier, J.C.; Gremillet, D.; Gonzalez-Solis, J.; Petry, M.V.; Phillips, R.A.; Wanless, R.M.; Paiva, V.H. Projected distributions of Southern Ocean albatrosses, petrels and fisheries as a consequence of climatic change. *Ecography* **2018**, *41*, 195–208. [CrossRef]
73. Harford, W.J.; Karnauskas, M.; Walter, J.F.; Liu, H. Non-parametric modeling reveals environmental effects on bluefin tuna recruitment in Atlantic, Pacific, and Southern Oceans. *Fish. Oceanogr.* **2017**, *26*, 396–412. [CrossRef]
74. Reguera-Rouzaud, N.; Diaz-Viloria, N.; Sanchez-Velasco, L.; Flores-Morales, A.L.; Pares-Sierra, A.; Aburto-Oropeza, O.; Munguia-Vega, A. Yellow snapper (*Lutjanus argentiventris*) connectivity in the Southern Gulf of California. *Marine Biodiversity* **2020**, *50*, 54. [CrossRef]
75. Ryan, J.P.; Green, J.R.; Espinoza, E.; Hearn, A. Association of whale sharks (Rhincodon typus) with thermo-biological frontal systems of the eastern tropical Pacific. *PLoS ONE* **2017**, *12*, e0182599. [CrossRef] [PubMed]
76. Taconet, M.; Kroodsma, D.; Fernandes, J.A. *Global Atlas of AIS-Based Fishing Activity—Challenges and Opportunities*; FAO: Rome, Italy, 2019.
77. Salazar, J.E.; Benavides, I.F.; Cabrera, C.V.P.; Guzman, A.I.; Selvaraj, J.J. Generalized additive models with delayed effects and spatial autocorrelation patterns to improve the spatiotemporal prediction of the skipjack (*Katsuwonus pelamis*) distribution in the Colombian Pacific Ocean. *Reg. Stud. Mar. Sci.* **2021**, *45*, 101829. [CrossRef]
78. Yu, W.; Chen, X.; Zhang, Y. Seasonal habitat patterns of jumbo flying squid *Dosidicus gigas* off Peruvian waters. *J. Mar. Syst.* **2019**, *194*, 41–51. [CrossRef]
79. Polovina, J.J.; Kleiber, P.; Kobayashi, D.R. Application of TOPEX-POSEIDON satellite altimetry to simulate transport dynamics of larvae of spiny lobster, *Panulirus marginatus*, in the Northwestern Hawaiian Islands, 1993–1996. *Fish. Bull.* **1999**, *97*, 132–143.
80. Griffin, D.A.; Wilkin, J.L.; Chubb, C.F.; Pearce, A.F.; Caputi, N. *Mesoscale Oceanographic Data Analysis and Data Assimilative Modelling with Application to Western Australian Fisheries*; Marine Research 2001 (Series: FRDC Project no. 1997/139); CSIRO: Canberra, Australia, 2001; pp. 1–52.
81. Scales, K.L.; Hazen, E.L.; Jacox, M.G.; Bograd, S.J. Fisheries bycatch risk to marine megafauna is intensified in Lagrangian coherent structures. *Proc. Natl. Acad. Sci. USA* **2018**, *115*, 7362–7367. [CrossRef]
82. Hitchman, S.M.; Reyns, N.B.; Thompson, A.R. Larvae define spawning habitat of bocaccio rockfish Sebastes paucispinis within and around a large southern California marine reserve. *Mar. Ecol. Prog. Ser.* **2012**, *465*, 227. [CrossRef]
83. Vestfals, C.D.; Mueter, F.J.; Hedstrom, K.S.; Laurel, B.J.; Petrik, C.M.; Duffy-Anderson, J.T.; Danielson, S.L. Modeling the dispersal of polar cod (*Boreogadus saida*) and saffron cod (*Eleginus gracilis*) early life stages in the Pacific Arctic using a biophysical transport model. *Prog. Oceanogr.* **2021**, *196*, 102571. [CrossRef]
84. Priya, R.K.S.; Balaguru, B.; Ramakrishnan, S. Improved exploration of fishery resources through the integration of remotely sensed merged sea level anomaly, chlorophyll concentration, and sea surface. In Proceedings of the Remote Sensing of the Ocean, Sea Ice, Coastal Waters, and Large Water Regions, Online, 23–26 September 2013; Volume 888, p. 888805. [CrossRef]
85. Balaguru, B.; Ramakrishnan, S.S.; Vidhya, R.; Thanabalan, P. A comparative study on utilization of multi-sensor satellite data to detect Potential Fishing Zone (PFZ). *Int. Arch. Photogramm. Remote Sens. Spat. Inf. Sci.* **2014**, *XL-8*, 1019–1026. [CrossRef]
86. Arur, A.; Krishnan, P.; Grinson, G.; Goutham Bharathi, M.P.; Kaliyamoorthy, M.; Hareef Baba Shaeb, K.; Suryavanshi, A.S.; Srinivasa Kumar, T.; Joshi, A.K. The influence of mesoscale eddies on a commercial fishery in the coastal waters of the Andaman and Nicobar Islands, India. *Int. J. Remote Sens.* **2014**, *35*, 6418–6443. [CrossRef]
87. Hsu, T.Y.; Chang, Y.; Lee, M.A.; Wu, R.F.; Hsiao, S.C. Predicting Skipjack Tuna Fishing Grounds in the Western and Central Pacific Ocean Based on High-Spatial-Temporal-Resolution Satellite Data. *Remote Sens.* **2021**, *13*, 861. [CrossRef]
88. Hazen, E.L.; Palacios, D.M.; Forney, K.A.; Howell, E.A.; Becker, E.; Hoover, A.L.; Irvine, L.; DeAngelis, M.; Bograd, S.J.; Mate, B.R.; et al. WhaleWatch: A dynamic management tool for predicting blue whale density in the California Current. *J. Appl. Ecol.* **2017**, *54*, 1415–1428. [CrossRef]
89. Hazen, E.L.; Scales, K.L.; Maxwell, S.M.; Briscoe, D.K.; Welch, H.; Bograd, S.J.; Bailey, H.; Benson, S.R.; Eguchi, T.; Dewar, H.; et al. A dynamic ocean management tool to reduce bycatch and support sustainable fisheries. *Sci. Adv.* **2018**, *4*, eaarar300. [CrossRef]

90. Salama, A.; Willis, J.; Srinivasan, M. Mapping Sea Level from Space. In *Climate. NATO Science for Peace and Security Series C: Environmental Security*; Linkov, I., Bridges, T., Eds.; Springer: Dordrecht, The Netherlands, 2011; pp. 419–431. [CrossRef]
91. Vignudelli, S.; Scozzari, A.; Abileah, R.; Gómez-Enri, J.; Benveniste, J.; Cipollini, P. Water surface elevation in coastal and inland waters using satellite radar altimetry. In *Extreme Hydroclimatic Events and Multivariate Hazards in a Changing Environment*; Maggioni, V., Massari, C., Eds.; Elsevier Press: Amsterdam, The Netherlands, 2019; pp. 87–127. [CrossRef]
92. Goni, G.; Demaria, M.; Knaff, J.; Sampson, C.; Price, J.; Mehra, A.; Ginis, I.; Lin, I.-I.; Sandery, P.; Buarque, S.; et al. The Ocean Observing System for Tropical Cyclone Intensification Forecasts and Studies. In Proceedings of the OceanObs' 09: Sustained Ocean Observations and Information for Society, Venice, Italy, 21–25 September 2009; p. 2.
93. Hossain, F.; Maswood, M.; Siddique-E-Akbor, A.; Wondmage, Y. A Promising Radar Altimetry Satellite System for Operational Flood Forecasting in Flood-Prone Bangladesh. *IEEE Geosci. Remote Sens. Mag.* **2014**, *2*, 27–36. [CrossRef]
94. Siddique-E-Akbor, A.; Hossain, F.; Lee, H.; Shum, C. Inter-comparison study of water level estimates derived from hydrodynamic–hydrologic model and satellite altimetry for a complex deltaic environment. *Remote Sens. Environ.* **2011**, *15*, 1522–1531. [CrossRef]
95. Climate Change Initiative Coastal Sea Level Team. Coastal sea level anomalies and associated trends from Jason satellite altimetry over 2002–2018. *Sci. Data* **2020**, *7*, 357. [CrossRef]
96. Woodworth, P.L.; Menéndez, M.; Gehrels, W.R. Evidence for century-timescale acceleration in mean sea levels and for recent changes in extreme sea levels. *Surv. Geophys.* **2014**, *35*, 1121–1151.
97. Cazenave, A.; Llovel, W. Contemporary sea level rise. *Annu. Rev. Mar. Sci.* **2010**, *2*, 145–173. [CrossRef]
98. International Altimetry Team. Altimetry for the future: Building on 25 years of progress. *Adv. Space Res.* **2021**, *68*, 319–363. [CrossRef]
99. Lin, I.-I.; Goni, G.J.; Knaff, J.A.; Forbes, C.; Ali, M.M. Ocean heat content for tropical cyclone intensity forecasting and its impact on storm surge. *Nat. Hazards* **2013**, *66*, 1481–1500. [CrossRef]
100. Chen, S.; Curcic, M. Ocean surface waves in Hurricane Ike (2008) and Superstorm Sandy (2012): Coupled model predictions and observations. *Ocean Model.* **2016**, *103*, 161–176. [CrossRef]
101. Song, Y.T.; Ji, C.; Fu, L.; Zlotnicki, V.; Shum, C.K.; Yi, Y.; Hjorleifsdottir, V. The 26 December 2004 tsunami source estimated from satellite radar altimetry and seismic waves. *Geophys. Res. Lett.* **2005**, *32*, 1–5. [CrossRef]
102. Smith, W.H.F.; Scharroo, R.; Titov, V.; Arcas, D.; Arbic, B. Satellite Altimeters Measure Tsunami. *Oceanography* **2015**, *18*, 11–13. [CrossRef]
103. Titov, V.; Song, Y.T.; Tang, L.; Bernard, E.N.; Bar-Sever, Y.; Wei, Y. Consistent Estimates of Tsunami Energy Show Promise for Improved Early Warning. In *Global Tsunami Science: Past and Future*; Geist, E.L., Fritz, H.M., Rabinovich, A.B., Tanioka, Y., Eds.; Pageoph Topical Volumes; Birkhäuser: Cham, Switzerland, 2016; Volume I. [CrossRef]
104. Hamlington, B.D.; Leben, R.R.; Godin, O.A.; Gica, E.; Titov, V.V.; Haines, B.J.; Desai, S.D. Could satellite altimetry have improved early detection and warning of the 2011 Tohoku tsunami? *Geophys. Res. Lett.* **2012**, *39*, L15605. [CrossRef]
105. Sandwell, D.T.; Smith, W.H.F.; Gille, S.; Kappel, E.; Jayne, S.; Soofi, K.; Coakley, B.; Géli, L. Bathymetry from space: Rationale and requirements for a new, high-resolution altimetric mission. *Comptes Rendus Geosci.* **2006**, *338*, 14–15. [CrossRef]
106. Gille, S.T.; Metzger, E.J.; Tokmakian, R. Seafloor topography and ocean circulation. *Oceanography* **2004**, *17*, 47–54. [CrossRef]
107. Vinogradova, N.; Hamlington, B. Sea level science and applications support coastal resilience. *Eos*, 29 June 2022; 103. [CrossRef]
108. Cazenave, A.; Gouzenes, Y.; Birol, F.; Leger, F.; Passaro, M.; Calafat, F.M.; Shaw, A.; Nino, F.; Legeais, J.F.; Oelsmann, J.; et al. Sea level along the world's coastlines can be measured by a network of virtual altimetry stations. *Commun. Earth Environ.* **2022**, *3*, 117. [CrossRef]
109. Benveniste, J.; Cazenave, A.; Vignudelli, S.; Fenoglio-Marc, L.; Shah, R.; Almar, R.; Andersen, O.; Birol, F.; Bonnefond, P.; Bouffard, J.; et al. Requirements for a Coastal Hazards Observing System. *Front. Mar. Sci.* **2019**, *6*, 348. [CrossRef]

remote sensing MDPI

Review

The Rising Concern for Sea Level Rise: Altimeter Record and Geo-Engineering Debate

Jim Gower [1] and Vittorio Barale [2,*]

[1] The Pillars of Hercules Foundation, Sidney, BC V8L 4B2, Canada; jimgower1940@gmail.com
[2] Tethys Research Institute, 20121 Milan, Italy
* Correspondence: vittorio_barale@yahoo.it

Abstract: The Oceans from Space V Symposium, held in Venice, Italy, on 24–27 October 2022, devoted special sessions to sea level rise, as described by a series of satellite altimeters, and to remediations of consequent calamities in vulnerable mediterranean seas. It emerged that various aspects of climate change can be modelled in time as a Single Exponential Event (SEE), with a similar trend (a 54–year e–folding time) for CO_2 concentration in the Earth's atmosphere, global average sea surface temperature, and global average sea level. The sea level rise record, combining tide gauges data starting in 1850, as well as more recent altimeter data, for the last 30 years, is already 25 cm above historical values. If the curve continues to follow the exponential growth of the simple SEE model, it will reach about 40 cm by the year 2050, 1 m by 2100, and 2.5 m by 2150. As a result, dramatic impacts would be expected for most coastal areas in the next century. Decisive remediations, based on geo-engineering at the basin scale, are possible for semi-enclosed seas, such as the Mediterranean and Black Seas. Damming the Strait of Gibraltar would provide an alternative to the conclusion that coastal sites such as the City of Venice are inevitably doomed.

Keywords: oceans from Space V; SEE model; sea level rise; satellite altimetry; geo-engineering; Mediterranean Sea; Gibraltar Dam; City of Venice; MOSE project

1. Introduction

> *"A friend recently told me her son had bought a water-front lot, but was worried about building a cabin on it. He had heard alarming news about future sea level rise. She asked for my advice. All I could say was that the only way of knowing the future for certain is to wait and see, but in this case, we can be fairly sure of what the immediate future holds. I suggested her son go ahead and build his cabin, and enjoy the coast (his lot is on Hornby Island, British Columbia, along the magnificent Pacific coast of Canada), but to be prepared for a global rise of almost half-a-metre in the first half of this century, and a full metre rise, or even more, by 2100. Hornby Island is far from flat, and unless he has bought a coastal marsh, he should be able to handle this. For other parts of the world, the danger is much more immediate, and will have vastly greater impact".*

Sea level rise is now a common topic for routine talks, in the wake of a growing public concern for the impending consequences of climate change, and for scientific conferences alike, in search of a holistic approach to its causes, effects, and remedies. The Oceans from Space V Symposium [1], held in Venice, Italy, on 24–27 October 2022, was no exception. The conference, the fifth in a series held at 10-year intervals since 1980 (with the latest delayed two years by the COVID-19 pandemic), aimed to review progress in the use of satellites to study the oceans [2]. Special sessions were devoted to the assessment of sea level rise by virtue of altimetric techniques [3], as well as to possible remediations to the expected damages on the coastal environment of marginal and enclosed seas [4] (a subject to be reviewed by Appendix A, in particular for the special case of the Mediterranean Sea and Black Sea). At the same time, as in previous editions, attendees had a chance to witness first hand the increasing impact of sea level variations on the City of Venice, as well as the city's response, most recently entrusted to the movable barriers of the MOSE (acronym of the Italian *MOdulo Sperimentale Elettromeccanico*, alluding in a secondary meaning to the biblical character Moses, or *Mosè* in Italian, who is credited for parting the sea), capable of sealing off the Venetian Lagoon from the Adriatic Sea, and now nearly operational [5].

Symposium participants were able to tour the MOSE control station, on the artificial island built in the middle of the Venetian Lagoon's main opening on the Adriatic Sea, from which the system is operated (see Appendix B for details on Venice, its periodic floodings known as *Acqua Alta*, and the MOSE project). Raising the movable barriers from the bottom platform, they were told, can provide temporary defence against short-term flooding events, which are due to the combined effects of tides and wind. However, the MOSE is not designed to protect against the long-term sea level rise due to climate change. Taking inspiration from the MOSE field trip, the conference sessions devoted to altimetry and sea level rise were somewhat expanded to discuss (a) impacts of sea level rise on semi-enclosed marine basins, with the Mediterranean Sea and Black Sea providing the most classical of examples; (b) the ecological danger of widespread coastal reinforcements, which might be adopted in the effort to prevent erosion and flooding; and (c) possible future actions that might be undertaken to protect coastal sites, infrastructures and urban settlements, of which Venice represents the archetype. The present review aims to provide a record of the main presentations delivered during these sessions and of the ensuing discussion themes as a contribution to the development of a public debate, advocated by several conference participants.

Global average sea level is indeed rising at an alarming rate [6]. In general, the media tend to over-emphasize immediate dangers of climate change, recently presenting the Thwaites Glacier in Antarctica, e.g., as "Doomsday Glacier" [7], possibly able to raise sea level in steps, causing sudden floods at uncertain times. But, in fact, observations of global sea level rise have, up to present, closely followed an unambiguous growth rate, pointing inexorably upwards into the future. A target rise of about 1 m by 2100 seems to be generally agreed upon in the literature [8], and such a value implies an exponential increase. This can be represented by the Single Exponential Event (SEE) model introduced during the Oceans from Space V Symposium (to be set out in the following section), which climate change indicators appear to be following precisely, and which shows no sign of sudden surges, or of slowing down. The parameters considered here, atmospheric concentration of carbon dioxide (CO_2), global mean Sea Surface Temperature (SST) and sea level can all be seen following the same trend (see Figures 1–3). This analysis is discussed in the following, where some of the key factors related to the changing level of the oceans [9] are reviewed. The sea level record, as defined in the last few decades by a number of altimeter space missions [10], is also reviewed. Finally, the main aspects of the Oceans from Space V discussion on impacts and defenses foreseeable (for the Mediterranean basin) is reported. More details about the envisioned geo-engineering possibilities and problems, and the particular case of Venice, will be deferred to the closing Appendices.

2. Single Exponential Event Model

It is becoming increasingly evident that the world is facing a major climate change disaster [11], driven by the ever-growing concentration of greenhouse gases in the Earth's atmosphere, attributable to anthropogenic causes [12]. The rate of change of atmospheric CO_2 concentration can be simulated by a simple mathematical model of the expected (exponential) form, which also provides a good approximation for a number of related parameters (i.e., global average SST and sea level, as will be seen later in this section), with a comparable e-folding time. The SEE-type trend has been ascribed to the impact of increasing anthropogenic pressure, as already debated in the scientific literature for almost a century [13]. The initial tendency was partly driven by the surge in human population, which was exponential up to about 1990, but has been significantly slower since then [14]. Industrial development and CO_2 production, however, have not slowed down. As a result, the Keeling Curve [15], plotted in Figure 1 as monthly averages of CO_2 concentration, with annual cycle removed, together with a best fit exponential for comparison, continues to show the same growing trend, in spite of all international efforts, led by the United Nations (UN), aiming at the reduction of human-caused carbon emissions [16].

Such efforts formally got under way with the establishment of the United Nations Framework Convention on Climate Change (UNFCCC), following the "Earth Summit"—in fact, the United Nations Conference on Environment and Development (UNCED)—held in Rio de Janeiro, Brazil, in 1992 [17], and continued with the 1997 Kyoto Protocol [18] and its 2012 Doha Amendment [19]. More recently, the Paris Agreement, reached by the parties to the UNFCCC at the 21st Conference of the Parties (COP), held in Paris, France, in 2015, originated a five-year cycle of "increasingly ambitious climate action" carried out by participating countries [20]. As a consequence, since 2020, countries have been submitting their first national climate action plans, or Nationally Determined Contributions (NDCs). However, given the UN's Intergovernamental Panel on Climate Change (IPCC) warning of "more severe climate change impacts, including more frequent and severe droughts, heatwaves and rainfall", should global warming cross the 1.5 °C threshold, the 27th COP, held in Sharm el-Sheikh, Egypt, in 2022, called for accelerated action and requested parties to "revisit and strengthen the 2030 targets in their NDCs, to align with the Paris Agreement temperature goal by the end of 2023" [21]. But, in the meantime, CO_2 concentrations shown by the Keeling Curve continue to rise, exhibiting the same exponential growth rate, with no effect apparent from these efforts.

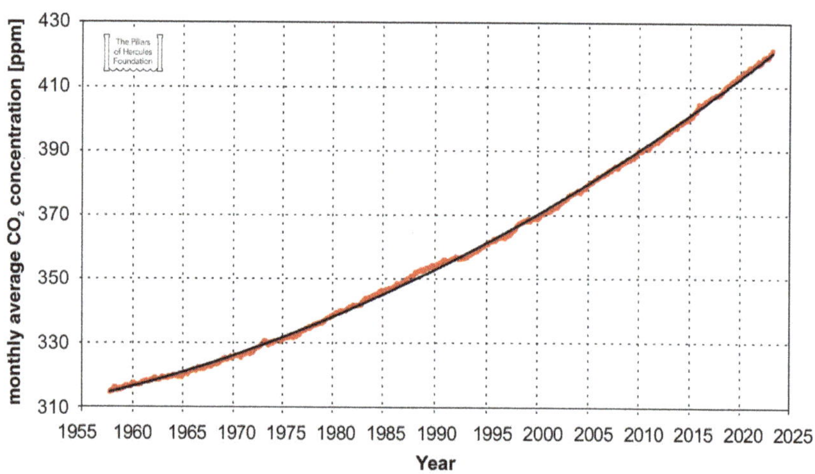

Figure 1. The Keeling Curve, showing the 1956–2022 evolution of CO_2 concentration [ppm] in the Earth's atmosphere (red line: CO_2 concentration monthly averages, seasonal cycle removed; public data obtained from [22]); compared to the best-fit exponential curve of the SEE model (black line).

2.1. The Keeling Curve

Data collection at the Mauna Loa Observatory—located on the north flank of the Mauna Loa Volcano, on the Big Island of Hawaii, at an elevation of 3397 m above sea level—on which the Keeling Curve is based, started in 1958 and continues through the present, albeit from a nearby station, with updates published regularly [23]. Actual measurements at the original facility stopped following the late 2022 eruption of Mauna Loa's Volcano, when a lava flow across the road leading to the site cut both access and power lines. Thanks to an emergency set-up, the same atmospheric sampling has been resumed at a temporary facility, using existing infrastructures on Maunakea, a dormant volcano located approximately 21 miles north of Mauna Loa [24]. Both observation points lie above the local marine temperature inversion layer, which separates the lower portions of the atmosphere from the above-standing cleaner troposphere. These remote locations, offering undisturbed conditions, with minimal influences of vegetation or human activity, are ideal for monitoring air constituents that can affect climate change. Thanks to the continuity of such measurements, well over 60 years of continuous CO_2 observations are now available, showing occasional minor dips and bumps due to volcanic eruptions and El Niño events, but staying depressingly close to the same exponential, in spite of the growing concern about the Earth's climate and the consequent UNFCCC-led actions recalled earlier in this section. So far, at least judging from the trend of the Keeling Curve, ongoing human activities apparently continue to generate the exponential greenhouse effect driving force [8]. The exponential best-fit, shown in Figure 1, gives an e-folding time of 54 years, with a baseline value of 269 ppm in the pre-industrial era, consistent with estimates of peak concentrations reached during warm interglacial periods in the last 800,000 years [25], based on Antarctic ice core data [26]. Concentrations have recently reached 420 ppm, 1.5 times the pre-industrial value. As of today, in 2023, the CO_2 amount is still growing exponentially, with no signs of any significant down-turn. By the year 2100, the SEE scenario reaches a value of 905 ppm, over three times the pre-industrial level.

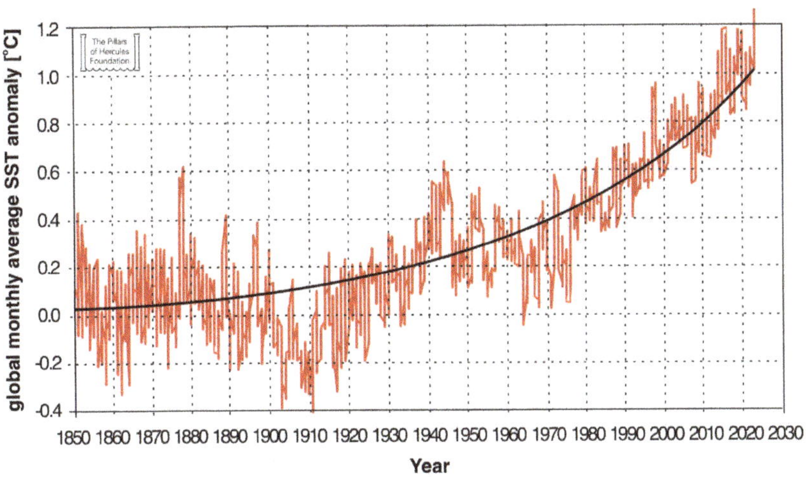

Figure 2. The 1850–2022 global monthly average SST anomaly [°C], as computed by the UK Hadley Centre, with 0.35 °C added to change anomaly reference date from the 1961 to 1990 period to the distant past (red line: SST anomaly monthly averages; HADSST4 public data set obtained from [27]); compared to the best-fit exponential curve, assuming the SEE 54-year e-folding time (black line).

2.2. *Global Average Sea Surface Temperature*

In comparison to the Keeling Curve for CO_2 concentration, the global average SST data from the UK Hadley Centre (HADSST4) shows more short-term and long-term variations [28], but with a longer data series, starting in 1850 (Figure 2). An exponential with an e-folding time of 54 years (also shown in Figure 2) again gives a good fit, with warming of 1.5 °C (the value that UN's IPCC reports [29] warn not to exceed) by 2046, 2 °C by 2062, and just over 4 °C by 2100. If the SEE model is taken as a prediction, this is clearly the most significant change of all those affecting sea level rise, given the most recent record temperatures reached in some basins of the northern hemisphere (e.g., in the summer of 2023 for the Mediterranean Sea [30]). Continued exponential growth, to 10 °C in 2150 and 26 °C in 2200, represents a run-away greenhouse effect that would eventually sterilize the Earth, in a Venus-like fashion [31]. Indeed, additional sources of CO_2, like the increasingly common wildfires raging in several regions of both the northern and southern hemispheres, must now be adding to the problem.

2.3. Global Average Sea Level

The global average sea level, as measured by satellite altimeters since the early 1990s [32], shows a similarly rising exponential curve (Figure 3), with the same e-folding time of about 54 years already introduced earlier. The more precise satellite data record is still rather short, so the e-folding time value is not yet well-determined, but the 54-year apparent figure is somewhat confirmed by the Reconstructed Sea Level [33], based on tide gauge data (also shown in Figure 3), going back to before 1900. The rise has amounted to 16 cm by the year 2000, and 24 cm by 2020; small amounts so far, but enough to start causing erosion and floods, locally and around the world [34]. The satellite altimeter data exhibit dips of up to 1 cm in years 2010 to 2016, which have been explained as large-area rain events, when water evaporated from the oceans has taken several months to flow back from land [35], but continues to follow rather precisely the SEE model. The difference between model and observed values for the period 1993 to 2022 shows a standard deviation of 2.5 mm, while the level has increased by 10 cm.

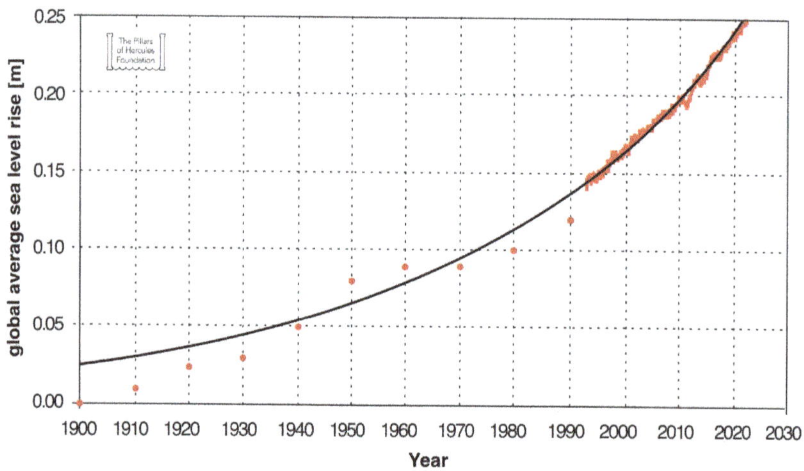

Figure 3. Global average sea level [m], as derived from tide gauge measurements before 1993 (red dots: reconstructed sea level based on [36], public data obtained from [37]), and by satellite altimeters after 1993 (red line: public data obtained from [38]), with 0.145 m added to change the height reference date from 1993; compared to the best-fit exponential curve (black line), assuming the SEE 54-year e-folding time.

These numbers are starting to attract widespread attention, though the rise is not yet large enough to cause the alarm that it deserves. The SEE model represents a continuing exponential rise, to 1 m in 2100, 2.6 m in 2150, even to 6.6 m in 2200. Of course, there's no way to be certain that the actual rise will continue to follow the model curve into the next century, but the present trend is clear and gives a good indication of what future to expect in the next few decades. If the model continues to be realistic, then sea level rise will have to be recognized as a really serious problem by the same national and international bodies mentioned above, which are now trying to reduce climate change, as well as by the general public at large (although global warming per se, rather than just rising sea levels, will probably continue to be the issue attracting the most attention).

3. Satellite Altimetry Data Record

Historical sea level changes cover a broad spectrum of space and time scales, resulting from complex processes taking place within the Earth system, as well as to external forcing due to both natural phenomena and anthropogenic factors [39]. In the last three decades, spaceborne instruments have detected accelerations in sea level rise, and in the evolution of other indicators of climate change, which are to be expected when the planet warms, based on current understanding of Earth's history and climate physics [40]. Most of the observed global sea level rise (in excess of ≈3.3 mm/year, in the satellite measurements record; see following paragraphs) is estimated to come from the melting of land-based ice, which adds to the ocean's volume (≈2.1 mm/year), and from the thermal expansion of ocean water as it warms (≈1.3 mm/year) [41]. Changes in land-water storage (e.g., due to the enlargement of surface reservoirs or the depletion of underground aquifers), as well as changes in global precipitation patterns, also make small contributions [42]. Vertical land motions, due to subsidence and/or to rebounding of the Earth's crust since the end of the Last Glacial Maximum, contribute to site-specific, regional variations in sea level [43]. The amount and speed of sea level rise also varies by location, and from near-coastal to open ocean areas, mainly due to the Earth's uneven distribution of mass, as well as to the changing physical properties of seawater and its dynamics [44]. In the present climate change context, both models and observations seem to agree that the dominant trend of sea level rise over most of the globe will follow the exponential increase exhibited by the global average curve (Figure 3). Prominent exceptions will occur wherever glacial isostatic adjustment is significant, either from the last ice age, as in Northern Europe, or from the ongoing land ice melting, over Greenland and Antarctica.

3.1. Relative Sea Level

Before the satellite altimetry era, tide gauges, which record the height of the water level with respect to a local land reference, were the only means to directly observe sea level variability [45]. Such measurements, limited to specific and unevenly distributed coastal sites, show relative changes due to the combination of absolute water level variations and local or regional vertical land movements. To derive absolute sea level changes from tide gauge records, these have to be corrected, in general by using global positioning systems [46]. Various methods are available to process tide gauge records over long periods and provide long-term mean values in which shorter-term variations are greatly reduced, so that local changes in sea level can be estimated to an accuracy at or below the cm level [47]. This renders the coverage of such in situ sensors invaluable, at least at their specific location, in following the evolution of average sea level, and of other coastal processes as well. The strong heterogeneity in tide gauge locations worldwide represents a main limitation at the global scale. Nevertheless, these data are also useful to validate numerical models and even to detect errors or drifts in altimeter datasets [48]. Reconstructed sea level data can also be obtained, combining the tide gauges data (and/or other in situ data) with satellite altimeters data, in order to estimate the pre-satellite era sea level. For example, an Empirical Orthogonal Function (EOF) can be fitted to the satellite record in order to generate an algorithm that uses tide gauge data back in time to extrapolate past global sea levels [49].

3.2. Geocentric Sea Level

Unlike tide gauges, satellite altimetry from the Earth's orbit can provide a measure of geocentric sea level variations [50]. Altimeters map the sea surface by measuring the satellite-to-surface return travel time of successive radar pulses in order to determine the height of the satellite above the water, i.e., its altimetric range. The difference between altimetric range and satellite altitude above a reference surface gives the height of the sea surface with respect to the same reference surface. The altimetric range must be corrected to mitigate effects caused by atmospheric refraction, sea state, and instrumental biases; a number of corrections due to different geophysical effects are also taken into account [51]. The proximity of land contaminates the return signal of the altimeter, so algorithms are used to provide corrections in coastal zones [52]. The measurements are collected along the ground tracks, the projection of the altimeter orbits on the Earth's surface, in any weather conditions and regardless of the time of overpass. Covering the oceans with repeated tracks, at repetition rates of some days, satellite altimeters can provide (*quasi*) global monitoring of sea surface height to an accuracy of 1 to 2 cm. Combining data from several satellites, by removing their respective biases, a continuous time series between mission can be obtained [53]. This provides unprecedented, detailed insights in the redistribution of water masses, thermal expansion and contraction, currents and eddies, thermohaline density variations, ocean–atmosphere coupling, and impact of fresh waters, including melting ice sheets and glaciers.

3.3. Satellite Altimeter Missions

For the last three decades, a string of diverse but coherent altimetry missions has been measuring routinely climate-related sea level dynamics, at both regional and global scales. A remarkable international group of space agencies is behind the realization of such a long-term series of satellite altimeters, and it includes the US National Aeronautics and Space Administration (NASA) and National Oceanographic and Atmospheric Administration (NOAA); the European Space Agency (ESA) and European Organisation for the Exploitation of Meteorological Satellites (EUMETSAT); the French Centre National d'Etudes Spatial (CNES); the Indian Space Research Organization (ISRO); the Canadian Space Agency (CSA); and the United Kingdom Space Agency (UKSA). Their early attempts to collect data on the structure of the sea surface, i.e., its elevation (with respect to the geoid) and roughness (as a function of winds, waves, wakes, slicks) started with the very first generation of ocean-viewing satellites, namely, Skylab in 1973 and Geos-3 in 1975 [54], carrying a suite of (active) microwave sensors, in particular altimeter and Synthetic Aperture Radar (SAR) prototypes. The 1978 SEASAT mission, also carrying a radar altimeter, as well as a number of other microwave sensors, explicitly aimed to validate the feasibility of global ocean monitoring by satellite, and to define the requirements of future operational satellite systems for ocean remote sensing [55].

More than a decade later, a series of follow-up *quasi*-operational altimeter missions begun to develop [56]. The sequence was initiated by the European ERS-1 and ERS-2, launched in 1991 and 1995, respectively, and by the American-French TOPEX/Poseidon, launched in 1992. American–European cooperative missions were continued by the Jason satellite series, i.e., Jason-1 in 2001, the Ocean Surface Topography Mission OSTM/Jason-2 in 2008, and Jason-3 in 2016. Next came the all-European Envisat, in 2002 [57]; the Indian-French altimetry mission SARAL/ALtiKa, launched in 2013 [58]; and, again, the European Sentinel 3A and Sentinel 3B, in 2016 and 2018, respectively; and Sentinel-6 *Michael Freilich* (S6MF) in 2020 [59]. Sentinel-3A and 3B (together with CryoSat-1, lost in 2005 due to a launch failure, and Cryosat-2, launched in 2010 [60]) were the first satellites to operate SAR (Delay-Doppler) mode altimeters, achieving major refinements, over both open ocean and coastal zone, improving the accuracy of measurements, mapping previously unresolvable features, and providing measurements closer to the coast than ever before. S6MF has ensured continuity of such observations. With the new digital altimeter technology and dedicated onboard processing of S6MF, more precise measurements of sea surface height are possible, as well as information on coastal ocean dynamics within a few kilometers of the coast.

Finally, the short-lived American–French, with Canadian and British contributions, Surface Water & Ocean Topography (SWOT) latest altimetry mission, with enhanced technologies for higher resolution data, was launched in late 2022 [61]. It will be followed by Sentinel-6B, scheduled for launch in 2025. SWOT, with new Ka band radar interferometer technology and a nadir altimeter, provided the first high-resolution measurements of both marine and other surface water features, obtained with a single satellite mission, resolving lakes of 250 m^2 in size, rivers 100 m wide, and sub-mesoscale ocean features.

The current time series of global mean sea level rise has been created essentially using Topex/Poseidon, the three Jason series, and S6MF. These have produced the unique, long-term, continuous time series of data, constituting the later, post-1992 part of the graph shown in Figure 3 [62]. While the historical data record indicates that sea level rose by about 15 cm on average, between 1900 and 1993, the altimetry-based global mean sea level has risen on average at a rate of 3.3 ± 0.3 mm *per* year, between the years of 1993 and 2022. This corresponds to a post-1990 average increase of almost 10 cm over the following 30 years, clearly showing acceleration of the rise. Moreover, the altimeter-derived sea level rise has been around 2.6 mm *per* year, between 1993 and 2008, but the rate of rise has grown to 4.2 mm *per* year, considering the period from 2007 to 2022 [35].

The SEE model curve, shown in Figure 3, displays an exponential increase, which has reached 25 cm above pre-1900 values in 2022 and will reach 40 cm by the year 2050, followed by 1 m by 2100 and 2.5 m by 2150. Institutions responsible for satellite altimetry in both the US and Europe have in the past suggested a much more moderate rise by publicizing only the linear best-fit trend of the satellite record. Until this year, 2023, websites of the US NOAA [63], the University of Colorado [64], and the French AVISO [65] all prominently displayed only a linear rise rate, close to 3.5 mm per year. This gives a much less alarming picture, but promotes a value from 15 years in the past (mean date of the satellite record is 2008) and suggests linear extrapolation for future prediction, to about 50 cm by 2100 and 85 cm in 2200, much less than the rise suggested by exponential growth. The most recent update of the NOAA website now emphasizes acceleration, and the University of Colorado website links to this. Only the French AVISO website continues to promote the linear rise. Presumably, this will be updated soon.

4. Coastal Areas: Impacts and Defenses

In the next 10 to 100 years, the dramatic effects of global sea level rise are expected to seriously affect most of the world's coastal regions (while in regions such as Greenland, or Antarctica, and nearby areas, the effects of significant glacial rebound will tend to prevail, as land ice melts so that loading plus gravitational attraction are removed). If the SEE scenario outlined above, showing exponential rates of sea level rise, continues to prove correct, then relatively soon, most of the world's coasts, which are not natural rock, will be either heavily reinforced artificially, or heavily eroded. One implication is that most of the present beaches are going to be lost [66]—a perspective that might be acceptable, as long as the coastline's changing structure were capable of evolving until it reaches a new equilibrium. But, of course, concern arises for the consequent anthropogenic stiffening of coastal environments, artificially reinforced or outright covered with concrete, to protect existing formations, which hinders the natural system's ability to adjust and adapt to the changes. In addition, there exist already large amounts of coastal infrastructures, along several stretches of coastline, worth retaining for environmental, cultural, or economic reasons (e.g., the City of Venice). A widespread discussion should be initiated on possible actions to be taken, including how to preserve some unchanged, natural coastline for future generations, as well as how to plan the future of presently modified coastline, which will have to be either abandoned or defended.

While sea level rise impacts and responses will need to be considered for all shorelines worldwide, the coastal areas of marginal and enclosed seas, especially those exhibiting high residential density, abundance of historical and cultural sites, and great economic value, due to local maritime activities such as tourism, commerce, transportation, aquaculture and fisheries, will represent special cases to be notably considered. The cluster of semi-enclosed basins in the Mediterranean region (Figure 4) embodies such a critical hotspot, where geography and history combine to create the cradle of (western) civilization [67]. At the same time, the basins' morphology offers an almost unique opportunity to study possible remediations to uncontrolled sea level rise, which near Gibraltar will be close to the global average [44], based on environmental engineering at the basin scale. A possible solution—proposed at the Oceans from Space V Symposium—to provide long-term, sustainable, and possibly affordable protection for all coastlines of the Mediterranean Sea, and Black Sea as well in fact, would be to close the Strait of Gibraltar with a dam, thereby regulating the water through-flow to and from the Atlantic Ocean and ultimately controlling sea level within the whole basin.

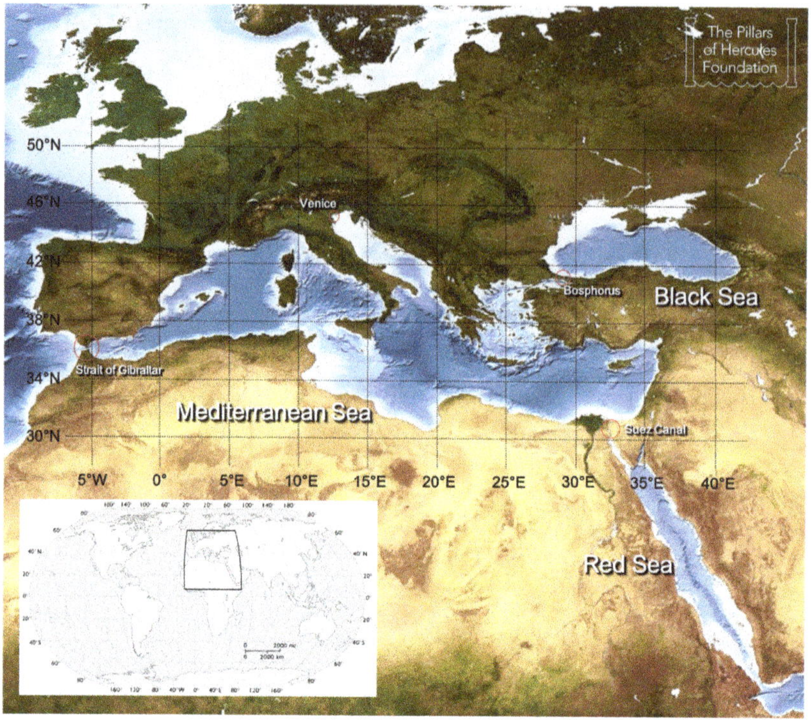

Figure 4. The semi-enclosed Mediterranean Sea, Black Sea, and Red Sea. The Mediterranean Sea is connected to the Atlantic Ocean by the Strait of Gibraltar; to the Black Sea by the Bosphorus; and to the Red Sea by the Suez Canal (all identifyed by red circles). The location of the Lagoon and City of Venice is also shown. The colour coding of sea areas indicates approximate bathymetry, from shallower waters (lighter blue tones) to deeper waters (darker blue tones), while that of land areas indicates vegetation cover (green tones) and desert areas (yellow and brown tones).

The notion of a Gibraltar Dam is not new. In the 1930s, for example, such a dam was proposed by Herman Sörgel as part of the Atlantropa Project, designed to lower the Mediterranean Sea level by up to 200 m, in order to gain new coastal lands and allow power generation from the resulting inflow at Gibraltar and the Bosphorous [68]. In 1958, Scientific American published an article by Henry Stommel, which mentioned the "entertaining fantasy" of controlling climate by building a Gibraltar Dam, blocking the deep outflow from the Mediterranean Sea and thus achieving climatic modifications that might possibly improve the planet for human habitation [69]. In 1977, a controversial paper by Roger Johnson, published in the EOS Transactions of the American Geophysical Union (AGU), claimed that damming the Strait of Gibraltar would prevent the onset of a new ice age, triggered by an increased salt outflow from the Mediterranean Sea, which in turn was a consequence of building the Aswan Dam on the river Nile [70]. More recently, in 2015, a precursor of the present review, published in Natural Hazards, revived the issue by advocating "a sea surface height control dam at the Strait of Gibraltar", aimed to stabilize the Mediterranean Sea and Black Sea at the present level, thus preventing the future loss of countless coastal sites of great economic and cultural value [71]. The debate was resumed once again at Oceans from Space V, during two ad hoc Sea Level Rise sessions, foreseen by the Symposium program for this very purpose, of which the present review constitutes a concise report.

The Strait of Gibraltar has been blocked in the process of continental drift, during the Messinian stage (the uppermost stage of the Miocene, in the geologic timescale), between 7.2 and 5.3 million years ago (Ma), with catastrophic consequences for the marine environment of the inner basins [72]. In the latter part of this stage, the connection between Atlantic Ocean and Mediterranean Sea was progressively restricted by tectonic movements, until its total closure during the Messinian salinity crisis, starting from 5.9 Ma [73]. Consequently, the Mediterranean basin went into a cycle of repeated partial desiccations, turning into a deep desert surrounding a number of residual extensive brine pools. After the Strait of Gibraltar precursor closed for the last time, around 5.6 Ma, the Mediterranean basin dried out almost completely within a thousand years [74]. The Messinian salinity crisis ended 5.3 Ma, when the Atlantic breached the closure once again and rapidly filled up the Mediterranean basin in what is known as the Zanclean flood [75].

A modern, artificial closure of the Strait of Gibraltar would require a careful evaluation of its effects, difficult international agreements, long realization plans, and relatively high costs (see [71] and references therein). If such a project could ever be realized, the long-term stability of water levels, as well as of the resulting new environmental equilibrium, would require an appropriate water flow management. Both positive and negative effects would result. On the positive side, the closure would allow an effective regulation of sea level and provide coastal protection throughout the inner basins, as well as a significant power generation potential and a land link between Europe and Africa. Adverse effects would include still unknown results of limiting the water exchange with the Atlantic Ocean; rising salinity in the Mediterranean Sea and possibly stronger stratification in the Black Sea; changes in nutrient cycling and ensuing ecological repercussions, including disrupted migration patterns of various species; and the need for maritime traffic to go through locks at both Gibraltar and Suez. No particular intervention would be required for the Black Sea sub-basin, where no alternative protection could be obtained by damming the Bosphorous, since the large freshwater inflow would eventually, and rather quickly, raise the water level. Appendix A offers a summary of this entire topic, as discussed during the Sea Level Rise sessions at Oceans from Space V, together with additional details and proper referencing for most of the issues above.

The general idea of responding to the effects of climate change with some form of "geo-engineering"—where a negative connotation might be given to the idea of solving a human-induced problem by tampering even more with nature, only to cause even worse side effects [76]—leads to important ethical considerations. Indeed, any plan for large-scale interference with Earth sub-systems implies the risk of unintentional disruptions of some natural balance and may result in a dilemma that such disruptions could be more damaging than the very damage they were meant to offset. Some geo-engineering proposals, in particular when related to climate change remediation, have been highly controversial, due to the large uncertainties about their effectiveness, their possible side effects, and just the outright possibility of unforeseen consequences [77]. However, the risks of adopting such interventions must be seen, eventually, in the context of the outcome of continuing climate change without them [78]. The proposed Gibraltar Dam does represent a major modification of environmental equilibria, but this is in response to the on-going global, albeit unplanned, "geo-engineering" that has so significantly increased the CO_2 concentration in the Earth's atmosphere and altered the climate in the first place.

A second ethical point to be noted is that a group of countries could unilaterally implement geo-engineering plans, which may be beneficial for themselves, but hurt others. Damming the Strait of Gibraltar to lower sea level in the Mediterranean basin would help riparian countries to safeguard their coastlines, and ease expected problems [79], but it would impose some level of hardship on low-lying countries in the rest of the world. In fact, should the Mediterranean Sea level stop rising in the future, the global ocean rise rate would increase by 0.8% (corresponding to the Mediterranean basin's share of the world's ocean area). This seems a relatively small fractional increase, but in principle it would add to the existing sea level problem worldwide (e.g., an 8 mm rise in global sea level for a drop of 1 m in Mediterranean Sea level). In this respect, the construction of a Gibraltar Dam has been labeled as "colonial" geo-engineering, in that Mediterranean countries would protect themselves at the expense of other countries, which would simply have to suffer the consequences of such imposition. This argument could be rejected on the basis of its relative scale: in fact, if even a microscopic rise is forbidden, then all flood protection becomes illegal, as in all cases water is being diverted to somewhere else, only to become somebody else's problem. Moreover, the Mediterranean basin is the only large marine area that could be dammed in this way (with the exception perhaps of the Red Sea and the Persian Gulf, although similar measures have been proposed also for the North Sea and Baltic Sea [80]), so there is little danger that similar solutions would proliferate, though major coastal defenses must be expected elsewhere, around the world. And, in the end, the whole climate change problem could be called "colonial", in that an initially small group of developed countries have already discharged their CO_2 waste into the Earth's atmosphere to further increase their own economic development, causing hardship to all (including themselves).

5. Concluding Remarks

The unfolding changes in the Earth's climate due to greenhouse gases emissions will have enduring consequences for the natural ecosystem—which, in some cases, may last up to one hundred thousand years [81]. They will also have immediate effects on human society. So far, commitments to cut emissions have proven unable to achieve the goals set by international agreements [82], so that various kinds of interim measures have been proposed [83]. While reducing emissions remains the foremost immediate priority, at least in the (very) long run, contingency plans would also be advisable to deal with the resulting changes that are already in progress, including sea level rise [84]. Instead of trying to modify the entire Earth's climate, though, as some of the geo-engineering plans recalled earlier would do, perhaps locally targeted interventions should be employed, aimed at specific high-leverage locations [85]. For sea level rise, future actions may encompass both traditional coastal protection and focused, regional, basin-scale geo-engineering.

As discussed at Oceans from Space V (more information on the conference and its contents can be obtained as indicated in the Supplementary Materials endnote), the simple SEE model proposes that climate change can be represented as a single exponential event in time, with a similar shape (a 54-year e-folding time) for indicators such as CO_2 concentration, average global SST, and global average sea level as well. An international suite of satellite altimeter missions has contributed a continuous 30-year time series to the long-term record of sea level observations. So far, the altimetry component indicates that global average sea level is rising at an increasing rate, and this suggests that the exponential increase, which has already reached 25 cm above historic values, will climb to about 40 cm by the year 2050, 1 m by 2100, and 2.5 m by 2150. Future satellite missions will help clarify the issue by extending the data record into the next decade and beyond [86]. Measurements at higher resolution and closer to coastlines, where the impact of level rise on coastal communities and infrastructures is expected to be the most significant, will also have relevance. But for now, the global altimeter measurements closely follow a growing exponential, indicative of a runaway greenhouse effect. Should this continue to be true, big changes are to be expected for the world's coastal regions [87]. To bring this situation back under control, at some point in the next two decades a concrete response will be required, in order for the curves of the various indicators to start bending down, below the present growth rate. This may include ambitious, possibly even daring, regional geo-engineering plans, to ensure that coastal icons such as the city of Venice may be spared.

Of course, managing sea level rise at its very source, mitigating and then reversing climate change by reducing the concentration of CO_2 in the Earth's atmosphere, offers the advantage of benefiting the entire planet, while a management strategy that aims just at local coastal protection represents an every-man-for-himself approach that may leave others behind. But still, local, targeted interventions, focusing on specific areas, do offer a softer alternative to global-scale geo-engineering. With enough research on all aspects of the proposal, it is plausible that damming the Strait of Gibraltar may prove itself as a solution that is both effective and achievable in order to protect in a single stroke the whole Mediterranean and Black Seas. A large amount of data collection, modeling, planning, testing, as well as of technological and logistical development, not to mention public discussion and political debate, will be needed before such a project is undertaken. The hypothesis that emerged from the Oceans from Space V debate, reported here (Appendices A and B) in some detail, can only be the starting point for a long incremental process of design upgrades that will be required, before the international community can agree on the merits of such a solution. Perhaps, after careful consideration, basin-scale geo-engineering may reveal to be not so viable, after all, while site-specific coastal protection, coupled to inland retreat where no protection measures would prove feasible or economical, may emerge as the proper answer. We shall SEE ...

Author Contributions: This paper is based on a series of private communications provided by J.G., later complemented with additional material and structured in the present form by V.B. All authors have read and agreed to the published version of the manuscript.

Funding: This research received no external funding.

Data Availability Statement: No new data were created or analyzed in this study. Data sharing is not applicable to this article.

Acknowledgments: Special thanks are due to A. Filipkowska, Technical Advisor, The Pillars of Hercules Foundation, for her invaluable help in the realization of the graphical part of this paper. Appreciation is expressed also to five anonymous reviewers for their contribution to the substantial improvement of the original manuscript. Finally, the authors are in debt with the "Oceans from Space V" Symposium attendees, who participated in the Sea Level Rise sessions and the ensuing discussions, from which this paper originates. "Oceans from Space V", scheduled for 2020, but delayed for 2 years due to the COVID-19 pandemic, and finally convened in Venice, Italy, on 24–27 October 2022, has been the latest event in a conference series held at 10-year intervals since 1980 to review uses of remote sensing from Earth's orbit in the study of the oceans.

Conflicts of Interest: The authors declare no conflicts of interest.

Appendix A

The Pillars of Hercules: Damming the Strait of Gibraltar

The Pillars of Hercules have stood for centuries at the separation of the Mediterranean Sea and the Atlantic Ocean [88]. They are commonly identified with the high mountain relief standing on both sides of the Strait of Gibraltar, and they are forever linked to the demigod Hercules—or Heracles, as he was known in the ancient Greek world—who ventured beyond the outer limits of the known world to perform one of his Twelve Labors [89]. According to Greek mythology, also shared by Etruscans and Romans, the Strait was opened by the hero himself in the mountain range that joined Africa and Europe. Both Seneca and later Pliny the Elder, in the first century AD, recall how Hercules decided not to go around the obstacle, but rather split the mountains with a blow from his sword, and then passed through the narrow strait. Conversely, in the preceding century, Diodorus Siculus, favoring an alternate school of thought on this matter, held that instead of smashing through an isthmus to create the Gibraltar opening, Hercules actually narrowed an existing, larger strait to prevent sea monsters of the Atlantic Ocean from entering the Mediterranean [90]. This second mythological hypothesis seems to point at a similar labor which faces mankind today: how to prevent the Atlantic Ocean itself from entering the semi-enclosed seas at the southern edges of the European continent, engulfing their coasts, submerging their riches, destroying their valuable infrastructures. The sea-level-rise monster is here, menacing to burst into *Mare Nostrum*, nullifying any effort to properly manage its present near-coastal environment.

The Pillars of Hercules Foundation—a vitual non-governamental organization born in the wake of the Oceans from Space V Symposium, held in Venice, Italy, in October 2022 [91]—was created to continue the debate started during the conference on climate change and its clear consequences for sites of the Venice class. Now that sea level rise is starting to be increasingly seen as a real and immediate danger, the advantages offered by regulating water flows at the Strait of Gibraltar (Figure A1), to provide protection for all coastlines of the Mediterranean and Black Seas, appear clearer than ever. The main outcome emerging from the 2022 Venice discussions is that the idea of countering the rising sea water level with a dam—for one of the few places in the world where a long-term, large-scale defense against sea level changes might indeed be practical—should be studied in greater detail. Planning and construction of a dam closing the Mediterranean Sea would be a major project, requiring complex negotiations and international agreements, not to mention significant resources. But most of all, the environmental impacts of controlling sea level in a semi-enclosed basin with a dam need to be carefully assessed. Certainly, such a project could not be undertaken any time soon, but discussions should start as early as possible in order to quantify uncertainties and impacts, benefits and drawbacks, alternatives and costs. The present paper should be seen as a contribution to this debate, aimed at highlighting what is known, or not known, about the southern European Seas, and what are the chances of mitigating problems foreseen in future sea level rise scenarios.

The Gibraltar Dam—a barrage, actually, meant to stabilize sea level with a relatively minor sea-level difference between its two sides, rather than a wall of concrete holding back water with a height difference of many metres—would be positioned along the Camarinal Sill, crossing the shallowest channel in the western part of the Strait. It should be designed to initially cause a height difference between its two sides of about 1 m, allowing inflow of Atlantic water to balance surface evaporation in the inner basin, and possibly a residual deep outflow from the Mediterranean itself. Such a design would need to accommodate the fact that height differences could increase significantly in future years, even up to 10 m within a century or so. The barrage would be built of loosely piled rock, dropped from ships or barges, or from the dam itself, with an angle of repose small enough to ensure stability. Since the structure would need to be designed so as to be stable in the event of at least moderate earthquakes, an angle of repose from 30° to 20° has been suggested [92], with the smaller angle giving increased dam stability. It could be on the order of 100 m wide at the top and about 10 m in height above the outer ocean level, to protect against high tides and storm surges. Being positioned west of the narrowest part of the Strait, it would be about 25 km long, following a shallow water line where the maximum depth is 284 m (rather than 800 m in the narrowest part of the Strait). Simple geometry gives an approximate required volume of rock, ranging from 1.23 km^3 at an angle of repose of 30°, to 1.45 km^3 at 25°, and to 1.78 km^3 at 20° (about the same order of magnitude as the whole Rock of Gibraltar).

A cost estimate can be derived by comparison with large-scale mining operations. For example, 1.5 km^3, or 5 billion tons, of rock is about half the volume of material mined from the 4.3 × 3.0 km^2 wide, 0.9 km deep Chuquicamata copper mine in northern Chile [94]. The total value of the copper extracted from this mine can be estimated at USD 50 billion. By comparison, if the dam is not built, the estimated total cost of coastal defenses could be considerably higher [95,96]. At the global scale, with about 200 million people living within coastal floodplains—and with 2 million km^2 of land and USD 1 trillion worth of assets lying less than 1 m above current sea level—sea-level rise is considered to be one of the major socio-economic hazards associated with global warming [97].

Power generation would also be possible at the dam, especially as the water level difference across the two sides increases in future years, offsetting over time the construction and maintenance cost of the infrastructure. A further major benefit of the dam would be a land link between Europe and Africa. Security for illegal immigration would need to be handled, but a major road and rail corridor between Spain and Morocco should help reduce the economic gap between Western Europe and North Africa. At present, both countries have royal commissions charged with designing an underground rail tunnel under the Strait [98,99] to improve this transport link. A tunnel would have to contend with the unstable geology of the Strait. The surface link provided by the dam (along the same route as the tunnel) would be simpler, safer, and of much higher capacity. The dam would face the same unstable geology, and could be subject to occasional slumping, but assuming it is built of loose rock, it should be relatively easily repaired.

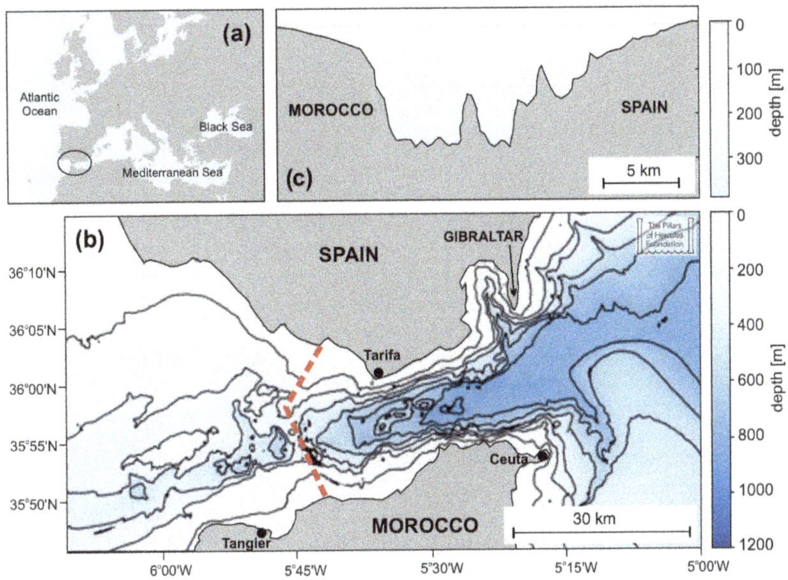

Figure A1. The Strait of Gibraltar, connecting Atlantic Ocean and Mediterranean Sea. Panel (**a**): location of the Strait at the westernmost end of the Mediterranean basin (black circle). Panel (**b**): bathymetry of the Strait and surrounding area, from [93]. Contours are shown at 150 m intervals. The red line indicates the approximate position for a barrage, along the Camarinal Sill, in the western part of the Strait. Panel (**c**): depth profile along the shallowest water line proposed for the barrage.

Unfavorable effects of the dam would include the disruption of present water flow patterns, rising salinity, changes in nutrient cycling, and still unclear biological impacts. Partially blocking the inflow of surface water from the Atlantic Ocean would cause sea level behind the dam to drop, because evaporation in the Mediterranean region exceeds the sum of river inflow and precipitation [100]. Sluices would be needed to allow and control water inflow to compensate for this drop. Evaporation would cause Mediterranean salinity to increase. Under present climate conditions, this average salinity increase is estimated at about 0.01 psu per year, or 1 psu every 100 years. This slow trend would eventually cause serious effects, but should give sufficient time (several hundred years) for climate change problems to be solved.

The outflow of deep, saline water into the Atlantic Ocean might also need to be preserved [101]. Numerical models of the global ocean, in which the exchange through the Strait of Gibraltar could be modulated at will, should be able to clarify this matter. Modeling of the Mediterranean itself would show the impact of changing stratification and vertical mixing patterns, affecting both oxygen ventilation of deeper layers, patterns of nutrients dynamics, and primary/secondary production. It remains to be seen whether fish and marine mammal migrations could be maintained via the dam openings (i.e., the sluices above). Ultimately, though, without a dam to control sea level rise, the trend to increase coastal reinforcements would lead to a catastrophic degeneration of a large part of the coastline, with dramatic impacts on the entire basin's flora and fauna [102].

Shipping would need to be regulated at the dam with locks, possibly built together with port facilities located to the north and to the south of the barrage, west of Tarifa, Spain, and east of Tangier, Morocco, for redundancy. Locks would also be needed for the Suez Canal. These could be installed in both the south and north sections of the waterway, allowing the central Bitter Lakes to return to the higher salinity that previously blocked invasive species from the Red Sea [103]. The locks would increase transit times for marine traffic, reducing the competitiveness of the Mediterranean shipping route. Moreover, undetected passage of submarines would become impossible, but that may prove not to be such a negative occurrence.

Appendix B

Saving the City of Venice

A barrage in the Strait of Gibraltar, regulating water exchanges with the Atlantic Ocean, together with proper locks in the Suez Canal, could protect all coastal environments and infrastructures throughout the Mediterranean Sea and Black Sea, eliminating the need for any other coastal defense against rising sea levels. A recent report by UNESCO [104] concluded that, due to the connection of the Mediterranean to the global ocean through the Strait of Gibraltar, the MOSE system of movable barriers is inadequate to defend the City of Venice on the long term [105]. The report states that "the sea will eventually rise to a level where even continuous closures will not be able to protect the city from flooding", and that "the question is not if this will happen, but only when".

Closing Gibraltar is an alternative to the conclusion that cities such as Venice are inevitably doomed. Clearly, this proposal needs to be studied and discussed. But it is also clear that a significant length of unchanged, natural Mediterranean coastline would be preserved if the dam were built. The dam would protect many historic cities and archeological sites which were designed for the low tidal range of the Mediterranean basin, and which have survived (with a few notable exceptions) thanks to the relatively stable sea level of the most recent centuries. The exponential increase in global sea level, which has already reached 25 cm above historic values, will reach 40 cm by the year 2050, 1 m by 2100, and 2.5 m by 2150. The present systems to prevent flooding in the City of Venice will find it hard to operate with an additional sea level rise of even 40 cm.

The *MOdulo Sperimentale Elettromeccanico* (MOSE), now nearly operational, is Venice's response to the recurring episodes of *Acqua Alta*, which combine storm surges and high astronomical tides (up to 187 cm, on 13 November 2019 [106], second highest only after the 194 cm record of 4 November 1966). These episodes cause repeated, picturesque, but devastating flooding in the city. The MOSE movable barriers rest on the bottom of the Venetian Lagoon openings (the three *Bocche di Porto*, i.e., *Lido*, *Malamocco*, and *Chioggia*) and can be raised from a control station on the artificial island built in the middle of the *Bocca di Lido* (Figure A2). Their closure seals off the Lagoon from the Adriatic Sea, preventing any water level rise in the city's historical centre, and in all other Lagoon sites. The MOSE, however, is designed to provide short-term defense from *Acqua Alta*, not from the long-term rise due to climate change. Ultimately, even if all Mediterranean coastlines were protected against the global sea level rise by a dam at Gibraltar, *Acqua Alta* would still occur in Venice, and the MOSE would still be required. The combination of MOSE and the dam would seem the best, and probably the only, long-term solution.

In the past 100 years, natural land subsidence lowered the City of Venice by 3 cm, and water pumping from underground aquifers caused additional lowering by 9 cm. In the same period, global sea level rise amounted to 13 cm. The total relative rise of about 25 cm is expected to grow by an additional 50 cm by the end of the century. By that time, a climate-change-driven sea level rise of 1 m and a possible 2 m *Acqua Alta* could give a grand total of 4 m! This would be well beyond the capability of the MOSE, which was designed to handle exceptional surges of up to 3 m, and to be closed only a few times per year. As of today, in 29 months of MOSE test operations, there have already been 44 closures, about 20 per year, a rate almost 10 times higher than expected [107].

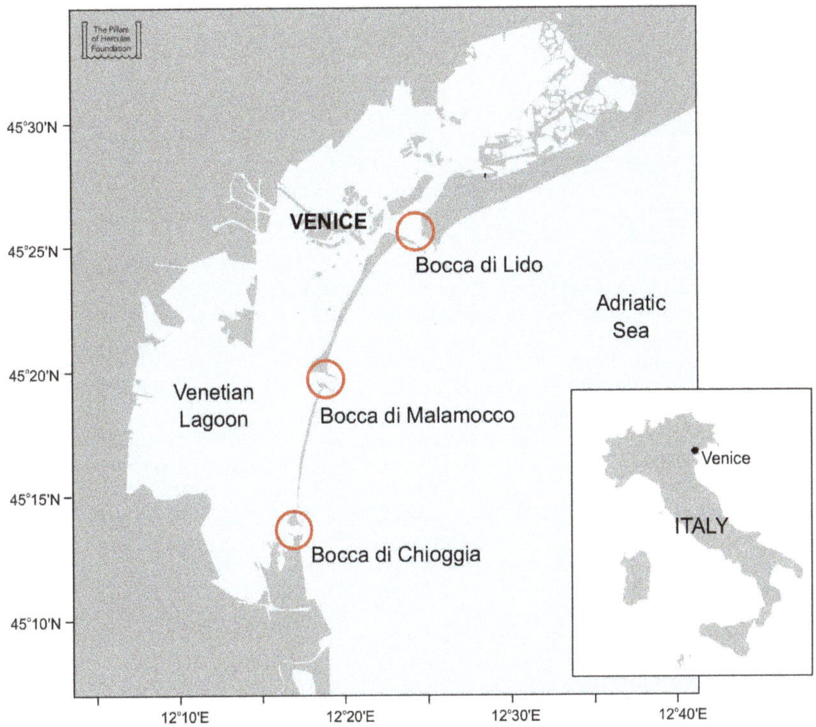

Figure A2. Location of the Venetian Lagoon's three main inlets, or *Bocca di Porto* (Lido, Malamocco, Chioggia), where the MOSE floodgates are located (red circles).

The book "Venice shall rise again", subtitled "Engineered uplift of Venice through seawater injection" [108], promotes the idea of raising by about 30 cm a circle of land ≈3.5 km in radius, centered under Venice, with lesser uplift out to a distance of ≈25 km. An obvious danger is that differential uplift could damage buildings, but the authors point out that the reverse procedure, i.e., the extraction of water for urban usage, has caused a drop by a significant fraction (about 30%) of the planned rise, without significant damage. Measurements of subsidence in the years 1950 to 1970 provide data on the differential movements due to extraction, and hence those to be expected during injection. The cost estimate appears to be modest, in comparison to that of MOSE, but, so far, the idea has not been pursued further.

Action to save Venice, if it is to include major projects like a Gibraltar Dam, requires active support by the entire Mediterranean community. Indeed, other countries are facing similar problems, due to the present and foreseen sea level rise. Egypt is a prime example, where the entire Nile delta is in danger of disappearing, while coastal cities, such as Alexandria, are struggling with the combined effects of sinking land and rising sea. It seems that an unprecedented level of collaboration would not be unlikely in the region. In the Mediterranean and Black Seas, the Gibraltar Dam would be a possible alternative to the massive intervention on coastal landscapes and infrastructures required by global sea level rise. But the present standard reaction to proposals to initiate a public debate on the subject appears to be quiet disbelief, probably directed more at the international cooperation needed for success, rather than at the reality of the coming sea level rise.

References

1. Barale, V.; Gower, J.F.R.; Alberotanza, L. (Eds.) *Proceedings of the "Oceans from Space" V, Venice 2022*; NSA GROUP: Rome, Italy, 2022; p. 252. [CrossRef]
2. Robinson, I.S. *Measuring the Oceans from Space. The Principles and Methods of Satellite Oceanography*; Springer-Praxis Books in Geophysical Sciences; Springer: Berlin/Heidelberg, Germany; New York, NY, USA, 2004.
3. Benveniste, J.; Bonnefond, P. 25 Years of Progress in Radar Altimetry. *Adv. Space Res.* **2021**, *68*, 317–1242. [CrossRef]
4. Zhou, D. Marginal Seas. In *Encyclopedia of Marine Geosciences*; Harff, J., Meschede, M., Petersen, S., Thiede, J., Eds.; Encyclopedia of Earth Sciences Series; Springer: Dordrecht, The Netherlands, 2016.
5. The Mobile Barriers for the Protection of Venice from High Tides. Available online: https://www.mosevenezia.eu/project/?lang=en (accessed on 17 May 2023).
6. Von Schuckmann, K.; Le Traon, P.Y.; Smith, N.; Pascual, A.; Djavidnia, S.; Gattuso, J.P.; Grégoire, M. Copernicus Marine Service Ocean State Report, Issue 5. *J. Oper. Oceanogr.* **2021**, *14* (Suppl. S1), s1–s185. [CrossRef]
7. Gramling, C. The 'Doomsday' Glacier May Soon Trigger a Dramatic Sea-Level Rise. Available online: https://www.snexplores.org/article/antarctica-thwaites-glacier-ice-shelf-collapse-climate-5-years (accessed on 2 April 2023).
8. Intergovernmental Panel on Climate Change (IPCC). *Climate Change 2023: Synthesis Report. Contribution of Working Groups I, II and III to the Sixth Assessment Report of the Intergovernmental Panel on Climate Change*; Core Writing Team, Lee, H., Romero, J., Eds.; IPCC: Geneva, Switzerland, 2023; p. 184. [CrossRef]
9. Forster, P.M.; Smith, C.J.; Walsh, T.; Lamb, W.F.; Lamboll, R.; Hauser, M.; Ribes, A.; Rosen, D.; Gillett, N.; Palmer, M.D.; et al. Indicators of Global Climate Change 2022: Annual update of large-scale indicators of the state of the climate system and human influence. *Earth Syst. Sci. Data* **2023**, *15*, 2295–2327. [CrossRef]
10. Cazenave, A.; Palanisamy, H.; Ablain, M. Contemporary sea level changes from satellite altimetry: What have we learned? What are the new challenges? *Adv. Space Res.* **2018**, *62*, 1639–1653. [CrossRef]
11. Pulhin, J.M.; Inoue, M.; Shaw, R. (Eds.) *Climate Change, Disaster Risks, and Human Security*; Book Series on Disaster Risk Reduction; Springer: Singapore, 2022; 450p.
12. Fox-Kemper, B.; Hewitt, H.T.; Xiao, C.; Aðalgeirsdóttir, G.; Drijfhout, S.S.; Edwards, T.L.; Golledge, N.R.; Hemer, M.; Kopp, R.E.; Krinner, G.; et al. Ocean, cryosphere, and sea level change. In *Climate Change 2021: The Physical Science Basis. Contribution of Working Group I to the Sixth Assessment Report of the Intergovernmental Panel on Climate Change*; Masson-Delmotte, V., Zhai, P., Pirani, A., Connors, S.L., Péan, C., Berger, S., Caud, N., Chen, Y., Goldfarb, L., Gomis, M.I., et al., Eds.; Cambridge University Press: Cambridge, UK, 2021.
13. Callendar, G.S.; Stewart, G. The artificial production of carbon dioxide and its influence on temperature. *Q. J. R. Meteorol. Soc.* **1938**, *64*, 223–240. [CrossRef]
14. Ritchie, H.; Rodés-Guirao, L.; Mathieu, E.; Gerber, M.; Ortiz-Ospina, E.; Hasell, J.; Roser, M. *Population Growth*; Our World in Data: Oxford, UK, 2023; Available online: https://ourworldindata.org/population-growth (accessed on 18 August 2023).
15. Keeling, R.F. Recording Earth's Vital Signs. *Science* **2008**, *319*, 1771–1772. [CrossRef] [PubMed]
16. United Nations Climate Change. Available online: https://unfccc.int (accessed on 14 August 2023).
17. Barcena, A. An Overview of the Oceans in *Agenda 21* of the 1992 United Nations Conference on Environment and Development. *Mar. Pollut. Bull.* **1992**, *25*, 107–111. [CrossRef]
18. Breidenich, C.; Magraw, D.; Anne Rowley, A.; Rubin, J.W. The Kyoto Protocol to the United Nations Framework Convention on Climate Change. *Am. J. Int. Law* **1998**, *92*, 315–331. [CrossRef]
19. Mayer, B. The Curious Fate of the Doha Amendment. Available online: https://www.ejiltalk.org/the-curious-fate-of-the-doha-amendment/ (accessed on 1 September 2023).
20. Schneider, L.; La Hoz Theuer, S. Environmental integrity of international carbon market mechanisms under the Paris Agreement. *Clim. Policy* **2019**, *19*, 386–400. [CrossRef]
21. Sharm El-Sheikh Climate Change Conference 2022. Available online: https://unfccc.int/cop27 (accessed on 17 May 2023).
22. NOAA. Global Monitoring Laboratory, Earth System Research Laboratories. Trends in Atmospheric Carbon Dioxide. Available online: https://gml.noaa.gov/ccgg/trends/data.html (accessed on 17 August 2023).
23. Thoning, K.W.; Crotwell, A.M.; Mund, J.W. Atmospheric Carbon Dioxide Dry Air Mole Fractions from continuous measurements at Mauna Loa, Hawaii, Barrow, Alaska, American Samoa and South Pole. 1973–2022 Version 2023-08-08 National Oceanic and Atmospheric Administration (NOAA), Global Monitoring Laboratory (GML), Boulder, Colorado, USA. Available online: https://doi.org/10.15138/yaf1-bk21 (accessed on 11 August 2023).
24. Gillespie, A. University of Hawaii, NOAA to Gather Climate Change Data Following Mauna Loa Eruption. Available online: https://www.noaa.gov/news-release/university-of-hawaii-noaa-to-gather-climate-change-data-following-mauna-loa-eruption (accessed on 17 August 2023).
25. NOAA. Climate.gov. Carbon Dioxide over 800,000 Years. Available online: https://www.climate.gov/ (accessed on 4 September 2023).
26. Lüthi, D.; Le Floch, M.; Bereiter, B.; Blunier, T.; Barnola, J.M.; Siegenthaler, U.; Raynaud, D.; Jouzel, J.; Fischer, H.; Kawamura, K.; et al. High-resolution carbon dioxide concentration record 650,000–800,000 years before present. *Nature* **2008**, *453*, 379–382. [CrossRef]
27. Met Office Hadley Centre Observations Datasets. Available online: https://www.metoffice.gov.uk/hadobs/hadsst4/data/download.html (accessed on 17 August 2023).

28. Kennedy, J.J.; Rayner, N.A.; Atkinson, C.P.; Killick, R.E. An ensemble data set of sea surface temperature change from 1850: The Met Office Hadley Centre HadSST.4.0.0.0 data set. *J. Geophyical. Res. Atmos.* **2019**, *124*, 7719–7763. [CrossRef]
29. Intergovernmental Panel on Climate Change (IPCC). Climate Change 2023, Synthesis Report, Summary for Policimakers. Available online: https://www.ipcc.ch/report/ar6/syr/downloads/report/IPCC_AR6_SYR_SPM.pdf (accessed on 17 May 2023).
30. Copernicus. OBSERVER: Record-Breaking Marine Heatwaves in the Mediterranean and Safeguarding Marine Ecosystems. Available online: https://www.copernicus.eu/en/news/news/observer-record-breaking-marine-heatwaves-mediterranean-and-safeguarding-marine (accessed on 1 September 2023).
31. Way, M.J.; Del Genio, A.D.; Aleinov, I.; Clune, T.L.; Kelley, M.; Kiang, N.Y. Climates of Warm Earth-like Planets. I. 3D Model Simulations. *Astrophys. J. Suppl. Ser.* **2018**, *239*, 24. [CrossRef]
32. Guerou, A.; Meyssignac, B.; Prandi, P.; Ablain, M.; Ribes, A.; Bignalet-Cazalet, F. Current observed global mean sea level rise and acceleration estimated from satellite altimetry and the associated uncertainty. *Ocean. Sci.* **2023**, *19*, 431–451. [CrossRef]
33. Church, J.A.; White, N.J. A 20th century acceleration in global sea level rise. *Geophys. Res. Lett.* **2006**, *33*, L01602. [CrossRef]
34. Grases, A.; Gracia, V.; García-León, M.; Lin-Ye, J.; Sierra, J.P. Coastal Flooding and Erosion under a Changing Climate: Implications at a Low-Lying Coast (Ebro Delta). *Water* **2020**, *12*, 346. [CrossRef]
35. Copernicus. Climate Indicators. Sea Level. Available online: https://climate.copernicus.eu/climate-indicators/sea-level (accessed on 5 September 2023).
36. Church, J.A.; White, N.J. Sea-level rise from the late 19th to the early 21st Century. *Surv. Geophys.* **2011**, *32*, 585–602. [CrossRef]
37. CSIRO. Sea Level Rise. Available online: https://www.cmar.csiro.au/sealevel/sl_data_cmar.html (accessed on 17 August 2023).
38. CNES. AVISO+, Satellite Altimetry Data. Data Access. Available online: https://www.aviso.altimetry.fr/en/data/data-access.html (accessed on 17 August 2023).
39. Church, J.A.; Clark, P.U.; Cazenave, A.; Gregory, J.M.; Jevrejeva, S.; Levermann, A.; Merrifield, M.A.; Milne, G.A.; Nerem, R.S.; Nunn, P.D.; et al. Sea Level Change. In *Climate Change 2013: The Physical Science Basis. Contribution of Working Group I to the Fifth Assessment Report of the Intergovernmental Panel on Climate Change*; Stocker, T.F., Qin, D., Plattner, G.K., Tignor, M., Allen, S.K., Boschung, J., Nauels, A., Xia, Y., Bex, V., Midgley, P.M., Eds.; Cambridge University Press: Cambridge, UK; New York, NY, USA, 2013.
40. Cazenave, A.; Moreira, L. Contemporary sea-level changes from global to local scales: A review. *Proc. R. Soc. A* **2022**, *478*, 20220049. [CrossRef]
41. NASA Earth Data. Sea Level Change. Observations from Space. Understanding Sea Level. Global Mean Sea Level. Available online: https://sealevel.nasa.gov/understanding-sea-level/global-sea-level/overview (accessed on 27 September 2023).
42. Pokhrel, Y.N.; Hanasaki, N.; Yeh, P.J.; Yamada, T.J.; Kanae, S.; Oki, T. Model estimates of sea-level change due to anthropogenic impacts on terrestrial water storage. *Nat. Geosci.* **2012**, *5*, 389–392. [CrossRef]
43. Moucha, R.; Forte, A.M.; Mitrovica, J.X.; Rowley, D.B.; Quéré, S.; Simmons, N.A.; Grand, S.P. Dynamic topography and long-term sea-level variations: There is no such thing as a stable continental platform. *Earth Planet. Sci. Lett.* **2008**, *271*, 101–108. [CrossRef]
44. Kopp, R.E.; Hay, C.C.; Little, C.M.; Mitrovica, J.X. Geographic Variability of Sea-Level Change. *Curr. Clim. Change Rep.* **2015**, *1*, 192–204. [CrossRef]
45. Woodworth, P.L.; Pugh, D.T.; Plater, A.J. Sea level measurements from tide gauges. In *Handbook of Sea-Level Research*, 1st ed.; Shennan, I., Long, A.J., Horton, B.P., Eds.; John Wiley & Sons Ltd.: Oxford, UK, 2015; pp. 557–574.
46. Boretti, A. Nonlinear absolute sea-level patterns in the long-term-trend tide gauges of the East Coast of North America. *Nonlinear Eng.* **2021**, *10*, 1–15. [CrossRef]
47. Gobron, K.; de Viron, O.; Wöppelmann, G.; Poirier, É.; Ballu, V.; Van Camp, M. Assessment of Tide Gauge Biases and Precision by the Combination of Multiple Collocated Time Series. *J. Atmos. Ocean. Technol.* **2019**, *36*, 1983–1996. [CrossRef]
48. Mitchum, G.T. Monitoring the Stability of Satellite Altimeters with Tide Gauges. *J. Atmos. Ocean. Technol.* **1998**, *15*, 721–730. [CrossRef]
49. Mu, D.; Yan, H.; Feng, W. Assessment of sea level variability derived by EOF reconstruction. *Geophys. J. Int.* **2018**, *214*, 79–87. [CrossRef]
50. Fu, L.L.; Cazenave, A. (Eds.) *Satellite Altimetry and Earth Sciences: A Handbook of Techniques and Applications*; International Geophysics Series; Academic Press: San Diego, CA, USA, 2001; Volume 69, 463p.
51. Nerem, R.S.; Mitchum, G.T. Chapter 6 Observation of sea level change from satellite altimetry. In *Sea Level Rise: History and Consequences*; Douglas, B.C., Kearney, M.S., Leatherman, S.P., Eds.; International Geophysics Series; Academic Press: San Diego, CA, USA, 2001; Volume 75, pp. 21–163. [CrossRef]
52. Vignudelli, S.; Birol, F.; Benveniste, J.; Fu, L.L.; Picot, N.; Raynal, M.; Roinard, H. Satellite Altimetry Measurements of Sea Level in the Coastal Zone. *Surv. Geophys.* **2019**, *40*, 1319–1349. [CrossRef]
53. Watson, C.; White, N.; Church, J.; King, M.A.; Butgette, R.J.; Legresy, B. Unabated global mean sea-level rise over the satellite altimeter era. *Nat. Clim. Change* **2015**, *5*, 565–568. [CrossRef]
54. Apel, J.R. Three Decades of Satellite Oceanography: The View from on High. In *Space Remote Sensing of Subtropical Oceans, Proceedings of the COSPAR Colloquium on Space Remote Sensing of Subtropical Oceans (SRSSO), Taiwan, 12–17 September 1995*; Liu, C.T., Ed.; COSPAR Colloquia Series; Pergamon: Oxford, UK, 1997; Volume 8, pp. 11–19. [CrossRef]
55. Born, G.H.; Dunne, J.A.; Lame, D.B. Seasat Mission Overview. *Science* **1979**, *204*, 1405–1406. [CrossRef]

56. Wilson, W.S.; Fellous, J.L.; Kawamura, H.; Mitnik, L.M. A History of Oceanography from Space. In *Remote Sensing of the Marine Environment*; Ryerson, R.A., Ed.; Manual of Remote Sensing; American Society for Photogrammetry and Remote Sensing: Bethesda, MD, USA, 2006; Volume 6, pp. 1–31.
57. ESA, Earth Online, ENVISAT. Available online: https://earth.esa.int/eogateway/missions/envisat (accessed on 25 May 2023).
58. Steunou, N.; Desjonquères, J.D.; Picot, N.; Sengenes, P.; Noubel, J.; Poisson, J.C. AltiKa Altimeter: Instrument Description and In-Flight Performance. *Mar. Geod.* **2015**, *38* (Suppl. S1), 22–42. [CrossRef]
59. ESA, Sentinel Online. Available online: https://sentinel.esa.int/web/sentinel/missions (accessed on 4 September 2023).
60. ESA, Earth Online, CryoSat. Available online: https://earth.esa.int/eogateway/missions/cryosat (accessed on 4 September 2023).
61. Fu, L.L.; Alsdorf, D.; Rodriguez, E.; Morrow, R.; Mognard, N.; Lambin, J.; Vaze, P.; Lafon, T. The SWOT (Surface Water and Ocean Topography) Mission: Spaceborne Radar Interferometry for Oceanographic and Hydrological Applications. In Proceedings of the OceanObs'09, Venice, Italy, 21–25 September 2009.
62. JPL, PODAAC, MEaSUREs—Integrated Multi-Mission Ocean Altimeter Data for Climate Research (MEaSUREs-SHH). Available online: https://podaac.jpl.nasa.gov/MEaSUREs-SSH (accessed on 30 August 2023).
63. Lindsey, R. Climate Change: Global Sea Level. NOAA Science & Information for a Climate-Smart Nation. Available online: https://www.climate.gov/news-features/understanding-climate/climate-change-global-sea-level (accessed on 25 May 2023).
64. University of Colorado, Sea Level Research Group. Available online: https://sealevel.colorado.edu/ (accessed on 25 May 2023).
65. CNES. AVISO+, Satellite Altimetry Data. Mean Sea Level. Available online: https://www.aviso.altimetry.fr/en/data/products/ocean-indicators-products/mean-sea-level.html (accessed on 25 May 2023).
66. Thiéblemont, R.; Le Cozannet, G.; Toimil, A.; Meyssignac, B.; Losada, I.J. Likely and High-End Impacts of Regional Sea-Level Rise on the Shoreline Change of European Sandy Coasts Under a High Greenhouse Gas Emissions Scenario. *Water* **2019**, *11*, 2607. [CrossRef]
67. Reimann, L.; Vafeidis, A.T.; Brown, S.; Hinkel, J.; Tol, R.S.J. Mediterranean UNESCO World Heritage at risk from coastal flooding and erosion due to sea-level rise. *Nat. Commun.* **2018**, *9*, 4161. [CrossRef]
68. Wikipedia. Atlantropa. Available online: https://en.wikipedia.org/wiki/Atlantropa (accessed on 6 September 2023).
69. Stommel, H. The circulation of the abyss. *Sci. Am.* **1958**, *199*, 85–90. Available online: https://sciam-cms.s3.amazonaws.com/sciam/cache/file/C39DA9C7-9B4C-4975-839909C607633730.pdf (accessed on 5 September 2023). [CrossRef]
70. Johnson, R.G. Climate control requires a dam in the Strait of Gibraltar. *Eos Trans. Am. Geophys. Union* **1997**, *78*, 277–282. [CrossRef]
71. Gower, J. A sea surface height control dam at the Strait of Gibraltar. *Nat. Hazards* **2015**, *78*, 2109–2120. [CrossRef]
72. Gradstein, F.M.; Ogg, J.G.; Smith, A.G. (Eds.) *A Geologic Time Scale 2004*; Cambridge University Press: Cambridge, UK; New York, NY, USA, 2005; p. 589.
73. Ruggieri, G.; Sprovieri, R. Messinian salinity crisis and its paleogeographical implications. *Palaeogeogr. Palaeoclimatol. Palaeoecol.* **1976**, *20*, 13–21. [CrossRef]
74. Hsü, K.J. *The Mediterranean Was a Desert*; Princeton University Press: Princeton, NJ, USA, 1983; p. 197.
75. Caruso, A.; Blanc-Valleron, M.M.; Da Prato, S.; Pierre, C.; Rouchy, J.M. The late Messinian "Lago-Mare" event and the Zanclean Reflooding in the Mediterranean Sea: New insights from the Cuevas del Almanzora section (Vera Basin, South-Eastern Spain). *Earth-Sci. Rev.* **2020**, *200*, 102993. [CrossRef]
76. Carlisle, D.P.; Feetham, P.M.; Wright, M.J.; Teagle, D.A.H. The public remain uninformed and wary of climate engineering. *Clim. Change* **2020**, *160*, 303–322. [CrossRef]
77. Kitchen, D. Geoengineering Sounds Like a Quick Climate Fix, But without More Research and Guardrails, It's a Costly Gamble—With Potentially Harmful Results. Available online: https://theconversation.com/geoengineering-sounds-like-a-quick-climate-fix-but-without-more-research-and-guardrails-its-a-costly-gamble-with-potentially-harmful-results-211705 (accessed on 7 September 2023).
78. Wagner, G. *Geoengineering: The Gamble*; Polity Press: Cambridge, UK; Oxford, UK; Boston, MA, USA; New York, NY, USA, 2021; p. 208.
79. Marcos, M.; Jorda, G.; Le Cozannet, G. Sea level rise and its impacts on the Mediterranean. In *The Mediterranean Region under Climate Change: A Scientific Update*; Moatti, J.P., Thiébault, S., Eds.; IRD Éditions: Marseille, France, 2016; pp. 265–275.
80. Groeskamp, S.; Kjellsson, J. NEED: The Northern European Enclosure Dam for if Climate Change Mitigation Fails. *Bull. Am. Meteorol. Soc.* **2020**, *101*, E1174–E1189. [CrossRef]
81. Archer, D. Fate of fossil fuel CO_2 in geologic time. *J. Geophys. Res. Oceans* **2005**, *110*, C09S05. [CrossRef]
82. INDEPENDENT. Climate. News. COP21: Paris Deal Far too Weak to Prevent Devastating Climate Change, Academics Warn, by T. Bawden. Available online: https://www.independent.co.uk/environment/climate-change/cop21-paris-deal-far-too-weak-to-prevent-devastating-climate-change-academics-warn-a6803096.html (accessed on 11 September 2023).
83. Shepherd, J.; Caldeira, K.; Cox, P.; Haigh, J.; Keith, D.; Launder, B.; Mace, G.; MacKerron, G.; Pyle, J.; Rayner, S.; et al. *Geoengineering the Climate: Science, Governance, and Uncertainty*; Royal Society Policy Document, 10/09; The Royal Society Publishing: London, UK, 2009; p. 82.
84. Arnell, N.W. The implications of climate change for emergency planning. *Int. J. Disaster Risk Reduct.* **2022**, *83*, 103425. [CrossRef]
85. Moore, J.C.; Gladstone, R.; Zwinger, T.; Wolovick, M. Geoengineer polar glaciers to slow sea-level rise. *Nature* **2018**, *555*, 303–305. [CrossRef]

86. International Altimetry Team; Verron, J.; Ryan, B.; Bonnefond, P.; Benveniste, J. Altimetry for the future: Building on 25 years of progress. *Adv. Space Res.* **2021**, *68*, 319–363. [CrossRef]
87. Nieves, V.; Radin, C.; Camps-Valls, G. Predicting regional coastal sea level changes with machine learning. *Nat. Sci. Rep.* **2021**, *11*, 7650. [CrossRef]
88. Wikipedia. Pillars of Hercules. Available online: https://en.wikipedia.org/wiki/Pillars_of_Hercules (accessed on 8 May 2023).
89. Wikipedia. Labours of Hercules. Available online: https://en.wikipedia.org/wiki/Labours_of_Hercules (accessed on 8 May 2023).
90. Siculus, D. *Library of History, Volume II: Books 2.35-4.58*; Oldfather, C.H., Translator; Loeb Classical Library No. 303; Harvard University Press: Cambridge, MA, USA, 1935.
91. Oceans from Space. Available online: https://www.oceansfromspacevenice2020.org (accessed on 11 September 2023).
92. Breeze, P. Chapter 3—Dams and Barrages. In *Hydropower*; Breeze, P., Ed.; Academic Press: Cambridge, MA, USA, 2018; pp. 23–33. [CrossRef]
93. EMODnet Bathymetry Consortium. *EMODnet Digital Bathymetry (DTM 2018)*; EMODnet High Resolution Seabed Mapping Project: Madeira, Portugal, 2018. [CrossRef]
94. USGS. Chuquicamata Mine, Chile. Available online: https://eros.usgs.gov/media-gallery/earthshot/chuquicamata-mine-chile (accessed on 11 September 2023).
95. Hallegatte, S.; Green, C.; Nicholls, R.J.; Corfee-Morlot, J. Future flood losses in major coastal cities. *Nat. Clim. Change* **2013**, *3*, 802–806. [CrossRef]
96. Hinkel, J.; Lincke, D.; Vafeidis, A.T.; Perrette, M.; Nicholls, R.J.; Tol, R.S.J.; Marzeion, B.; Fettweis, X.; Ionescu, C.; Levermann, A. Coastal flood damage and adaptation costs under 21st century sea level rise. *Proc. Natl. Acad. Sci. USA* **2014**, *111*, 9–3292. [CrossRef] [PubMed]
97. Stern, N. *The Economics of Climate Change: The Stern Review*; Cambridge University Press: Cambridge, UK, 2007. [CrossRef]
98. Sociedad Espanola de Estudios Para la Communicion Fija a Traves del Estrecho de Gibraltar (SECEGSA), Agenda 2030. Available online: https://www.secegsa.gob.es/secegsa/lang_castellano/ (accessed on 25 May 2023).
99. Ste Nationale d'Etudes du Detroit de Gibraltar (SNEDG). Available online: https://www.charika.ma/societe-ste-nationale-d-etudes-du-detroit-de-gibraltar-17101 (accessed on 25 May 2023).
100. Mariotti, A.; Struglia, M.V.; Zeng, N.; Lau, K.M. The hydrological cycle in the Mediterranean region and implications for the water budget of the Mediterranean Sea. *J. Clim.* **2002**, *15*, 1674–1690. [CrossRef]
101. Rahmstorf, S. Influence of Mediterranean outflow on climate. *Eos Trans. Am. Geophys. Union* **1998**, *79*, 281–282. [CrossRef]
102. Meinesz, A. *Proteger la Biodiversite Marine*; Odile Jacob: Nice, France, 2021; 304p.
103. Galil, B.S. Taking stock: Inventory of alien species in the Mediterranean Sea. *Biol. Invasions* **2009**, *11*, 359–372. [CrossRef]
104. Umgiesser, G. *From Global to Regional: Local Sea Level Rise Scenarios, Focus on the Mediterranean Sea and the Adriatic Sea. The Future of Venice and Its Lagoon in the Context of Climate Change*; UNESCO Office Venice and Regional Bureau for Science and Culture in Europe: Venice, Italy, 2011; Available online: https://unesdoc.unesco.org/ark:/48223/pf0000215105?posInSet=8&queryId=92c0667e-e72d-4198-bf42-dc33119acd56 (accessed on 17 May 2023).
105. Bras, R.L.; Harleman, D.R.F.; Rinaldo, A.; Rizzoli, P. Obsolete? No. Necessary? Yes. The gates will save Venice. *EOS Trans.* **2002**, *83*, 217–224. [CrossRef]
106. Harlan, C.; Pitrelli, S. Venice Submerged by Highest Tides in Half a Century. Washington Post, 13 November 2019. Available online: https://www.washingtonpost.com/world/europe/venice-partly-submerged-by-highest-tides-in-half-a-century/2019/11/13/fa36566e-05fa-11ea-8292-c46ee8cb3dce_story.html (accessed on 16 August 2023).
107. Umgiesser, G. The impact of operating the mobile barriers in Venice (MOSE) under climate change. *J. Nat. Conserv.* **2020**, *54*, 125783. [CrossRef]
108. Gambolati, G.; Teatini, P. *Venice Shall Rise Again*, 1st ed.; Elsevier Insights: London, UK, 2013; 100p.

MDPI AG
Grosspeteranlage 5
4052 Basel
Switzerland
Tel.: +41 61 683 77 34

Remote Sensing Editorial Office
E-mail: remotesensing@mdpi.com
www.mdpi.com/journal/remotesensing

www.ingramcontent.com/pod-product-compliance
Lightning Source LLC
Chambersburg PA
CBHW040819250926
51850CB00007BA/1210
* 9 7 8 3 7 2 5 8 7 6 3 1 0 *